Agricultural Survey Methods

AGRICULTURAL SURVEY METHODS

Edited by

Roberto Benedetti
'G. d'Annunzio' University, Chieti-Pescara, Italy

Marco Bee
University of Trento, Italy

Giuseppe Espa
University of Trento, Italy

Federica Piersimoni
National Institute of Statistics (ISTAT), Rome, Italy

WILEY

A John Wiley and Sons, Ltd., Publication

This edition first published 2010

John Wiley & Sons Ltd, The Atrium, Southern Gate, Chichester, West Sussex, PO19 8SQ, United Kingdom

Library of Congress Cataloguing-in-Publication Data

Benedetti, Roberto, 1964-
Agricultural survey methods / Roberto Benedetti, Marco Bee, Giuseppe Espa, Federica Piersimoni.
p. cm.
Based on papers presented at the 1998, 2001, 2004 and 2007 International Conferences on Agricultural Statistics.
Includes bibliographical references and index.
ISBN 978-0-470-74371-3 (cloth)
1. Agriculture–Statistical methods. I. Bee, Marco. II. Piersimoni, Federica. III. Title.
S566.55.B46 2010
338.1021–dc22

2010000132

A catalogue record for this book is available from the British Library.

ISBN: 978-0-470-74371-3 (H/B)

Typeset in 10/12 Times-Roman by Laserwords Private Limited, Chennai, India
Printed and bound in the United Kingdom by CPI Antony Rowe, Chippenham, Wiltshire.

To Agnese and Giulio
Roberto Benedetti

To Chiara and Annalisa
Marco Bee

To Massimo, Guido and Maria Rita
Giuseppe Espa

To Renata and Oscar
Federica Piersimoni

Contents

List of Contributors

Luca Alinovi
Agricultural Development Economics Division (ESA)
FAO of the United Nations
Rome, Italy.
Luca.Alinovi@fao.org

Dale Atkinson
Research Division
National Agricultural Statistics Service (NASS)/USDA
Fairfax, VA, USA.
Dale.Atkinson@usda.gov

Bettina Baruth
IPSC-MARS, JRC
Ispra, Italy.
Bettina.baruth@jrc.ec.europa.eu

Marco Bee
Department of Economics
University of Trento
Italy.
marco.bee@unitn.it

Roberto Benedetti
Department of Business, Statistical, Technological and Environmental Sciences (DASTA)
University 'G. d'Annunzio' of Chieti-Pescara
Pescara, Italy.
benedett@unich.it

Gianni Betti
Department of Quantitative Methods
University of Siena
Italy.
betti2@unisi.it

Andrea Carfagna
Department of Statistical Sciences
University of Bologna
Italy.
andreacarfagna@virgilio.it

Elisabetta Carfagna
Department of Statistical Sciences
University of Bologna
Italy.
elisabetta.carfagna@unibo.it

Raymond L. Chambers
School of Mathematics and Applied Statistics
University of Wollongong
Australia.
ray@uow.edu.au

Denis Chartrand
Agriculture Division
Statistics Canada
Ottawa, Canada.
Denis.Chartrand@statcan.gc.ca

Arijit Chaudhuri
Applied Statistics Unit
Indian Statistical Institute
Kolkata, India.
arijitchaudhuri@rediffmail.com

Nhu Che
Australian Bureau of Agriculture and Research Economics (ABARE)
Canberra, Australia.
NChe@abare.gov.au

Jim Cotter
National Agricultural Statistics Service (NASS)/USDA, retired.

Carrie Davies
Area Frame Section of National Agricultural Statistics Service (NASS)/USDA
Fairfax, VA, USA.
Carrie_Davis@nass.usda.gov

Jacques Delincé
Institute for Prospective Technological Studies
Seville, Spain.
jacques.delince@ec.europa.eu

Marcelle Dion
Agriculture, Technology and Transportation Statistics Branch
Statistics Canada
Ottawa, Canada.
Marcelle.Dion@statcan.gc.ca

Giuseppe Espa
Department of Economics
University of Trento
Italy.
giuseppe.espa@economia.unitn.it

Pieter Everaers
Eurostat, Directorate D
External Cooperation, Communication and Key Indicators
Luxembourg.
Pieter.EVERAERS@ec.europa.eu

Danila Filipponi
Istat
National Institute of Statistics
Rome, Italy.
danila.filipponi@istat.it

Javier Gallego
IPSC-MARS, JRC
Ispra, Italy.
javier.gallego@jrc.ec.europa.eu

Giulio Ghellini
Department of Quantitative Methods
University of Siena
Italy.
ghellini@unisi.it

Antonio Giusti
Department of Statistics
University of Florence
Firenze, Italy.
giusti@ds.unifi.it

Carol C. House
Agricultural Statistics Board
National Agricultural Statistics Service
Washington, DC, USA.
Carol_House@nass.usda.gov

Philip N. Kokic
Commonwealth Scientific and Industrial Research Organisation (CSIRO)
Canberra, Australia.
phil.kokic@csiro.au

Ulf Jorner
Statistics Sweden
Stockholm, Sweden.
Ulf.Jorner@scb.se

Claude Julien
Business Survey Methods Division
Statistics Canada
Ottawa, Canada.
Claude.Julien@statcan.gc.ca

Erdgin Mane
Agricultural Development Economics Division (ESA)
FAO of the United Nations
Rome, Italy.
Erdgin.Mane@fao.org

Paul Murray
Agriculture Division
Statistics Canada
Ottawa, Canada.
Paul.Murray@statcan.gc.ca

Jack Nealon
National Agricultural Statistics Service (NASS)/USDA
Fairfax, VA, USA.
Jack.Nealon@usda.gov

Jeroen Pannekoek
Department of Methodology
Statistics Netherlands
The Hague, Netherlands.
j.pannekoek@cbs.nl (and jpnk@cbs.nl)

Federica Piersimoni
Istat
National Institute of Statistics
Rome, Italy.
piersimo@istat.it

Paolo Postiglione
Department of Business, Statistical, Technological and Environmental Sciences (DASTA)
University 'G. d'Annunzio' of Chieti-Pescara
Pescara, Italy.
postigli@unich.it

J.N.K. Rao
School of Mathematics and Statistics
Carleton University
Ottawa, Canada.
jrao@math.carleton.ca

Ray Roberts
National Agricultural Statistics Service (NASS)/USDA
Fairfax, VA, USA.
Ray.Roberts@usda.gov

Donato Romano
Department of Agricultural and Resource Economics
University of Florence
Firenze, Italy.
donato.romano@unifi.it

Vijay Verma
Department of Quantitative Methods
University of Siena
Italy.
verma@unisi.it

Frederic A. Vogel
The World Bank
Washington, DC, USA.
fvogel@worldbank.org

Ton de Waal
Department of Methodology
Statistics Netherlands
The Hague, Netherlands.
t.dewaal@cbs.nl

Anders Wallgren
Örebro University and BA Statistiksystem AB
Vintrosa, Sweden.
ba.statistik@telia.com

Britt Wallgren
Örebro University and BA Statistiksystem AB
Vintrosa, Sweden.
ba.statistik@telia.com

Introduction

The importance of comprehensive, reliable and timely information on agricultural resources is now more than ever recognized in various practical situations arising in economic, social and environmental studies. These renewable, dynamic natural resources are necessary for many countries where the growing population pressure implies the need for increased agricultural production. To improve the management of these resources it is necessary to know at least their quality, quantity and location.

In Western countries, agriculture is assuming a more and more marginal economic role in terms of its percentage contribution to GNP, but recent radical economic and social transformations have caused a renewed interest in this sector. Such interest is due not only to economic factors but also to issues related to the quality of life and to the protection of the public health. Today the food industry suffers from phenomena that originate in the primary sector, such as diseases on farms or the production of genetically modified organisms. The growing attention of consumers to the quality of food products has strongly reinforced the need to look at a single agro-food product as the result of a chain of processes linked together. In this approach, agriculture represents not only an economic sector but also the origin of the food chain, and because of this role it deserves special attention. These aspects, together with the related recovery and protection of the environment, have led to deep modifications in the data provided in this sector.

Agricultural surveys are thus conducted all over the world in order to gather a large amount of information on the classic crops, yields, livestock and other related agricultural resources. As a result, the statistics produced are so strongly conditioned by this largely diversified demand that many countries, in order to be able to comply with these requests, began to set up a complex system of surveys based on a harmonized and integrated set of information whose design, implementation and maintenance require a strong methodological effort.

Apart from the difficulties typical of business data, such as the quantitative nature of many variables and their high concentration, agricultural surveys are indeed characterized by some additional peculiarities that often make it impossible or inefficient to make use of classical solutions proposed in the literature. In particular we refer to the following:

(a) The definition of the statistical units to be surveyed is neither obvious nor unique, because the list of possible options is quite large (family, agricultural holding, household, parcel of land, point, etc.) and its choice depends not only on the phenomenon for which we are interested in collecting the data, but also on the availability of a frame of units of sufficient quality.

(b) Typological classifications of the statistical units are very important tools to define the estimation domains and to design an efficient survey. However, harmonized hierarchical nomenclatures are usually not available for a certain definition of statistical unit, or they do exist but are so subjective that they cannot be considered as standard.

(c) The concentration of many variables is often even higher than in other business surveys.

(d) In many countries the use of the so-called 'census frames' is considered an ordinal procedure, with the obvious consequence that the list of sampling units is not updated and is very far from the target population. This has evident implications in terms of non-sampling errors due to under- or, less dangerous, over-coverage of the list.

(e) When designing a sample, the theory suggests two classical ways of using a size measure existing in the frame: a scheme with inclusion probabilities proportional to the size of the unit, and a stratification obtained through the definition of a set of threshold levels on the size variable. Nonetheless, a typical frame of agricultural units has a large amount of auxiliaries and the size of each unit is usually multivariate. In this case the solutions proposed by the methodological literature are much more complex.

(f) At least when using geographically referred units, there often exists a particular auxiliary variable requiring ad hoc procedures to be used in a sampling environment: the remotely sensed data. Remote sensing is nothing but a tool to get information about an object without being in physical contact with the object itself, and it is usually represented by digital images sensed from a satellite or an aircraft.

As far as we know, in the current literature there exists no comprehensive source of information regarding the use of modern survey methods adapted to these distinctive features of agricultural surveys.

However, the successful series of conferences on agricultural statistics, known as ICAS (International Conference on Agricultural Statistics), demonstrates that there is a broad and recognizable interest in methods and techniques for collecting and processing agricultural data. In our opinion, the remarkable number of high-quality methodological papers presented in these conferences may serve to fill this gap.

This book originates from a selection of the methodological papers presented at this set of conferences held in Washington, DC (1998), Rome (2001), Cancún (2004) and Beijing (2007). The declared aim was to develop an information network of individuals and institutions involved in the use and production of agricultural statistics.

These conferences were organized by the national statistical offices of the hosting countries – the National Agricultural Statistics Service (NASS) and United States Department of Agriculture (USDA); Italian National Statistical Institute (Istat), Agri-food and Fishery Information Service, Mexico (SIAP) and National Bureau of Statistics of China (NBS), in collaboration with international institutes such as Eurostat, FAO, OECD, UN/ECE, and ISI.

This book is an attempt to bring together the competences of academics and of experts from national statistical offices to increase the dissemination of the most recent survey methods in the agricultural sector. With this ambition in mind, the authors were asked to extend and update their research and the project was completed by some chapters on specialized topics.

Although the present book can serve as a supplementary text in graduate seminars in survey methodology, the primary audience is constituted by researchers having at least some prior training in sampling methods. Since it contains a number of review chapters on several specific themes in survey research, it will be useful to researchers actively engaged in organizing, managing and conducting agricultural surveys who are looking for an introduction to advanced techniques from both a practical and a methodological perspective.

Another aim of this book is to stimulate research in this field and, for this reason, we are aware that it cannot be considered as a comprehensive and definitive reference on the methods that can be used in agricultural surveys, since many topics were intentionally omitted. However, it reflects, to the best of our judgement, the state of the art on several crucial issues.

The volume contains 22 chapters of which the first one can be considered as an introductory chapter reviewing the current status of agricultural statistics, and the remaining 21 are divided into five parts:

I. Census, frames, registers and administrative data (Chapters 2–5). These chapters provide an overview of the basic tools used in agricultural surveys, including some practical and theoretical considerations regarding the definitions of the statistical units. Attention is then focused on the use of administrative data that in the last few years have evolved from a simple backup source to the main element in ensuring the coverage of a list of units. The opportunity to reduce census and survey costs implies growing interest, among statistical agencies, in the use of administrative registers for statistical purposes. However it requires attitudes, methods and terms that are not yet typical in the statistical tradition. The keyword is the harmonization of the registers in such a way that information from different sources and observed data should be consistent and coherent. In particular, the expensive agricultural census activities conducted periodically in every country of the world should benefit from such a radical innovation.

II. Sample design, weighting and estimation (Chapters 6–9). These chapters review advanced methods and techniques recently developed in the sampling literature as applied to agricultural units, in the attempt to address the distinctive features (c)–(e) described above. Some interesting proposals arise from the field of small-area estimation, which has received a lot of attention in recent years due to the growing demand for reliable small-area statistics needed for formulating policies and programmes. An appraisal is provided of indirect estimates, both traditional and model-based, that are used because direct area-specific estimates may not be reliable due to small-area-specific sample sizes.

III. GIS and remote sensing (Chapters 10–13). These chapters describe the use of the Geographic Information System technology as a tool to manage area- and point-based surveys. These devices are applied to carry out a wide range of operations on spatial information retrieved from many kinds of mixed sources. They provide a detailed description of the procedures currently used in the European Union and United States to develop and sample area frames for agricultural surveys. Additionally, the usefulness of remotely sensed data as the main auxiliary variables for geographically coded units is assessed through empirical evidence, and some techniques to increase the performance of their use are proposed.

IV. Data editing and quality assurance (Chapters 14–17). These chapters deal with the classical problem of error handling, localization and correction. This is strictly connected with the issue of guaranteeing data quality, which obviously plays a central role within any statistical institute, both in strategic decisions and in daily operations. In this framework, it is evident that quality is not as much concerned with the individual data sets as with the whole set of procedures used. Some approaches to ensure data quality in collecting, compiling, analysing and disseminating agriculture data are described.

V. Data dissemination and survey data analysis (Chapters 18–22). These chapters examine some experiences in the use of statistical methods to analyse agricultural survey data. In particular, regression analysis (or some of its generalizations) is quite often applied to survey microdata to estimate, validate or forecast models formalized within agricultural economics theory. The purpose is to take into account the nature of the data analysed, as observed through complex sampling designs, and to consider how, when and if statistical methods may be formulated and used appropriately to model agricultural survey data. Web tools and techniques to assist the users to access statistical figures online are then described, for complete, safe and adequate remote statistical analyses and dissemination.

We would like to thank Daniel Berze of the International Statistical Institute for useful advice and suggestions during the starting phase of the project. Thanks are also due to Susan Barclay, Heather Kay and Richard Davies of John Wiley & Sons, Ltd for editorial assistance, and to Alistair Smith of Sunrise Setting Ltd for assistance with LaTeX.

Finally, we are grateful to the chapter authors for their diligence and support for the goal of providing an overview of such an active research field. We are confident that their competence will lead to new insights into the dynamics of agricultural surveys methods, to new techniques to increase the efficiency of the estimates, and to innovative tools to improve the timeliness and comprehensiveness of agricultural statistics.

Roberto Benedetti
Marco Bee
Giuseppe Espa
Federica Piersimoni
Pescara, Trento and Rome

November 2009

1

The present state of agricultural statistics in developed countries: situation and challenges

Pieter Everaers
EUROSTAT, Directorate D – External Cooperation, Communication and Key Indicators, Luxembourg

1.1 Introduction

Agricultural statistics in the UN Economic Commission for Europe (UNECE) region are well advanced. Information collected by farm structure surveys on the characteristics of farms, farmers' households and holdings is for example combined with a variety of information on the production of animal products, crops, etc. Agricultural accounts are produced on a regular basis and a large variety of indicators on agri-economic issues is available. At the level of the European Union as a whole the harmonization – forced by a stringent regulatory framework – of the collection, validation and analysis of agricultural information has led to a settled system of statistics. Recent developments in methodologies for data collection (e.g. using hand-held computers, registers and administrative sources, advanced sampling techniques and remote sensing) are shaking up the development of agricultural statistics. There is a lack of uniformity in the pace at which different countries in the UNECE region are introducing new methodologies and techniques. The need to reduce the burden on farmers and to organize the collection and analysis of data more efficiently creates a set of challenges for the countries.

Agricultural Survey Methods Edited by Roberto Benedetti, Marco Bee, Giuseppe Espa and Federica Piersimoni

This chapter covers the situation for the majority of UNECE countries. At the time of writing the UNECE comprises the whole of Europe as well as the USA, Canada, New Zealand, Australia and Brazil. The work of Eurostat (the statistical office of the EU) covers in full the 27 member states and the four countries of the European Free Trade Association. The requirement for the acceding and pre-acceding countries to comply with the regulations at the moment of accession means that the western Balkan countries and Turkey are still developing their standards; in these countries the situation in both agriculture and agricultural statistics is more traditional, and at different stages en route to the EU model. For the EU's other neighbours in Europe, the model is more based on the Commonwealth of Independent States (CIS) approach to statistics, but also strongly harmonized and converging to the EU model. The contribution of the USA can be considered partly valid for Canada, Australia and New Zealand. Nevertheless, especially for these countries, some specific circumstances might not be fully covered. Finally, the situation in Brazil is that of a specific country with a strong development in new technologies for statistics and a very specific agricultural situation.

The term 'agricultural statistics' is here taken to include statistics on forestry and fisheries. It implicitly also includes statistics on trade in agricultural products (including forest and fishery products) as well as issues related to food safety. The definition of agricultural statistics is based on three conditions, all of which have to be met. In this definition, agriculture consists of the use of land, the culture of a living organism through more than one life cycle, and ownership. Land is used for many purposes, ranging from mining to recreation. Agricultural land supports the culture of living organisms and their ownership. This separates aquaculture from capture fishing and tree farming from forestry. Agriculture includes the management of water, the feeding and raising of organisms through several growth stages. Tree farming includes the management of the soil, fertilization, and pest management as the trees or other ornamental plants are raised through varies stages of growth. In both cases, farmers can choose to use the land for other purposes than aquaculture or raising tree crops.

The raising of awareness of the effects of globalization and the impact of climate change have led in the UNECE region to a greater understanding by statisticians as well as the politicians of the need to analyse different societal developments in relation to each other rather than in isolation. However, the interrelatedness of agriculture with, for example, land use and rural development, and also with environmental sustainability and overall well-being, is considered to be not yet fully reflected in available statistical information.

Agriculture in the UNECE region is in general characterized by the use of highly advanced technologies. Machinery, new production methods, fertilizers, pesticides and all kinds of supporting instruments have created a sector that is more businesslike than some traditional primary industries. At the same time, the recent emphasis on sustainability, environmental protection and ownership has led to more attention being given to the important role of rural areas. The increased use of modern technologies and the increase of scale has created a farming sector with relatively strong relations to other sectors of society, both at the level of the sector as a whole as well as at the level of individual farmers and their households, for example, with regard to employment and time use. The paradox of increased efficiency on the one hand and more emphasis on sustainability and environmental protection on the other is considered one of the main challenges for current agricultural statistics that justifies a reflection on their future.

The burden on farmers and decreasing response rates have forced the statistical institutes and other agricultural data collectors to enhance their efforts to deploy new data collection techniques and to make more use of other types of information, for example, from administrative sources. Growth in the availability of advanced IT tools for data collection, analysis and dissemination techniques is also forcing the statistical institutes to review the methodologies applied in agricultural statistics. Furthermore, the pressure to lower the administrative burden by simplifying regulations in agricultural statistics has created an emphasis on changes in the fundamental legal and methodological bases for agricultural statistics.

The enlargement of the EU in 2004 and 2007 and enhanced cooperation with future acceding countries and other neighbouring countries has visibly increased the impact of decisions on the organization and content of agricultural statistics in the EU. The greater variety in crops and methods demands a review of the existing statistics, specifically in the EU context.

In agricultural statistics in the UNECE region, the number of international and supranational organizations involved is currently quite limited. In the UNECE context, only the Food and Agriculture Organization (FAO) in Rome and Eurostat in Luxembourg play a role of any significance. The UNECE secretariat in Geneva and the Organisation for Economic Co-operation and Development (OECD) in Paris are no longer heavily involved. At a global level, the number of organizations involved is also very limited, this being the main reason for the decision of the UN Security Council in 2006 to close down the Inter-Secretariat Working Group on Agricultural Statistics. Both in Northern America and in Europe, however, there are many other organizations outside statistics involved in agricultural statistics and information. Traditionally, many of the agricultural organizations as well as the agricultural ministries are involved – as part of, for example, the Common Agricultural Policy – in collecting and using information on agriculture. This chapter has been written from the viewpoint of the national statistical offices, related governmental bodies as well the international and supranational organizations mentioned above. However, for a complete overview of ongoing work in agricultural statistics, this has to be complemented with information from other international and branch organizations.

The chapter is structured as follows. In Section 1.2 the current state and political and methodological context of agricultural statistics in the UNECE region is described. The main items discussed are the infrastructure for agricultural statistics, the information systems for collecting structural information, the statistics on production, the monetary elements, the added value in the production of agricultural statistics, other important sources, the relations with other statistics and the use of administrative data. Fishery and forestry statistics are also briefly discussed. In Section 1.3 governance and horizontal issues are discussed in more detail. In Section 1.4 some developments in the demand for agricultural statistics and some challenges are discussed. Finally, Section 1.5 focuses on the main recommendations for agricultural statistics in the UNECE region, regarding both content and governance.

1.2 Current state and political and methodological context

1.2.1 General

Agricultural statistics in the UNECE region have a long history at national level, especially in the European Union. The harmonized and integrated European system has evolved over the past five decades into a sophisticated and effective system. The priority attached to agricultural statistics in earlier years (especially from the middle of the last century till the early 1990s) reflected the need for this statistical information for the implementation and evaluation of the agreed Common Agricultural Policy[1] (and later also the Common Fisheries Policy) and the share of agriculture in the economy, employment and land use, both in EU and in national budgets. Nevertheless, resources for agricultural statistics have for the last decade or so been constant or diminishing and, compared to other areas of statistical development and innovations in agricultural statistics, are rather limited, thus also showing that resources have been shrinking. As a result, in many countries the current priority of this domain of statistics does not reflect the attention it deserves, considering the important position it has meanwhile assumed, for example, in the conservation of nature and the impact of climate change. Only in the last few years has an increase in attention become apparent, mainly as a result of the recognition of the important relation agriculture has with environmental issues such as climate change and the emphasis on developments in small areas.

In recent years, several initiatives have been taken to review the effectiveness of the current system of agricultural statistics, especially in the EU. This is partly the result of the emphasis on better regulations but also a direct result of changes in the Common Agricultural Policy, shifting its main objectives from purely directing the market to a position where developments are followed more indirectly. As agricultural and rural development support mechanisms have also been changed, instruments to monitor developments also need to be renewed. A set of recommendations on renewing agricultural statistics were developed in 2004. Part of these initiatives related to restructuring, and part related to simplification.

The developments described are in principle valid for the whole of the UNECE region. However, the regions in Europe outside the EU and the (pre-)acceding countries are still characterized by different degrees of a more traditional set of agricultural statistics. The main differences and changes that can be observed between the different systems relate to the main lines of resolution – geographical, temporal and subject-oriented resolution for agricultural statistics.

For the CIS countries there is clearly a general understanding of the need for further improvement of agricultural statistics on the basis of the international standards. It should be emphasized that agriculture statistics in the CIS countries have undergone significant changes since the inception of the CIS in 1993. The changes in statistics reflect to a considerable extent the transformation of centrally planned economies into market-oriented economies in all economic sectors, especially agriculture. As a result of

[1] The creation of a Common Agricultural Policy was proposed in 1960 by the European Commission. It followed the signing of the Treaty of Rome in 1957, which established the Common Market. The six member states were individually responsible for their own agricultural sectors, in particular with regard to what was produced, maintaining prices for goods and how farming was organized. This intervention posed an obstacle to free trade in goods while the rules continued to differ from state to state, since freedom of trade would interfere with the intervention policies.

economic reforms in agriculture a number of large agricultural enterprises (with state or collective ownership) were liquidated and significant number of small and medium size private enterprises and farms were created in their place. This development resulted in considerable problems with the collection of primary data needed for compilation of agricultural statistics relating to various aspects of economic process in this sector of economy. Under these conditions a system of sample surveys had to be introduced in order to obtain data on the activities of numerous small private farms and personal plots of households which were to supplement the data from the reports submitted to statistical authorities by agricultural organizations (relatively large enterprises with different type of ownership).

An example of this transition in the CIS countries is the transition from the Material Product System (MPS) to the System of National Accounts (SNA). This transition required the introduction of new concepts of output and intermediate consumption in order to ensure compilation of production accounts for agriculture consistent with the SNA requirements. The CIS countries were assisted in this endeavour by CIS-STAT in the context of more general work associated with the implementation of SNA 1993 by the CIS countries. The issue which requires attention in this context is the treatment of work in progress as prescribed in the SNA, treatment of different types of subsidies, and adjustment of figures on seasonality. Using the above concepts in practice required considerable work associated with collection of primary data both on output and input and achieving consistency of the estimates based on the data from different sources. Recall that in the former USSR an essential element of agricultural statistics used for compilation of the most important tables of the MPS was a wide and detailed system of supply and use tables compiled for major groupings of agricultural products (both in physical and value terms) by the major types of agricultural enterprises (state farms, collective farms, personal plots of members of collective farms, personal plots of employees, etc.). Data from this system of tables were used for computation of agricultural output, major elements of input and some items of disposition of agricultural goods (final consumption, increase in stocks, etc.). Unfortunately, during the market economy transition this system of tables was considerably reduced and the structure of tables was simplified. As a result agricultural statisticians face serious problems associated with provision of data for compilation of major SNA accounts in strict compliance with the adopted definitions and classifications. For example, the structure of supply and use tables currently used by many CIS countries does not make it possible to isolate the consumption from own production which is to be valued in basic prices as the SNA requires.

In the rest of this section, the developments in agricultural statistics are described along the lines of the main statistical infrastructures available. The emphasis is on the national statistical institutes as the data providers, or, in the context of North America, on the National Agricultural Statistics Service (NASS) and United States Department if Agriculture (USDA). In many countries, however, in this domain there is a range of other governmental organizations collecting information on farms and the farming industry, for administrative reasons but also for statistical purposes. The situation in the USA is a strong example of such wide involvement. While the NASS collects agricultural data primarily through direct contact with farmers or farm-related businesses, other parts of the US government also provide statistics relevant to American agriculture. The Census Bureau collects international trade statistics for all products and works with the USDA to develop relevant product definitions for food and agriculture. The Agricultural Marketing Service collects market prices for a wide range of agricultural products. The Natural Resource and Conservation Service collects statistics on land use, soil quality and other environmental indicators. The Economic Research Service (ERS) in the USDA has worked with the Census Bureau and the Department of Health and Human Services to add modules to surveys of consumer behaviour and health status to support economic analysis of food security and food choices. ERS has also purchased private data to support research into and analysis of consumer food choices.

1.2.2 Specific agricultural statistics in the UNECE region

The description in this section is restricted mainly to those sources that are managed by the statistical offices. Administrative use of farm data, especially for the bookkeeping of subsidies, premiums, etc., is an important issue in this basic and essential sector for society. These sources are very different in the countries concerned and require a different approach. For reasons of simplicity, only those that are reflected at the regional level are discussed. A more detailed analysis in future should be used to shed light on the still relatively many undeveloped possibilities for integrating these other sources into the compilation of official agricultural statistics. Such an endeavour might fit very well with the aim of making as much use as possible of existing data sources.

Another issue which is difficult to avoid touching on in a discussion on agricultural statistics is the traditional relations with trade and customs statistics – food and agricultural products being very strongly related to agricultural production and for many countries and regions an important source for indirect income via taxes and levies. In this chapter this issue will only receive very brief attention.

Farm register

The availability of an up-to-date register of farms and farm holdings is considered an important feature of a good infrastructure for agricultural statistics in developed countries and is seen as the basis for a coherent system and also, if coordinated with national business registers, a tool contributing to the integration of agricultural information with that of other sectors. The fact that farm registers are often not included in business registers, or are kept separately, poses problems when constructing the sample frames for the surveys. The European Union has experienced technical and coordination problems with updating EU-level farm registers and protection of individual data. Several of the countries in the UNECE region, in relation to the agricultural census, have developed a farm register. A farm register provides a basic tool as a frame for sampling and, provided that appropriate information is included, it may permit effective sample design with stratification by size, type and location. It could also call into question the cost-effectiveness of full agricultural censuses. However, the coverage of a farm register should be carefully analysed, otherwise the costs for keeping it up to date could be too high. A possibility would be to improve household statistics to contain data on subsistence farming, i.e. small farm holdings not producing for the market, but merely or mainly for their own consumption.

Most recent experiences show that the overall support for such a register at EU level – preparing the way for EU sampling and EU surveys – is not yet sufficient. This way of substantially reducing the burden and allowing a linking of sources has until now been possible only in a limited number of countries. The development of farm registers with at least a minimum of common coverage (e.g. containing only the market-oriented farms as described above) could be regarded as an ideal situation and as an effective response to the future. In the EU, it is not (yet) possible to discuss a regulation including farm registers because of specificities in the agricultural statistical systems of the member states. The fact that a common approach to farm registers is not yet possible can be considered a serious problem but also one of the most important challenges for further development, especially given the need for effectiveness and the desire to reduce the burden on farmers. For future strategic planning in this domain, an overview of the countries that have and those that do not have a farm register would be useful. Where a farm register is available its characteristics should be described, and when no farm register is available an indication should be given of alternatives on which the census of agriculture is based.

Farm structure surveys

The farm structure survey (FSS) is considered in UNECE countries to be the backbone of the agricultural statistics system. Together with agricultural censuses, FSSs make it possible to undertake policy and economic analysis at a detailed geographical level. This type of analysis at regular time intervals is considered essential. In the EU, several simplifications have been made in recent years. From 2010 the frequency of FSSs will be reduced from every two to every three years. The decennial agricultural census carried out within the FAO framework will take place in most UNECE countries by 2010. Furthermore, not all the variables are subject to detailed geographical or temporal analysis. This allows the regular FSSs to focus on a set of core variables and to be combined with specific modules with less geographical detail and eventually more subject detail. In the coming years, such a system with a base FSS and a set of specific modules on, for example, use of fertilizers and production methods will be developed. This method is considered to deliver an important contribution to reducing the response burden. In opposition to this development, however, is the increased pressure to add new variables and items to the questionnaire. These new demands stem from the new developments mentioned above – production methods, water usage, etc.

For the EU countries, the design and basis content of the FSS is regulated by European law. For many member states, the survey instrument is an ideal tool to which can be added some country-specific questions. This so-called 'gold plating' is a topic in many of the discussions on the burden of statistics, but also an issue that implicitly generates a more effective use of the survey instrument. In light of this, further extensions of the scope of surveys are in principle not recommended. Furthermore, decision-makers should be informed on the substantial costs of agricultural surveys, especially when no administrative data are available.

In Brazil, the situation is in principle similar but in its implementation is more advanced than in the EU. The integration of the National Address List for Statistical Purposes with the Registers of the Census of Agriculture has allowed the Brazilian Institute of Geography and Statistics (IBGE) to construct the first list frame of productive units in completely computerized form. This list will gather data on all of the country's 5.2 million agricultural producers, with their respective geographical coordinates. On the other hand, the rural area which encompasses all the sectors surveyed by the Census of Agriculture will form the Area Frame including all the information surveyed. Both list and area frame will be able to function as a source for the selection of the agricultural holdings to be researched by agricultural surveys based on probability sampling. These surveys, combined with the current surveys, will make up the National Brazilian Agriculture Statistics System which is presently being developed.

An issue that has recently attracted much discussion in the context of the new EU Farm Structure Survey Regulation and of the preparations for the new regulations on crops and on meat and livestock is the reference to geographic entities. From the descriptions above – and also from knowledge of the Brazilian and US situation – it is clear that there is increased demand for small-area estimates and for data that allow the description of land use and rural development on a small scale. Such detail is also required especially for agri-environmental indicators. Geocoding or the reference to small geographical entities is, however, an issue that is discussed with respect to both confidentiality and increased burden.

The FSS in the EU is in the census years upgraded to cover all farms and holdings. For the EU and most of its neighboring countries this will be held in 2010. For example, in Armenia, Belarus, Moldova, Tajikistan, Turkmenistan, Uzbekistan and Ukraine the agricultural censuses are intended to be carried out in the foreseeable future. In recent years a number of CIS countries have already carried out agricultural censuses: Kyrgyzstan in 2002, Georgia in 2004, Azerbaijan in 2005, Kazakhstan in 2006, and Russia in 2006. As a result of these censuses valuable information on the state and development of agriculture (both nationally and regionally) was obtained. For example, data on a number of agricultural enterprises were updated and can be used for planning and organizing different sample surveys.

Farm typology

In the EU, closely related to the farm structure surveys is the management of the community farm typology. This typology plays an important role in the classification of holdings by economic size and type. It functions as a bridge between the Farm Accounts Data Network and the farm structure surveys. Recently this typology has been updated to better take into account the recent changes in the Common Agricultural Policy to decoupled support.

Farm Accounts Data Network

The Farm Accounts Data Network (FADN) is a specific EU instrument, developed and managed by the Directorate-General for Agriculture. The FADN is an important source for micro-economic data relating to commercial holdings. For purposes of aggregation, the FADN sample results are linked to population results derived from the FSS using groupings based on the community typology. The creation of unique identifiers in the context of the agricultural register would enhance this linkage and, if privacy and confidentiality concerns could be dealt with satisfactorily, would permit more complex analysis, at least if the FADN were a subsample of the FSSs. The current status of confidentiality in both the FSS and the FADN does not, however, allow the combination of these two very rich surveys.

For the USA, a similar situation obtains for the Agricultural Resource Management Survey (ARMS). Policy issues facing agriculture have also become increasingly complex in the USA. In the past 20 years, government support for farmers has changed from being based primarily on supporting market prices to policies that include direct payments to farmers, government support for crop and revenue insurance, payments for environmental practices on working farm lands, and payments for not farming environmentally sensitive land. Increasingly complex agricultural policies require new types of data and at lower geographical scales. For example, land conservation programmes require information about land qualities or services provided as well as the value of alternative uses of land. In the USA, statistics on rental rates have been mandated in the recent Farm Bill for very small geographical areas. In addition, government support for risk management requires information to determine farmers' eligibility for insurance payments. The types of statistics required to support economic analysis of the new suite of farm programmes extend beyond those required for programme management. Farmers participate voluntarily in US government programmes, and data required to analyse the effects of programmes start with information that affects programme participation, including participation in off-farm employment and demographic characteristics. Other information needs include statistics on production decisions, technology choices, and farm financial outcomes. The ARMS provides the main source of farm business and farm finance data for US agriculture and is jointly conducted by the NASS and ERS. ARMS data support indicators of farm sector health such as income and expenditures. Equally important, the micro-data from the ARMS serves as the basis for research on programme outcomes, including informing ongoing international policy debates about the degree to which different types of payments are linked to distortions in agricultural markets.

Area frame sampling

A recent and very important development in agricultural statistics is the use of remote sensing and aerial photographs in combination with in-situ observations. In the UNECE region, these methods for collecting information on rural areas have developed into a strong pillar of agricultural and land use statistics. Similar examples from USDA and IBGE illustrate this. The Land Use and Cover by Area Frame Sampling (LUCAS) survey of the EU is conceived as an area-based sample, designed to provide timely estimates of the area of the principal crops with high precision and a relatively low level of geographical resolution with the advantage of a low response burden. Nevertheless for small or very heterogeneous countries the reduction of response burden by using LUCAS or other EU sample surveys could be smaller than expected, as they might need to be completed in order to have any usefulness at national level. The LUCAS survey has demonstrated its usefulness and versatility. In the recent past, several LUCAS surveys have been carried out and analysed as a set of pilot studies and the next survey, covering all EU member states, will be carried out in 2009.

The continuation of LUCAS is currently under consideration. For its original objective of calculating early estimates of cultivated areas, the LUCAS surveys had to compete with the more structural inventories, perhaps not with more geographical detail but with an expected higher level of accuracy as these are based on farmers' detailed information about their own parcels. Based on the evaluation of the potential use of LUCAS, a wider use of this survey is foreseen, not solely serving agriculture and changes in land use but focusing more on non-agricultural applications, such as environmental (soil, land use and ecosystem) issues. The possibility of combining aerial interpretation with georeferencing and observations on the spot allows the combined analysis of data on agriculture, environment, and more general land use issues. The use of a fixed sample, with observation points stable over more than one survey period, allows the construction of panel data and monitoring of important developments in land use to a high level of statistical significance.

As the EU Common Agricultural Policy has changed over the years, funding has focused more on developing rural areas than on price support for farmers. In order to monitor the changes, rural development statistics are needed. These statistics are related not only to agriculture, but to all entrepreneurial activities in rural areas, as well as other socio-economic issues. However, as there are several methods used to define rurality, the problem has been to decide the regional level at which the data should be collected. The solution chosen so far has been to collect the data at the lowest available regional level, and then to flag these regions/districts as rural, semi-urban or urban, depending on the methodology chosen for the analysis at hand.

An issue deserving special mention in the context of rural development is the work of the Wye Group. This group prepared the handbook on *Rural Households' Livelihood and Well-Being* (United Nations, 2007) which gives an excellent overview of possible statistics on rural households.

The statistics described above are based on surveys on a more ad hoc basis or on regularly collected financial data based on farmers' administrative obligations. However, most of the agricultural statistics are not based on ad hoc surveys but on a very well-established and traditionally organized system of counting on a regular basis the stocks and changes thereto as well as the number of transactions (transport and slaughter) and specific points in the chain from crop and product to food. Statistics on meat and livestock, and on milk and crops, are examples characterized by a high frequency of data collection. Statistics on vineyards and orchards are normally collected less frequently. Most countries have seen an increased use of information technology in the reporting by farmers and from farmers to the statistical institutes. Automated systems support the modern farmer in monitoring the milk yield per cow, the food consumption of the animals, the use of specific extra nutrition, pesticides and fertilizers but also the use of water. EU member states' statistical collection systems and also those of the other countries in the region are increasingly based on the use of internet and related systems to collect this regular information from the farms and holdings. However, this situation is by no means universal. The use of ad hoc surveys is also still considered an important instrument for collecting regular information on the flows in the production cycle. These statistics are now briefly described with reference mainly to the situation in the EU.

Meat, livestock and egg statistics

These traditional animal and poultry product statistics – resulting from traditional regular livestock surveys as well as meat, milk and eggs statistics – still play a key role in the design, implementation and monitoring of the EU Common Agricultural Policy and also contribute to ensuring food and feed safety in the EU. European statistics on animals and animal products are regulated by specific EU legislation. Member states are obliged to send monthly, annual and multi-annual data to the European Commission within pre-defined deadlines. In addition, for several meat products and eggs, the supply balance sheets provide the major type of resources and uses.

The first chronological animal data series in the EU were created for bovine animals in 1959, followed by series for sheep and goats in 1960, monthly meat production in 1964 and pigs and livestock in 1969. The statistical system was progressively improved and enlarged, and now Eurostat receives statistical information from the 27 member states broken down into roughly over 700 individual values per country, some of which are multiplied by 12 for monthly data or by 4 or 2 for quarterly and half-yearly data, respectively.

For these traditional statistics, recent years have witnessed a substantial effort on both the methodology applied by the countries and the improvement of the procedures for data transmission, in particular by using standard formats in the telecommunication net. For example, to achieve this goal, the EU's new eDAMIS[2]/Web Forms application has to considerably improve the data transmission from the member states to Eurostat and has allowed for an improvement not only of work efficiency but also of efficacy for both Eurostat and EU member states. As a result, data quality has improved in parallel with the simplification of data treatment operations.

Milk statistics

Milk statistics relate to milk produced by cows, ewes, goats and buffaloes. For the EU they are concerned with milk collected by dairies (monthly and annually) at national and regional level, milk produced in agricultural holdings (farms), the protein content and the supply balance sheets. Triennial statistics provide information on the structure of the dairies. Data collection and pre-validation are carried out through, for example, the use of the Web Forms system which ensures the management of deadlines and monitors the data traffic.

Statistics on crop production

The traditional statistics on crop production correspond in general to four families of data. First, the Early Estimates for Crop Products (EECP) provide, for cereals and certain other crops, data on area, yield and production before the harvest. Second, the current crop statistics provide at national level, for a given product, the area, yield and production harvested during the crop year. For some products data are requested at regional level. Third, the supply balance sheets give, for a product or group of products, the major type of resources and uses. Finally, the structural data for vineyards and orchards and the inter-annual changes for vines of wine grape varieties give information about age, density and variety of the different species. Crop product statistics cover cereal production, products deriving from field crops, fruits and vegetables, and the supply balance sheets for a large number of crop products. They also include two specialized surveys: one on vineyards with a basic survey every 10 years and an annual one to monitor changes that have occurred; and another one on fruit tree plantations every 5 years.

In the USA, NASS collects and publishes, based on annual or monthly surveys, data on crop production, livestock inventories, livestock products, farm finances, sector demographics, chemical usage, and other key industry information. In contrast to statistical programmes in other countries, government statistical agencies in the USA are focused specifically on data needs relative to the agency, department or cabinet area in which they reside.

Eurostat transmits to EU member states the forecasts (for the currenr year) of the Eurostat Agromet model (obtained by extrapolation of the statistical trend) for area, yield

[2] Electronic Data Files Administration and Management Information System.

and production. From February to October, the member states react to these proposed data and transmit their own new estimates back to Eurostat. The objective is to obtain data on area and production for the main crops before the harvest. The EECP is one of the main inputs used by the Directorate-General for Agriculture for its short-term forecasts and analysis of the agricultural markets on the commodities considered.

Vineyards

Surveys on vineyards are intended to collect information on vines and wine production in the EU member states at different geographic levels (nationally and regionally) and over time. Thus, they provide basic information to enable evaluation of the economy of the sector at production level and, in particular, they permit evaluation and measurement of the impact of the implementation of the common market organization for wine. Member states on whose territory the total area of vines cultivated in the open air is more than 500 hectares have to do a survey on these areas. They have to conduct their surveys within a fixed time-frame and have to ensure high-quality results. The scope of the basic survey is the area under vines, while the survey unit is the agricultural holding. In the case of the intermediate survey, only areas under vines for wine are surveyed.

Survey on plantations of certain species of fruit trees

Basic surveys (apple, pear, peach, apricot, orange, lemon and small-fruited citrus trees) are carried out in the EU every five years, to determine the production potential of plantations from which fruit produced is intended for the market. Data are collected on the areas under fruit trees broken down by region (production zone), species, variety, density (number of trees/ha) and age of the trees.

The chronological crop data series start with data from the early 1960s. The statistical system has been progressively improved and enlarged, and now Eurostat receives and publishes harmonized statistical information from the 27 member states broken down into several thousand individual values per country, some of which are multiplied by several (1 to 8) waves for the different updates taking place every year. As for meat and livestock statistics and animal products, a substantial effort has also been made in this area to improve the methodology applied by the member states and candidate countries, as well as the procedures for data transmission, in particular by using standard formats and electronic data transmission.

Although validation procedures have recently improved, mainly because they were introduced into the data treatment process, there is still considerable room for further improvement, especially in advanced validation.

Agricultural monetary statistics

Collection and validation of these include the economic accounts for agriculture and forestry, the agricultural labour input statistics, and the agricultural price in absolute terms and in indices. The agricultural accounts data at the national level are regulated by legislation which prescribes the methodological concepts and definitions as well as data delivery. The accounts data at regional level and the agricultural price statistics are transmitted on the basis of gentlemen's agreements. Data for agricultural accounts are provided annually, while price statistics are transmitted by the EU member states on a quarterly and annual basis. In the CIS countries, due to a considerable reduction in the number of supply-and-use tables and the simplification of their structure, there are problems with the integration of agricultural statistics with national accounts.

Fisheries statistics

The programme of fisheries statistics in the EU provides statistical information on fisheries needed for the management of the Common Fisheries Policy. The programme comprises the following elements: catch statistics, landing statistics, aquaculture production statistics, supply balance sheets for fisheries products, fishing fleet statistics, employment statistics, socio-economic data, and structural and sustainability indicators. The programme of work is designed primarily to provide statistical support for the management of the Common Fisheries Policy and to meet the EU's commitments to international bodies of which the EU is a contracting party. Apart from meeting numerous ad hoc requests for data from EU institutions, national and international organizations, and public and private organizations and individuals, Eurostat meets routine requests for data from the FAO: fishing fleet statistics, thereby removing the obligation of EU member states, the Northwest Atlantic Fisheries Organisation and other regional international organizations to supply data; catch statistics to meet the EU's obligations as a contracting party of these organizations and the International Council for the Exploration of the Sea (ICES); and catch statistics, under the terms of the Eurostat/ICES Partnership Agreement.

For the EU region Eurostat is planning a redesign of the fisheries statistical database. The new design moves away from a global system of management of fisheries statistics to a system which is more focused on the needs expressed by users and of higher quality. With revised needs and uses for fisheries statistics and fewer resources available, there is a need for a more efficient and higher-quality data environment and a decrease in the workload for data providers. An important consideration for this redesign is a decrease in the overlap of data systems with those of other international organizations, such as the FAO.

Forestry statistics

The Forest Action Programme of the European Communities, adopted in 1989, and more specifically Regulation (EEC) No. 1615/89 establishing a European Forestry Information and Communication System, are the basis for the collection of EU forestry statistics, not only on the present situation of woodlands and their structure and the production and consumption of wood, but also on developments in the afforestation of agricultural land, the forestry situation in the various regions of the Community, and the exploitation, processing and marketing of forest products.

Cooperation between Eurostat, the Directorate-General for Agriculture, UNECE, FAO and the International Tropical Timber Organisation (ITTO) take places through the Inter-Secretariat Working Group on Forest Sector Statistics (IWG), in which the OECD also initially participated.[3] The aim of the IWG is the optimization of the use of scarce resources, so that each piece of information is collected only once from each country and there would be only one entry for each transaction in all the international data sets. Together, the partners created the Joint Forest Sector Questionnaire (JFSQ) and its harmonized definitions in 1999. For each country, the JFSQ produces core data on

[3] Eurostat document 'IWG Mandate Jan 1996' of 20 October 2003.

harvesting of roundwood from the forest (by type of wood), production and overall trade of primary wood products (by quantity and value), and overall trade in secondary processed wood and paper products (by value). These production data are very reliable when available directly from the countries. When not available, they are estimated from (as a minimum) export figures (which is unsatisfactory) or from other sources, such as industrial associations, company news on the Internet (which is very time-consuming). Agreement must be obtained from the country's correspondent to be able to publish the estimates.

As a consequence of enterprises reducing activities in some countries and/or enterprise mergers, some of the production data can no longer be reported due to confidentiality rules. It would be possible to produce different kinds of aggregates if countries could be persuaded to supply the relevant confidential data.

Some countries are experiencing difficulty in obtaining data on wood produced in private forests, so the total for those countries may be considerably underestimated. Another source of underestimation is likely to be the non-reporting of household use of roundwood or the direct sale of wood by forest owners to private households, mainly for heating purposes. It is clear that the reliability of the data produced could be improved by wood balances for each member state and the EU as a whole, as was shown by the recent Joint Wood Energy Enquiry of UNECE.[4] Such data would also be a valuable source of information for energy statistics.

Data on trade has declined in quality ever since the introduction of simplified customs procedures for intra-EU trade in 1993. From then on, data was collected directly from companies, and threshold values-below which nothing has to be reported-were applied. As of 2006, further simplification allows Member States to drop the net mass of a product if a supplementary unit of measurement is reported. Several Member States have chosen to apply this option. As of 2008, only dispatches are foreseen to be reported, doing away with the data on arrivals. The possibilities for cross-checking anything are rapidly diminishing.

Integrated Environmental and Economic Accounting for Forests uses an exhaustive questionnaire. Data have been collected once as a test and a second time in 2007. The proposal is therefore to further simplify the questionnaire and to collect these data every 5 years, which would be adequate for the slow rate of change in forestry. The purely economic data for forestry and logging (output, intermediate consumption, net value added, entrepreneurial income, labour input, etc.) covered in one of the tables could be collected yearly.

Agri-environmental indicators

The requirement to include environmental assessments in all policy areas has led to the collection in the EU of a set of 28 agri-environmental indicators; these have been selected from a group of 75 indicators that are usually collected. Many relate to other environmental statistics already collected, except that they are broken down by the agricultural sector. Some of them relate to specific policy actions and are therefore available from administrative sources, where other indicators have been developed specifically for the

---

[4] http://www.unece.org/trade/timber/docs/stats-sessions/stats-29/english/re-port-conclusions-2007-03.pdf.

purpose. The basic principle in the EU is that already available data should be used wherever possible, and that new data collection should be used only when really necessary. The agri-environmental indicator data collection system is still under construction, partly because some of the indicators are still under development, partly because the required data are not collected and proxies have to be used.

Rural development statistics

These are a relatively new domain and can be seen as a consequence of the reform of the Common Agricultural Policy, which accords great importance to rural development. Eurostat has started collecting indicators for a wide range of subjects – such as demography (migration), economy (human capital), accessibility to services (infrastructure), social well-being – from almost all member states at regional level. Most of the indicators are not of a technical agricultural nature. Data collected cover predominantly rural, significantly rural and predominantly urban areas according to the OECD typology.

The UN Committee of Experts on Environmental-Economic Accounting, in cooperation with the London Group, is preparing the revision of the System of Economic and Environmental Accounting. Many UNECE countries are involved in this process. The objective is to provide a framework that allows the development of indicators for monitoring and directing policy decisions where economy and environment are interlinked. Agricultural and forest accounting are strongly related to this system and very well established. This development of integrated accounting is considered one of the most useful developments towards a single consistent statistical framework for agriculture and the environment. Sustainability of ecosystems is also related to these topics. The developments in this domain, mainly initiated by the European Environmental Agency, are at an early stage of development.

The overview given in this section is far from exhaustive. In general, however, it reflects the main agricultural statistics in the UNECE region. Smaller data collections on endangered species, home farming and ornamental aquaculture – which might be useful in a full description of the situation in a specific country – have not been included.

1.3 Governance and horizontal issues

1.3.1 The governance of agricultural statistics

The success of statistics depends to a large extent on issues of governance. The governance of the statistical work chain traditionally in the UNECE countries covers a well-developed system of agricultural statistics as reflected in the overview above. Data collection and analysis are done via well-established and documented procedures and good cooperation between the national institutes and other governmental organizations. In the EU context an important part of procedures for collecting and providing data is set in stone by the EU regulations. In the neighbouring domains of trade and employment statistics this is also the case. However, in the relatively young domain of environmental statistics and ecosystems these statistical procedures are far from being well established. The agencies involved are still developing their methods and the availability and accessibility of many data sources is still not appropriately documented.

In the USA there is a great variety of organizations involved in the collection and analysis of data on agricultural and related issues. This holds also true for Europe. In

2005 the European Environmental Agency, the Joint Research Centre, Eurostat and the Directorate-General for Environment agreed to work together on the development of data centres on environmental and related issues. Ten such data centres are planned (e.g. on land use, forests, and water). The objective of each data centre is to function as a portal for all the available information in that specific field and to create a virtual knowledge centre.

The recent EU communication on the Shared Environmental Information System (SEIS) even goes a step further, at both European and member state level, in promoting the enhancement of the exchange of available data sources, avoidance of overlapping data collection methods and the use of administrative sources and non-official statistics to supplement official statistics. This development is considered an important way forward in an efficient use of all the available information and will be an important asset in the work on combing agricultural statistics with several domains of environmental statistics.

International coordination

The number of international and supranational organizations involved in agricultural statistics is rather limited. The FAO and Eurostat are the main international organizations involved. The OECD and UNECE were more involved, especially via the Inter-Secretariat Working Group on Agriculture. However, the activities of these organizations are currently limited and there is presently no forum to discuss issues concerned with agricultural statistics at the UNECE level (except for forestry statistics).

In the early 1990s cooperation among states led to a set of so-called joint questionnaires which facilitate the efficient coordination of data collection in some areas. Changing demands of course require a regular updating of these joint questionnaires. This regular updating, however, is not easy to organize, even if only because of natural divergences in the use of the data by the organizations as the years pass.

The FAO regularly collects information on agriculture that can be compared to the information Eurostat collects for the EU member states. In coordination with the FAO, Eurostat tries to limit the burden for the member states as much as possible – one way of doing this is to avoid asking questions that overlap with those asked by the FAO. It can be concluded that cooperation, especially on the traditional agricultural statistics, needs to be improved. At a recent EU–FAO meeting (in Brussels in December 2008) the need for closer cooperation was emphasized. For fisheries and forestry, relations are considered to be good.

Compared to other fields of statistics, the international global cooperation in agricultural statistics has not resulted in many overarching groups such as city groups. The only relevant city group is the Wye Group as mentioned earlier. Perhaps more important for the functioning of the global governance structure for agricultural statistics is the network around the International Conferences on Agricultural Statistics (ICAS) meeting and the network of regional conferences initiated by the FAO.

1.3.2 Horizontal issues in the methodology of agricultural statistics

From the description in Section 1.2, a distinction can be made between the regular inventories on products and stocks, the ad hoc surveys and the special data collections by more advanced techniques such as remote sensing, etc. For the regular data collections, well-established systems have been developed and these do not need to be discussed. Several specific problems, however, occur in the field of agricultural surveys. These problems are to an important extent not typical of agricultural statistics but also characterize data collection experiences in social and business statistics. These problems are summarized below. The examples below are mainly taken from IBGE and NASS/USDA, but are also valid for the rest of the countries.

Respondent reluctance: privacy and burden concerns

Although the agricultural sector is somewhat unique and not directly aligned with the general population on a number of levels, concerns regarding personal security and privacy of information are similar across most population subgroups in the USA, Brazil, and Europe. Due to incidences of personal information being released by businesses and government agencies, respondents now have one more reason for not responding to surveys. While this is not the only reason for increasing non-response levels on surveys, it represents a huge challenge for future data collection efforts. The strong protection afforded to respondents by law is sometimes not enough, particularly considered alongside the other challenge faced by statistics of reducing respondent burden. With trends showing that fewer farms increasingly represent a larger proportion of UNECE agricultural production, respondents are being contacted multiple times within a sampling year.

The NASS, in an effort to mitigate the reluctance of respondents, employs a variety of practices intended not only to encourage response to a specific survey, but also to demonstrate the value of good data. These strategies include personal contact by interviewers familiar with the respondents, small monetary incentives, a sampling methodology that factors burden into the probability of selection, flexible use of multiple data collection modes, and public relations efforts demonstrating the uses of data. Over the past few years, the NASS has directed resources specifically towards increasing response rates on the agency's two largest projects, the Census of Agriculture and the ARMS. Although resulting in some short-term increases in response, the efforts have not shown an overall increase in survey response rates or their concern for non-response bias. The NASS is putting extra effort into better understanding the characteristics of non-respondents so as to be able to describe them, make appropriate data adjustments, and better understand the potential magnitude of bias introduced. One could also surmise that changes in response rates are directly tied to the changing face of agriculture.

In the EU, the experiences with the FSS are similar, with member states reporting increasing problems with response rates. This is the most cited reason for not wanting to increase the number of variables to be collected. Some countries have addressed the problem by making response a legal obligation, but most have decided that this is an inappropriate solution. Perhaps the most common approach is to try to make use of administrative sources, making the respondents aware that data are collected only where it is really necessary. Other countries have reformed collection methods, combining, for example, computer-aided telephone interviews with prefilled questionnaires sent in advance. Obviously, there is no unique solution available; problems like these must be solved based on the cultural situation for each respondent group, with, for example, bigger enterprises being treated in a different way than the small part-time farmer.

The increased demand from users for FSS micro-data also creates a problem in the EU. Confidentiality rules do not allow these data to be disseminated, even within the group of restricted users under the EU confidentiality regulations. The process of making FSS micro-data available to researchers has not yet been approved by the member states, and as some have indicated that they will use their right of veto to protect their data, this clearly hinders the further increase in the use of such data for advanced analysis and thus also their credibility.

Small and diversified farm operations

For agricultural statistics in general, and for the EU and the NASS survey programmes in particular, the coverage (for the different crops such as acres of corn or the number of cattle represented by the farms on the frame) is a very important issue. In the regulations used for EU statistics, the desired accuracy and coverage are described in detail. Furthermore, countries are requested to provide detailed metadata and quality information.

In general, active records eligible for survey samples account for 80–95% of total US production for most major items. Medium to large size operations are typically sampled at higher rates as they represent a greater proportion of production being measured. This is adequate for the survey programme where the major focus is agricultural totals at the federal and state levels. For the census programme, the focus is county-level data on farm numbers by type and size, demographics, acreage, production, inventory, sales, labour, and other agricultural census items. Consequently, adequate coverage of all types and sizes of farms is needed to ensure reliable census results.

Even though the NASS publishes coverage-adjusted census data, a specific issue for the USA is the need for adequate list frame coverage for all types and sizes of farms to ensure reliable county-level data for all census items. Although coverage goals are established to generate increased agency attention to list-building needs, coverage of the total number of farms has been decreasing over the last few censuses. These decreases are due primarily to the increasing number of small farms which are difficult to locate through traditional list-building approaches. Also, they are difficult to properly maintain on the list frame due to their borderline 'farming/not farming' status. Small farms routinely enter and exit at a faster pace than larger, more commercial size farms. To keep coverage high for farm numbers, the NASS must keep rebuilding its lists. Additionally, before conducting the 2007 Census of Agriculture, the NASS recognized the extensive interest in minority farm numbers and specialty commodity farms and attempted to improve the reliability of these data through extensive list-building efforts.

Estimates for small domains and areas

An issue already reflected on earlier in this chapter is the increasing demand for data for small domains. In agriculture, these small domains could be geographical areas or unique commodities. Legislators are more frequently seeking data at lower levels of aggregation. In order for survey-based estimates to be reliable, the sample sizes would be required to increase beyond the organization's capacity to pay. The NASS's approach has been to augment probability-based survey estimates with non-probability-based survey data. Much effort is put into investigating statistical methods for small-area estimation that use models borrowing strength from other data sources such as administrative data or other areas. This procedure allows estimates that can have a proven measure of error.

Uses of data for unintended purposes

For many years, the NASS has estimated crop and livestock production at the federal, state, and in some instances county level. NASS stakeholders have utilized published estimates for marketing and production decisions, agricultural research, legislative and policy decisions, and implementation of farm programmes. Data needs have evolved over the past several years, resulting in uses of NASS information to establish USDA farm programme payment levels and calculate the USDA Risk Management Agency (RMA) insurance indemnity payments to farmers.

The RMA has provided group risk insurance products, Group Risk Income Protection (GRIP) and Group Risk Plan (GRP), to farmers for a number of years. These policies were designed as risk management tools to insure against widespread loss of production of the insured crop in a county. NASS county yields for insured crops are currently used in determination of payments to farmers. The NASS county estimates were not originally designed for such use. The estimates for a 'small area' (such as a county) are often not as precise as one would desire as the basis for insurance policies. However, the NASS estimates are the only source of data at the county level available to the RMA.

Programme content and stakeholder input

National statistical offices work very hard to understand the needs of the data user community, although the future cannot always be anticipated. As the primary statistical agency for the USDA, the NASS services the data needs of many agencies inside and outside the Department. Partnerships have been in place with state departments of agriculture and land-grant universities through cooperative agreements since 1917 to ensure statistical services meet federal, state, and local needs without duplication of effort. This coordination maximizes benefits while minimizing respondent burden and costs to the taxpayer. The NASS also considers the thousands of voluntary data suppliers as partners in the important task of monitoring the nation's agricultural output, facilitating orderly and efficient markets, and measuring the economic health of those in agriculture.

The NASS uses numerous forums to obtain programme content and customer service feedback. For many years, NASS has sponsored data user meetings which are a primary source of customer input that keeps the NASS agricultural statistics programme on track with the needs of the user community. Data user responses have played a vital role in shaping the agency's annual and long-range planning activities.

For the EU, the Standing Committee on Agricultural Statistics (CPSA), along with several other committees, functions as the sounding board for initiatives in the field of agricultural statistics. Most of the initiatives come from coordination meetings at expert level, often generated by policy debates in the EU Council and Parliament.

Funding for agricultural statistics

Agricultural statistics and especially the farm structure surveys are an expensive method of data collection. In the EU, the European Commission co-finances the data collection work of the FSSs and also finances the LUCAS survey. For the 2010–2013 round of the FSSs, the European Commission has reserved a budget of around €100 million. However, an important part of the work has to be funded by the countries themselves.

The funding situation for the NASS as a national statistical institute responsible for agricultural statistics is different. As the need for data continues to grow, so does the NASS budget. From its inception as an agency in 1961, the NASS appropriated budget has grown from under \$10 million annually to its current level of about \$140 million. In addition to appropriated funding, NASS receives approximately \$15-\$20 million annually through reimbursable work for other federal agencies, state governments, and agricultural commodity groups. NASS funding level increases have come about primarily due to a corresponding increase in workload. However, the NASS continues to find ways to become more efficient and currently employs fewer personnel than it did in its early years as an agency.

Legal procedures

An issue specific to the situation in the EU is the procedure for the agreement of the EU Council and Parliament on new regulations. The organization of this process is complex and time-consuming; however, in the context of the necessary legal basis for statistics, it is very necessary. The preparations begin with task forces and working groups in the member states before progressing to the level of the CPSA or the Statistical Programming Committee who then agree to submit the proposal for discussion with the other services of the Commission and then the Council and Parliament.

The way the regulations are organized ensures that the requirements for the statistics to be collected and delivered by the member states are described in detail. Changing or adding to these requirements, or actively integrating new developments into the data collection process, is therefore almost impossible. This means that the instruments are well developed but somewhat inflexible. It also allows member states, via so-called 'gold plating', to take the initiative of adding questions or variables to the questionnaires or making changes to existing ones for their own use.

1.4 Development in the demand for agricultural statistics

Agricultural statistics (including fisheries and forestry) have a long history. The subject has an extensive literature. A major reason for the review on which this chapter is based is the recognition of the need for a reorientation of agricultural statistics in order to integrate them into the wider system of statistics. New demands on environmental impact and ownership of rural areas, water and energy use, etc. have been signalled and need to be included. Recent conferences have concluded that globalization and issues such as climate change demand a different approach to statistics, given the important role of agriculture in the global economy, the sustainability of the global economy and modern society more generally, and this clearly includes agricultural statistics. More information is needed on the demand side and the non-food use of agricultural products. Furthermore, these conferences have concluded that, especially in developing countries, the capacity to produce agricultural statistics has decreased. The main developments are described below.

Small-area statistics

There is increasing demand for information on small areas, and the interplay between rural and agricultural issues as well as issues of territorial cohesion has become important in many countries. Coastal areas, small island economies, urban and rural areas all require a specific set of indicators that reflect the integration/cohesion and development of these areas. There is an increasing demand for indicators for these types of areas. The need for spatial information combined with socio-economic information and environmental data is, for example, expressed in several communications from the European Commission. Agricultural statistics will be pushed to deliver small-area information. Surveys based on samples are an important instrument (with a sufficient coverage, of course). Multi-purpose surveys with a georeference are seen as important sources for data to be complemented with spatial agricultural information. Next to this approach, aggregated municipal or regional information is also considered important information on this level.

Integrated economic and environmental accounting

At the aggregated level, sound indicators that give a good insight into the mechanism of agricultural society in relation to the economy and environment are needed. The integration of agricultural statistics with other statistics is a process that is tackled especially from the viewpoint of integrated economic and environmental accounting. The UNECE region is actively participating in preparations for the revision of the System of National Accounts (dating from 2008) where the relevance of satellite accounts is emphasized. The related revision of the System of Environmental and Economic Accounting is on track with the work of the London Group. Agriculture is well represented in the discussions. The process of building these integrated systems, the extraction of valid sets of indicators and the adjustment of the basic statistics to these more systematic approaches remains a medium-term project.

Farm register and area frame

Farm structure surveys are the backbone of agricultural statistics, delivering micro-information that allows the detailed analysis of mechanisms on individual farmers' and farms' behaviour. The response burden described in Section 1.3 forces investment in the use of more efficient instruments for collecting data. Linking sources is a way forward in combination with a permanently updated farm register and area frame. Such frames facilitate sampling, but in themselves can already supply a lot of basic information. In many countries these farm registers have been built or are under development.

New data collection tools

Modern technologies for data collection for agricultural and land use statistics are being implemented in many countries. As in many surveys, the use of data collection via computer-assisted personal or telephone interviewing has become the rule rather than the exception. The NASS and many EU countries have used the Blaise software for such interviewing for many years. A more recent development is the use of Internet questionnaires mainly for annual and monthly inventories. Both NASS/USDA and IBGE have accumulated considerable experience in using modern technologies in data collecting. For IBGE, there is the experience in electronic collection of the 2007 Census of Agriculture, integrated with the population count and with the construction of a National Address List for Statistical Purposes. This operation covered the entire 8.5 million square kilometres of the national territory, collecting information from 5.2 million agricultural establishments, in 5564 municipalities, and from 110 million persons, in 28 million households located in 5435 municipalities. In the censuses, the integration of these surveys was facilitated by the use of a hand-held computer, the personal digital assistant (PDA), equipped with GPS, in the stage of field operation.

The use of this technology enabled the construction of a more consistent rural address list. For the first time, Brazil conducted an operation of this magnitude using only digital collectors (PDAs), which allowed better control of the quality of data obtained in the fieldwork, both at the stage of collection and in the supervision by the central bureau. This operation required the used of 82 000 PDAs with GPS and the participation of 90 000 persons. The electronic transmission of the data directly from the PDA of the census takers to the central computer of the IBGE reduced the data processing time and contributed a significant saving of resources, since it eliminated the stages of transportation, storage and digitization of the data, essential when paper questionnaires are used.

The use of PDAs in combination with other remote sensing tools constituted a unique combination for data collecting. The PDAs were equipped with GPS and helped to associate the collected agricultural data with the geographic coordinates of the 5.2 million rural units visited. Each agricultural holding could be visualized by means of Google Earth images, combined with the grid of rural census sectors. This procedure allowed the IBGE to monitor the evolution of the entire data collection operation more closely.

Georeferencing

Information on the positioning (georeferencing) of agricultural holdings creates new possibilities for dissemination of information from the Census of Agriculture, such as the publication of agriculture maps, with the description of the process of occupation of the national territory, according to the diverse products, agricultural techniques, areas of forest reserves and biomes, hydrographic basins, Indian lands, and several other instances of georeferenced information. For the future design of LUCAS in the EU, the design used by IBGE is an important precedent. The NASS has a long history of using geographic information system (GIS) techniques to assist in fulfilling its mission. In the NASS approach, some recent developments are described in great detail. It is evident that the methods described above are an important addition to the development of good statistics on land use. The need for detailed spatial information requires the use of these types of new tools.

Small-area estimates

An important development is the need for up-to-date and accurate small-area estimates. The demand for early estimates for advance warning on crops, and for results for small domains, continues to increase. In agriculture these small domains could be geographical areas or unique commodities. Statistical methods are being used for small-area estimation that use models and modelling techniques borrowing strength from other data sources such as administrative data or other areas.

The overview of recent developments is not complete without mention of the permanent need to update the existing list of products, goods, etc.: crops for bio-fuels and genetically modified products, organic production methods, etc.

1.5 Conclusions

From the foregoing description there are clear priorities for developments in agricultural statistics. The developments of course are in the substance of the individual statistics.

They are described in the paragraphs above and will not be repeated here. However, we will focus here on some of the more horizontal issues of governance and dissemination of information.

In general, cooperation between countries and international organizations requires greater priority so that maximum use can be made of the global statistical infrastructure in order to improve agricultural statistics. In the international cooperation an intensified cooperation in Joint Questionnaires but also in the sharing of experiences has been relatively weak in the last decade. With respect to the increased need for high-quality agricultural statistics, stronger cooperation and leadership are needed, as indeed they are in relation to the need to link with other areas of statistics.

With respect to governance at the national level, close cooperation with the main stakeholders active in agricultural statistics at the national level is essential. Many governmental organizations are involved in the collection and use of agricultural statistics. In this schema national statistical institutes play an important role as data providers, but also a reference for data quality issues. For efficient use of available information, good coordination is needed, both at the level of data collection and analysis and in describing the need for statistics and the feasibility of collecting certain types of data.

Agricultural statistics can still be characterized as a rather traditional sector in statistics, and it has only recently been recognized that linkages with other fields such as the environment and socio-economic issues are relevant. The agricultural statistical system is somewhat inflexible. This is partly due to the way the system is constructed (many regulations) but can also be related to the relatively low priority given in recent years to modernization. Recent developments clearly indicate a need to liaise closely with environmental and spatial statistics and, in the context of rural development strategies, a stronger interrelation with social and other economic statistics.

The access to micro-data for researchers is an important development in increasing the value and credibility of agricultural statistics. Solutions for the issue of confidentiality have to found both in information technology and in legal structures.

Acknowledgements

This chapter is based on the in-depth review of agricultural statistics in the UNECE region prepared for the Conference on European Statistics (CES). In its third meeting of 2007/2008, the CES Bureau decided on an in-depth review of this topic. It was requested that the review took into account recent developments such as the increase in food prices and the impact of climate change, and incorporated the final conclusions and recommendations reached at the fourth International Conference of Agricultural Statistics (ICAS IV, Beijing, November 2007) on the situation of agricultural statistics. The in-depth review was discussed in the October 2008 CES Bureau meeting in Washington, DC, again discussed and approved at its February 2009 meeting and finally presented to the CES Plenary meeting in June 2009 in Geneva. The review was also updated following the written January 2009 consultation of the member countries. In general the countries were very positive about the review and comments that referred to issues not yet covered have been included in the final version. The review also takes into account the results of the Expert Meeting on Agricultural Statistics held in Washington on 22–23 October 2008. The CES recognized the timeliness of this review because of the food crisis, increases in food prices, climate change, etc., and stressed that the current crisis might help to raise the profile and emphasize the importance of agricultural statistics. The in-depth review was based on preparatory contributions from Eurostat, the IBGE and the NASS. To compile the review, use was made of a variety of information sources on the current state and future challenges of agricultural statistics, an important input being the policy reviews on agricultural statistics issues of Eurostat, the Wye Group Handbook (United Nations, 2007) and the results of the 26th CEIES seminar 'European Agricultural Statistics – Europe First or Europe Only' (Brussels, September 2004). Furthermore, the review benefited greatly from the input from the CES Bureau members, input from experts from the UNECE member states and especially from CIS countries. This chapter was written with contributions from the United States Department of Agriculture and the National Agricultural Statistics Service (Mary Bohman and Norman Bennet), the Brazilian Institute of Geography and Statistics (Eduardo Nunes Pereira) and the Interstate Statistical Committee of the Commonwealth of Independent States (Michael Korolev). However, the final responsibility for this chapter rests with the author.

Reference

United Nations (2007) *Rural Households Livelihood and Well-being: Statistics on Rural Development and Agriculture Household Income*. New York: United Nations.

Part I

CENSUS, FRAMES, REGISTERS AND ADMINISTRATIVE DATA

2

Using administrative registers for agricultural statistics

Anders Wallgren and Britt Wallgren
Örebro University and BA Statistiksystem AB, Vintrosa, Sweden

2.1 Introduction

The official statistics of the Nordic countries are today based upon a comprehensive use of registers that build on administrative data; many other countries are also increasing their use of administrative sources. Using data from several nation-wide administrative systems, a national statistical agency can produce a number of statistical registers. These statistical registers are used on the one hand as frames for sample surveys or censuses and to complement survey data with register information, and on the other hand for pure register-based statistics. Data from the tax authorities, for example, can be used to produce demographic statistics, turnover statistics and payroll statistics.

Within agricultural statistics, too, administrative registers should be used to create registers for statistical purposes. A system for official statistics on agriculture can be based on the following sources:

(a) agricultural sample surveys based on register-based sampling or area sampling;

(b) censuses of agricultural enterprises that can be the basis of a farm register;

(c) administrative sources about farms, such as the Integrated Administrative and Control System (IACS) and Cattle Database (CDB) registers;

(d) other administrative registers that can be linked to the units in the farm register.

Agricultural Survey Methods Edited by Roberto Benedetti, Marco Bee, Giuseppe Espa and Federica Piersimoni

Sources (a) and (b) are the traditional ones, but for the farm structure survey many countries also use data from sources of type (c). In the future, we expect that more extensive use of administrative data from many integrated registers (source (d)) will be established practice.

2.2 Registers, register systems and methodological issues

A *register* is a complete list of the objects belonging to a defined object set. The objects in the register are identified by *identification variables*. This makes it possible to update or match the register against other sources.

A *system of statistical registers* consists of a number of registers that can be linked to each other. To make this *exact linkage* between records in different registers possible, the registers in the system must contain reference numbers or other identification variables. The definitions of the objects and variables in the system must be harmonized so that data from different registers can be used together. Reference times must also be consistent. Microdata from different sources can then be integrated to form a new register. Figure 2.1 illustrates how the system of statistical registers is created and used for statistical purposes.

To design a *register-based survey* is a different statistical task than designing a *sample survey*. In the case of sample surveys the first step is to determine the population and which parameters are to be estimated for which domains of study. This in turn determines the character of the survey with regard to sampling design and estimation. Thus the definition of population and parameters comes first, then data collection. As a rule one survey at a time is considered, with a limited number of parameters.

In the case of a register-based survey a different approach is taken, since data have already been collected and are available in different administrative registers that are not tailored for a particular statistical application. With the aid of available registers a selection is made of objects and variables that are relevant to the issue addressed by the register-based survey. It may be that, on the basis of available registers, new

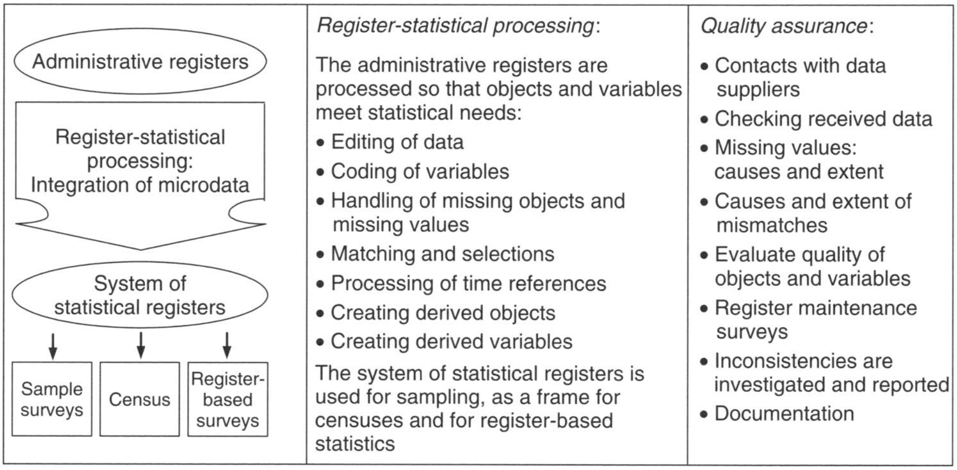

Figure 2.1 From administrative registers to a system of statistical registers.

variables – and possibly new objects as well – have to be derived. Thus, the data come first, and then the determination of population, parameters and domains of study. Sample errors do not restrict the possibilities of selecting domains of study for the coming analysis and reporting of results.

In designing statistical registers and register systems there is a desire to make them flexible and thus widely applicable. Therefore, an extremely important part of the register-statistical methodological work consists of structuring and improving the whole, that is, finding the best design for the register system. Included in this is the long-term work of monitoring and influencing the access to administrative data for statistical purposes.

So even though much of the statistical methodology is the same for sample survey statistics and register-based statistics – for example, all non-sampling errors and all problems of analysis and presentation – the ways of thinking are different, since sampling errors and the design problems arising from them are so central to sample survey statistics. In register-based statistics the system approach is fundamental: to improve the quality you cannot look at one register at a time, you have to consider the system as a whole and pay special attention to identification variables used for linking purposes.

2.3 Using registers for agricultural statistics

Making greater use of administrative sources is one way to reduce costs and response burden. Also, by making more efficient and flexible use of existing data, new demands can be met. Integrating different administrative sources and data from sample surveys and censuses will also create new opportunities to improve quality factors such as coverage, consistence and coherence. How should the statistical system be designed so as to allow efficient use of these administrative sources? There are important distinctions between three different ways of using administrative data, in this case for agricultural statistics.

2.3.1 One source

One way is to use one administrative source almost as it is, for example IACS data to obtain aggregated crop area statistics. For the reliable crop categories in IACS, this simple way of using data will give aggregated statistics of good quality. The objects in the administrative register need not be of high statistical quality as it is not necessary to link these objects with other objects at the microdata level. The identification variables are now not important and can therefore be of low quality.

This way of thinking focuses on one specific source at a time – how can this source alone be used for agricultural statistics? This way of using administrative data often gives rise to undercoverage errors that can escape notice if the source is not compared with data from other sources.

Example 1. During the years when farmers got area-based subsidies for some crops but not for others, the arable land could be divided into land used for crops for which the IACS alone gave reliable estimates, and crops for which the IACS alone did not give reliable estimates. In Table 2.1 IACS data and data from the census-based Swedish Farm Register are compared. For the unreliable crop areas the bias is substantial, about 20%, but for crops with area-based subsidies the quality of the IACS data is high.

Table 2.1 Crop areas in the IACS and the Farm Register (FR), in thousands of hectares.

	Reliable areas			Unreliable areas		
Year	IACS	FR	IACS/FR	IACS	FR	IACS/FR
1995	1641	1634	1.004	912	1133	0.805
1996	1706	1710	0.998	888	1102	0.806
1997	1705	1715	0.994	919	1083	0.848

2.3.2 Use in a farm register system

The second way of using the IACS register is as one source when creating a new kind of farm register, which can be used in conjunction with other administrative sources such as the CDB. The objects in this register must be of high quality at the microdata level. They must be well defined and the identification variables must be of good quality. This farm register can then be used as a sampling frame and also as a frame for an agricultural census. Figure 2.2 illustrates this kind of system.

To achieve a high-quality farm register it may be necessary to do *register maintenance surveys* where questionnaires are sent to units with uncertain activity.

This way of thinking focuses on the possibilities for agricultural statistics only – which sources can be used and combined to produce traditional agricultural statistics? If this system is mainly based on *one* source (e.g. the IACS register), then the coverage errors and changes of the IACS system will result in statistics of low quality.

2.3.3 Use in a system for agricultural statistics linked with the business register

The third approach is to consider the entire system of surveys at the statistical agency. Figure 2.3 shows the possibilities the statistical offices in the Nordic countries have to combine data in different statistical registers.

How does the system of agricultural registers and surveys fit and interplay with other parts of the system? Are agricultural statistics consistent and coherent with national accounts statistics, labour market statistics and regional statistics? If the entire system is well coordinated and designed, then those who work with agricultural statistics can use all relevant sources in the system, those who work with other parts of the system can use agricultural data for their purposes, and the statistical quality of these integrated data sets will be good.

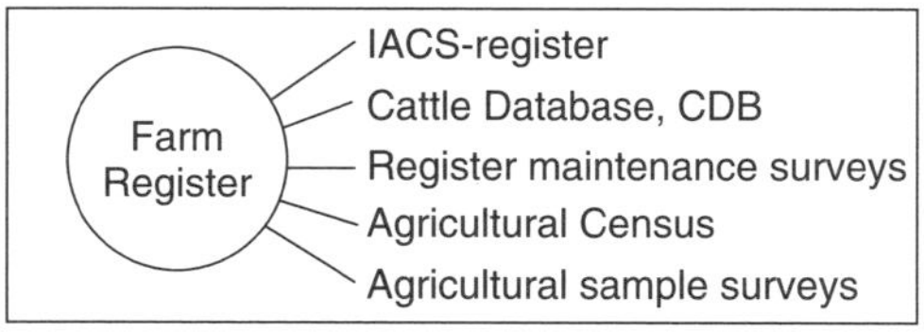

Figure 2.2 An agricultural system based on the farm register.

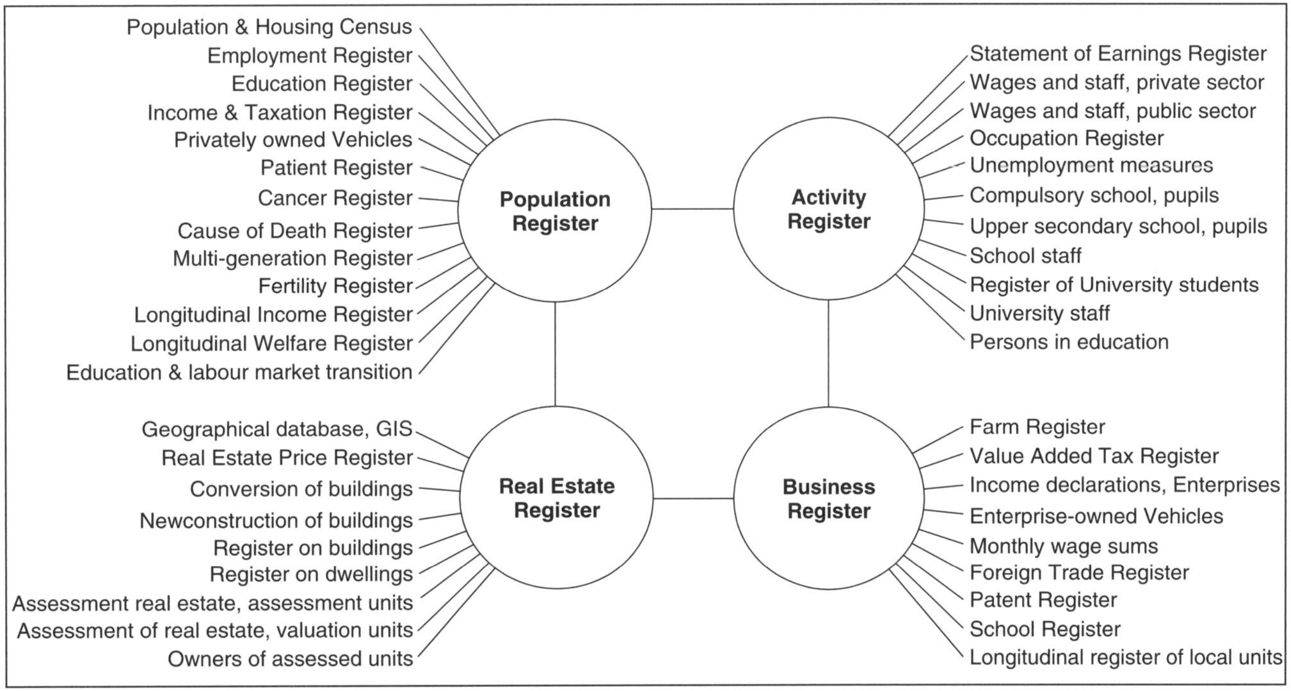

Figure 2.3 The entire system of statistical registers by object type and subject field. *Source*: Wallgren and Wallgren (2007, p. 30).

Systems of registers and register-statistical methods are discussed in Wallgren and Wallgren (2007). A statistical register system is made up of the following parts:

1. *Base registers*: Statistical registers of *objects/statistical units* of fundamental importance for the system. The four circles in Figure 2.3 are the base registers.
2. *Other statistical registers*: Registers with *statistical variables*. In Figure 2.3 there are 42 other statistical registers with a large number of statistical variables describing the populations in the base registers.
3. *Linkages* between the objects/statistical units in different base registers and between base registers and statistical registers. The lines in Figure 2.3 are these linkages.
4. *Standardized variables*: Classifications of fundamental importance for the system.
5. *Metadata*: Definitions of objects, object sets and statistical variables, information about quality and comparability over time should be easily accessible.
6. *Register-statistical methods* including routines for quality assurance.
7. Routines for the protection of integrity.

This third way of creating a system for agricultural statistics aims to integrate the farm register with the business register and thus with the entire system. The farm register here is not only based on IACS and CDB data but also on data from the Business register coming from the tax authorities. This new kind of farm register can then be used in conjunction with many other administrative sources. The objects/units in this register must also be of high quality at the microdata level and the identification variables must be of good quality. This register can also be used as a sampling frame but can also be linked to objects in other statistical registers to produce data sets with many interesting variables describing well-defined populations.

If the aim is to study 'wide agriculture', IACS variables must be combined with other economic, demographic or social variables in other registers to describe farming families or rural enterprises. If it is possible to link objects in this register over time it can also be used for longitudinal studies.

Example 2. In Figure 2.4 a number of registers in the entire system have been selected. All variables in these statistical registers, which are linked to the business register, can be used to analyse the agricultural sector. If the holdings in the farm register can be linked to local units (establishments) in the business register, then it is possible to combine data in the farm register with economic data describing the enterprises connected with these local units. Yearly income declarations including balance sheets and profit and loss statements for the agricultural sector can then be combined with IACS data, and value-added tax (VAT) data can be used in the same way. The vehicle register contains register data about vehicles owned by firms and persons that can be linked to holders or enterprises. This means that all these registers contain data concerning the agricultural sector.

When we use administrative registers for statistical purposes, we become dependent on changes in the administrative system that generates the data. For example, the IACS system has been used both for updating the population in the Farm Register and for crop variables. If the IACS system is changed in the future, these two ways of using IACS

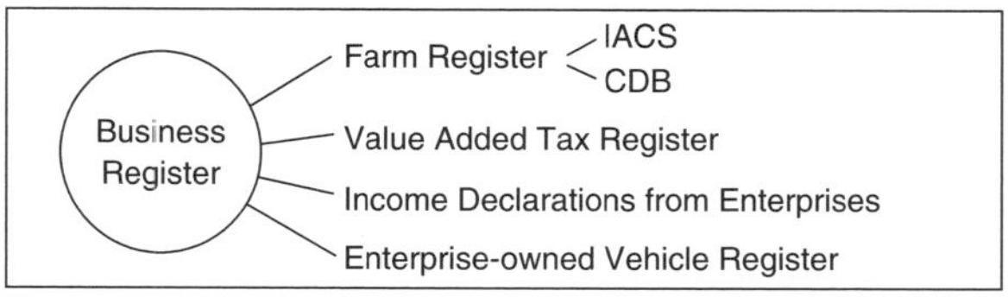

Figure 2.4 An Agricultural Register System linked to the Business Register.

data may become impossible. But if the Farm Register is well harmonized with the entire register system, other administrative registers regarding enterprises together with register maintenance surveys can be used to update the population in the Farm Register via the Business Register instead of using the IACS register.

This third way of thinking requires new statistical methods to be developed, and new quality issues such as coverage, consistence and coherence to be dealt with in a systematic way. Integration of many sets of data at the micro level will require new statistical methods, but the payoff in the way of improved quality and new possibilities will be great. In our opinion, this will be the future of advanced official statistics.

The Business Register and Farm Register at Statistics Sweden are not harmonized today. The populations of agricultural enterprises and establishments differ, the objects/statistical units differ and the activity codes in the Business Register are often of low quality for agricultural enterprises. Selander (2008) gives a description of the differences between these registers. Coverage and activity classification in the Business Register will, however, be improved in the years to come. In this way the agricultural part of the Business Register will correspond better to the population in the Farm Register.

Even if it is not possible to achieve perfect correspondence due to different definitions, the objects in the two registers should be linked to each other, if necessary with many-to-one linkages. The Farm Register and the agricultural part of the Real Estate Register are also not harmonized. This depends on the fact that ownership of agricultural real estate is not necessarily related with agricultural production. These two causes of lack of harmonization mean that the Swedish system of statistical registers at present is not used for agricultural statistics in the same way as it is used for social and economic statistics.

There are also different international/European groups that decide on guidelines and regulations for the farm structure survey, the structural business statistics survey, the statistical farm register, the business register and the national accounts. All these surveys and statistical products are used for the production of agricultural statistics, and if these rules and regulations are not harmonized it will be difficult to achieve consistency and coherence in the entire statistical system.

If we want to use administrative data for statistical purposes in the most efficient way, important changes from the present situation are necessary:

- *New attitudes*: From the present one-survey-at-a-time thinking, or one-sector-at-a-time thinking, we must develop our ability to think about the entire statistical system.
- *New statistical methods and terms*: Traditional sample survey theory and methods are based on one-survey-at-a-time thinking, and the statistical theory is based on probability and inference theory. For the work with administrative data we will

also need terms and statistical theory regarding registers and systems of registers. The survey errors we discuss today are the errors that arise in surveys with their own data collection. For register-based surveys we will also need quality concepts that describe the errors that arise when we use and integrate different administrative sources.

In the following sections we will outline the methodological issues that will be important when we develop agricultural registers that fit in a harmonized system of statistical registers. In a harmonized system both register data and sample survey data can be combined to produce statistics that are consistent and coherent and use administrative data in the most efficient way.

2.4 Creating a farm register: the population

When a statistical register is created, *all* relevant sources should be used so that the coverage will be as good as possible. When microdata from different sources are integrated many quality issues become clear that otherwise would have escaped notice. However, a common way of working is to use only one administrative source at a time. Figure 2.5 shows the consequences of this. The Swedish Business Register has been based on only one source, the Business Register of the Swedish Tax Board. Between 1996 and 1997 the rules for VAT registration were changed. Earlier enterprises with a turnover smaller than 200 000 SEK were not registered for VAT, but from 1997 these are included in the Tax Board's register. Quality improvements can explain the change between 2005 and 2006; those who work with the Business Register now also use typology data from the Farm Register. The definition of NACE 01 in the Business Register contains more activities than the definition used for the population in the Farm Register, which can explain the difference during 2007; if detailed NACE is used this

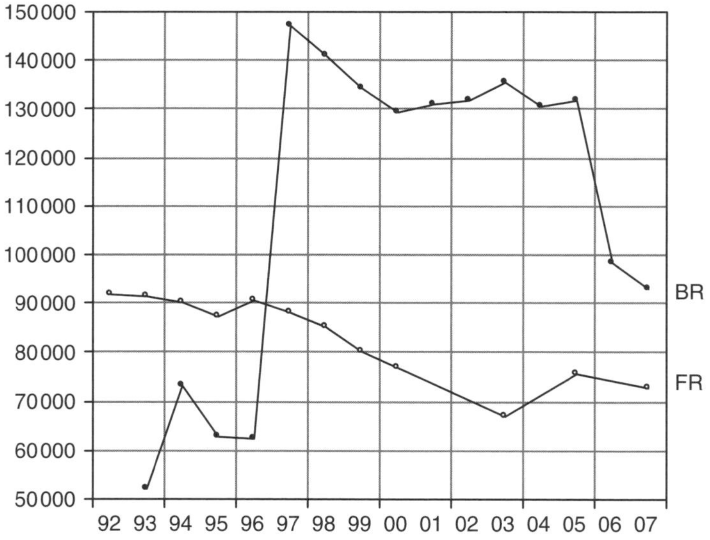

Figure 2.5 Number of agricultural enterprises in Sweden according to the Business Register (BR) and the Farm Register (FR).

difference can be reduced. But it is clear that the Business Register series is of low quality – the Business Register has both undercoverage and overcoverage, the quality of NACE for agricultural enterprises has been low and administrative changes disturb the time series pattern.

The errors in the Farm Register series are not so dramatic, but still they can be seen in Figure 2.5. The traditional Farm Register was replaced by a register mainly based on IACS during 2003. Due to undercoverage (not all farmers could apply for subsidies) the value for 2003 may be too low. During 2005 the system for subsides was changed and the undercoverage was reduced. The IACS register may have undercoverage and here, too, we see that changes in the IACS system will disturb the time series. There are two kinds of administrative systems – some systems concern all (e.g. everyone has to report and pay tax), while other systems concern only those who apply or use the benefit or system (e.g. only those who benefit from subsidies are found in the IACS register). This means that different administrative registers can be more or less sensitive to undercoverage.

To be able to create a Farm Register harmonized with the Business Register the following problems must be solved:

1. A Farm Register that is not disturbed by undercoverage and changes in the administrative system should be created. This is discussed below.

2. The under- and overcoverage of agricultural enterprises in the Business Register that we have today must be eliminated. This includes the harmonization of NACE in the Business Register with the typology in the Farm Register. This is discussed below.

3. The administrative units in taxation data, the administrative units in IACS applications and the statistical units in the Farm Register and Business Register must be linked so that different sources can be integrated. This is discussed in Section 2.5.

4. All enterprises with some agricultural production should be included in the register population of the Farm Register with a link to the Business Register, and the importance of agriculture should be incorporated in the registers. The aim should be to have the importance of all NACE activities measured in percentage terms for each enterprise and establishment. This is discussed in Section 2.5.

We will illustrate these issues with examples from Statistics Sweden. We think that similar problems also exist in most other countries and the methods to solve them will be the same.

We begin with a discussion of the farm register design. Today the Swedish Farm Register is of high quality, updated yearly with IACS data and every second or third year with the Farm Structure Survey that contains questions aimed at improving register quality with regard to the statistical units. However, Business Register data are not used to update and improve the Farm Register and IACS data are not used to update the population in the Business Register. This means that available data are not used in an optimal way today and also that these two registers are not coordinated and that statistics based on them are not as consistent and coherent as they could be. The quality of the Farm Register is today mainly dependent on the administrative IACS system with regard to coverage. This is a potential risk – in the future the quality of this administrative source may change.

Within the entire statistical system, the main task of the Business Register is to keep track of the population of all kinds of enterprises and organizations. So the population of agricultural enterprises should be defined by the Business Register and all sources available to do this should be used by the Business Register. An important source is the weekly updates of the Tax Board's Business Register that are sent to Statistics Sweden's Business Register that are not used for the Farm Register today.

From a theoretical point of view, the Business Register should also be based on the sources used for the Farm Register, and the Farm Register population should also be based on the sources used for the Business Register. Then data would be used in the most efficient way, we would get the best possible quality and agricultural statistics would be consistent and coherent with other kinds of economic statistics.

Since 2006, we have created a calendar year version of the Business Register that consists of all enterprises and organizations active for some part of a specific calendar year. For this register we use all available sources and the register describing year t can be created during January of year $t+2$. Activity is defined as non-zero values in at least one administrative source regarding year t. This calendar year register has very good coverage and we plan to use it as a standardized population for the economic statistics produced for the yearly National Accounts. We have also used the calendar year register to study coverage errors in our present economic statistics, and it is planned to use our findings to improve quality. This project is described in Wallgren and Wallgren (2008).

The IACS and the Farm Register should also be used when this calendar year register is created – the population and NACE for the agricultural part of the register will then be as good as possible. But this calendar year register cannot be used as frame for samples or censuses. This means that frames must be created almost as they are today, but with better sources. Figure 2.6 illustrates the production processes for 2007. The points in time when the frames for the Farm Structure Survey and the Structural Business Survey were created are shown above the time axis. Based on all administrative sources below the time axis, the calendar year register for 2007 was created during January 2009. With this calendar year register the preliminary estimates for the Farm Structure Survey and the Structural Business Survey can be revised. The revisions give information about the frame errors in preliminary estimates. This knowledge is the starting-point for the work to improve the Farm Register and the Business Register. We expect that the frames produced with these

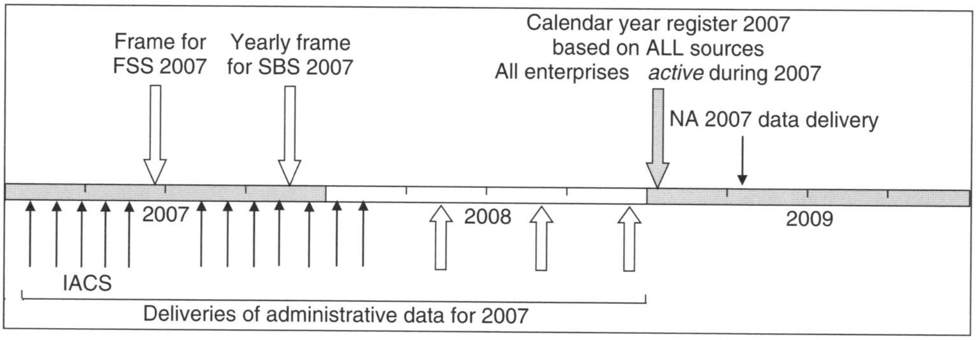

Figure 2.6 Frames used for the Farm Structure Survey (FSS) and Structural Business Survey (SBS) and the corresponding calendar year register.

Table 2.2 Coverage errors in the Business Register (BR) compared with the Farm Register (FR) data for 2004 and 2005.

	Number of units in			Coverage errors in		
	FR	BR	Activity	SBS	SEK millions	
1. NACE 01 in BR not active		6 787	No	Overc. production	1110	1.8%
2. NACE 01 in BR not in FR		64 498	Yes, not in 01			
3. In FR and BR	67 112	67 112	Yes, in 01			
4. In FR not in BR	8 696		Yes, in 01	Underc. turnover	4425	6.8%
Total number:	75 808	138 397				

registers will gradually improve. As a result, more and more Business Register data will be used for the Farm Register and the dependence of IACS will become smaller.

We now turn to under- and overcoverage in the Business Register. Selander (2008) describes the undercoverage in the Business Register with regard to agricultural enterprises. Statistics Sweden (2007) describes the corresponding overcoverage. In Table 2.2 the population in the Farm Register has been matched against the units in the Business Register.

In the Farm Register there are 75 808 units and in the Business Register there are 138 397 units coded as active in NACE 01. Of these 138 397 units 6787 (line 1) were not found in any administrative source for the reference year, thus they represent overcoverage in the Business Register. In the Structural Business Survey (SBS) these inactive enterprises were treated as non-response and positive values were imputed. This gave rise to an overcoverage error of 1110 million SEK, or 1.8% of the total production value for NACE 01.

A further 64 498 units in the Business Register (line 2) were coded as active within NACE 01. However, as they did not belong to the Farm Register, we suspect that the NACE classifications in the Business Register of these enterprises in most cases are wrong – they are not active in NACE 01 but in another branch of industry instead. These enterprises thus represent overcoverage with regard to NACE 01 and undercoverage with regard to other unknown branches of industry.

A Further 8696 units in the Farm Register were not found in the Business Register (line 4). We can classify these as undercoverage in the Business Register, and due to this undercoverage the estimate of turnover in the Structural Business Survey was 4425 million SEK too low, or 6.8% of the total turnover value for NACE 01. We estimated this coverage error with information in the Value Added Tax Register during a project in which we analysed the quality of the Swedish Business Register. This project is reported in Statistics Sweden (2007) and in Wallgren and Wallgren (2008).

The conclusion is that the Swedish Business Register must be redesigned. More administrative sources, such as VAT, monthly payroll reports to the tax authorities and yearly income declarations from enterprises must be used to improve the coverage. The IACS and CDB registers should also be used to improve the agricultural part of the Business Register.

The typology in the Farm Register should be used to improve the NACE codes in the Business Register. The enterprises in the Business Register, today coded as NACE 01 but missing in the Farm Register, should get new NACE codes after a register-maintenance survey.

After these improvements to the Business Register it will be possible to integrate the Farm Register with the Business Register and use a large number of administrative sources for agricultural statistics in the same way as we today use many administrative registers to produce social and economic statistics.

2.5 Creating a farm register: the statistical units

The administrative units in taxation data, the administrative units in IACS applications and the statistical units in the Farm and Business Registers must be linked so that different sources can be integrated. The methodology for this linkage requires that different kinds of *derived statistical units* are created. The following administrative and statistical units should be considered here:

- Agricultural holding (AH): A single unit which has single management and which undertakes agricultural activities either as its primary or secondary activity. The target population of the Farm Register consists of this kind of units.
- Legal unit (LeU): A unit that are responsible for reporting to tax authorities or other authorities. Almost all administrative data we use for statistical purposes come from such legal units. Each legal unit has a unique identity number used for taxation purposes.
- For the Business Register a number of statistical units are created: enterprise units (EU), local units (LU), kind of activity units (KAU) and local kind of activity units (LKAU).

The structure and relations between different legal units can often be complex. This must be considered when we use administrative data. Table 2.3, with data from the Structural Business Survey, illustrates that data from related legal units sometimes must be aggregated into data describing an enterprise unit.

The relations between different kinds of units in the Business Register are illustrated in Figure 2.7. Agricultural holdings almost correspond to local kind of activity units the agricultural part of a local unit.

Table 2.3 One enterprise unit consisting of four legal units.

		Turnover, SEK millions		Wage sum, SEK millions	
		Source 1	Source 2	Source 1	Source 3
EU 1	LeU 1				0.1
EU 1	LeU 2	8.6		1.3	
EU 1	LeU 3		0.2		
EU 1	LeU 4		0.6		1.2
EU 1	Sum:	8.6	8.8	1.3	1.3

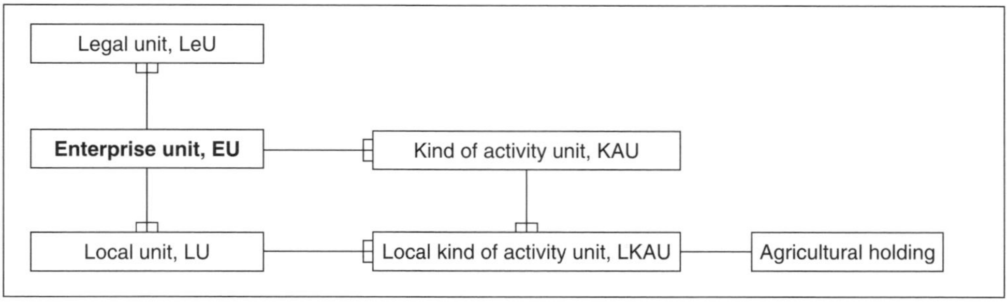

Figure 2.7 Statistical units in the Business Register and Farm Register.

If we wish to integrate data from two administrative sources (e.g. turnover from the VAT Register and wheat area from the IACS Register), we must consider the relations between different kinds of units and have access to a Business Register and a Farm Register where these relations are stored. The example in Figure 2.8 illustrates how the two administrative registers should be matched and also how the statistical estimates depend on the matching method.

The first record in Figure 2.8 is perhaps the most common case – there is a one-to-one relation between all kinds of units.

The record with LeU 2 and LeU 3 could be one holding where husband and wife both report income as self-employed but only one of them applies for subsidies. They are then registered as two different legal units by the tax authorities.

The record with LeU 4 and LeU 5 could be one holding where husband and wife both report income as self-employed and both apply for subsidies for different parts of the agricultural activities.

The record with LeU 6 describes a case where one enterprise has two local units and two holdings. The enterprise sends in one VAT report but applications for subsidies are sent in for each holding by different holders.

The last record with LeU 8 describes a case with one local unit and one holding, but agriculture is the secondary activity. The local unit is the divided into two local kind of activity units. As a rule, we have information in our Swedish Business Register describing the proportions of each activity, here 60% forestry and 40% agriculture. In Table 2.4, we illustrate how data from the two registers should be matched and how the estimates are influenced by the matching method.

The correct way of integrating these two administrative registers is shown in columns 1 and 2. For each holding we add the values that belong to it, e.g. $75 = 45 + 30$ for AH 2.

For holdings AH 4 and AH 5 we have only one common VAT report. The turnover value of 50 can be divided between the two holdings by a model that describes turnover as proportional to some measure of the production on each holding.

For holding AH 6 we have one VAT report for the forestry and agricultural parts together. With the information that the agricultural proportion is 40% we estimate the agricultural part of the turnover as $0.40 \cdot 100$.

Columns 3–8 in Table 2.4 illustrate the consequences of matching the two administrative registers directly with the legal unit identities. Due to the fact that we do not use the correct statistical units we get mismatch in two cases. This gives us two missing

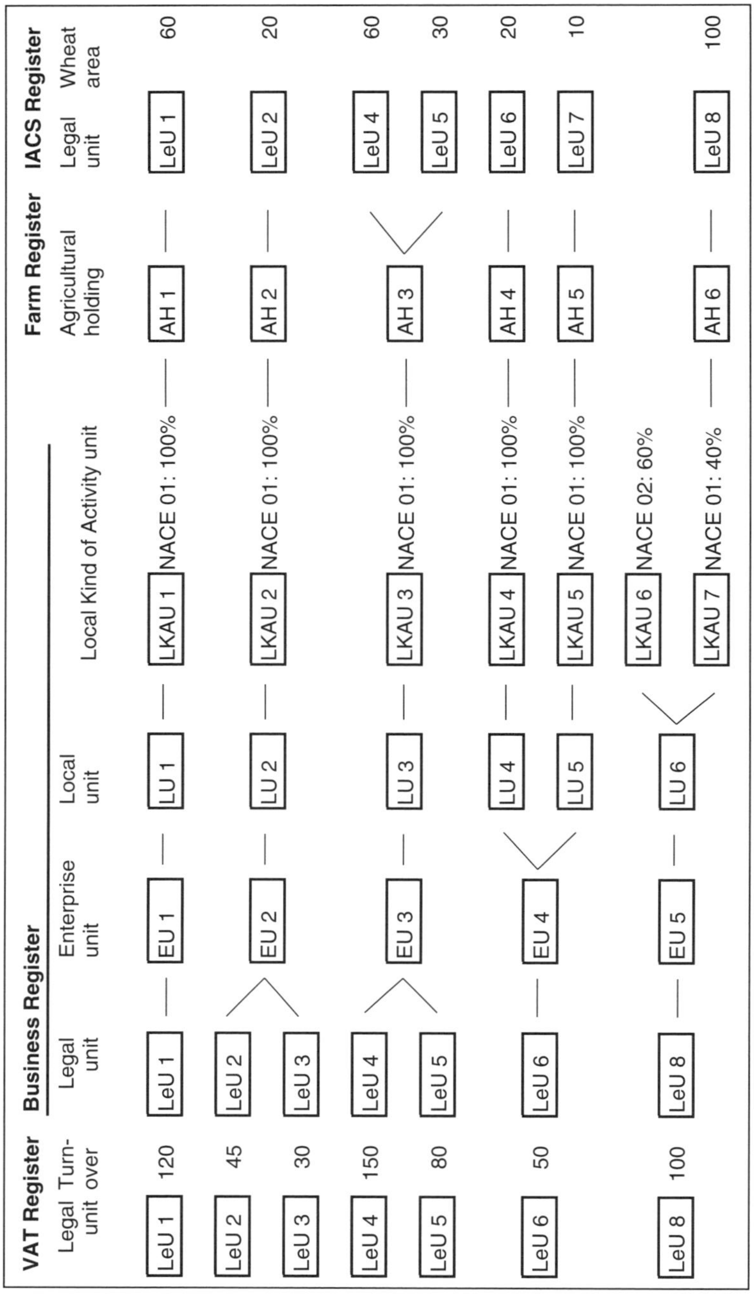

Figure 2.8 Statistical units in the VAT, Business, Farm and IACS registers.

Table 2.4 Integrating the VAT Register and the IACS.

						Model for imputation		Imputed values	
	VAT Turnover (1)	IACS Wheat area (2)		VAT Turnover (3)	IACS Wheat area (4)	VAT Turnover (5)	IACS Wheat area (6)	VAT Turnover (7)	IACS Wheat area (8)
AH1	120	60	LeU1	120	60	120	60	120.0	60.0
AH2	75	20	LeU2	45	20	45	20	45.0	20.0
AH3	230	90	LeU3	30	No hit		No hit	30.0	16.0
AH4	$50 \cdot p$	20	LeU4	150	60	150	60	15.0	60.0
AH5	$50 \cdot (1 - p)$	10	LeU5	80	30	80	30	80.0	30.0
AH6	$0.40 \cdot 100$	100	LeU6	50	20	50	20	50.0	20.0
			LeU7	No hit	10	No hit		18.8	10.0
			LeU8	100	100	100	100	100.0	100.0
Sum	515	300		575	300	545	290	593.8	316.0
						1.88	*0.53*		

values that can be handled in different ways. If we use the six records in columns 5 and 6 we can use the relation between turnover and wheat area (1.88) and the corresponding relation between wheat area and turnover (0.53) to impute values in columns 7 and 8. The errors created by wrong matching method are then 593.8 – 515 for the turnover estimate and 316 – 300 for the wheat area estimate.

In a calendar year register, which is discussed in the previous section, units should be combined into complex units that follow a holding during the calendar year. If, for example, holder A takes over a holding from holder B during the year, two legal units must be combined into one unit. Administrative data from these two legal units should be combined to describe the new enterprise unit.

2.6 Creating a farm register: the variables

When we create statistical registers, we usually integrate microdata from different sources. As a consequence of this there is a fundamental difference between microdata from a sample survey and microdata in a statistical register. In Figure 2.9 the production process in surveys with their own data collection (sample surveys and censuses) is compared with the production process in a register-based survey, in this example based on four different sources.

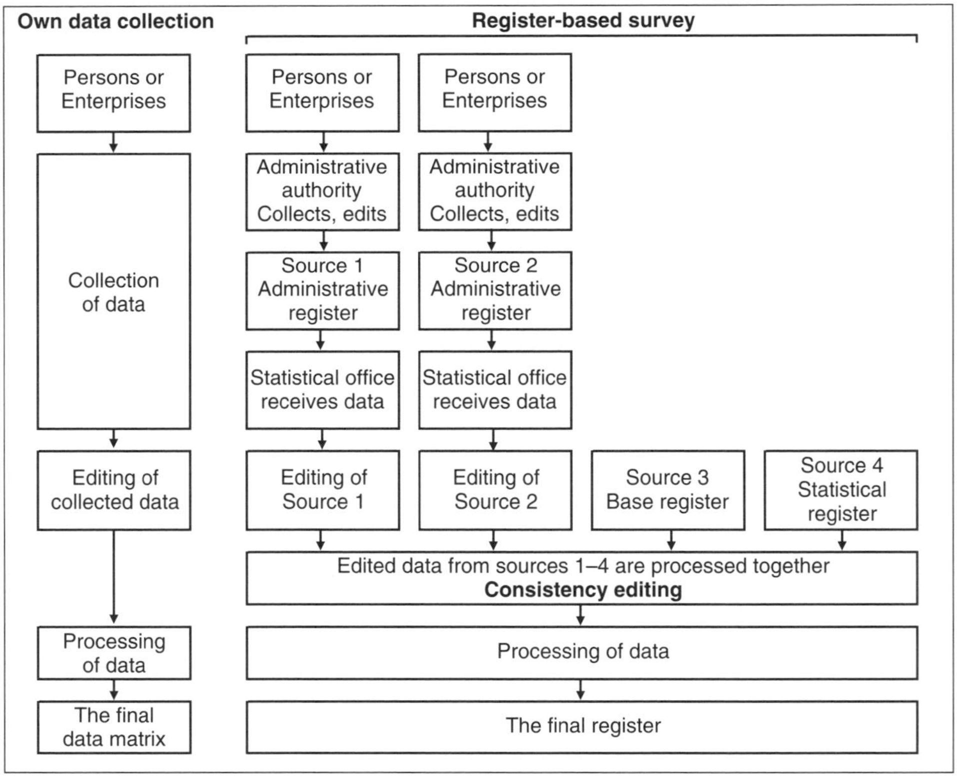

Figure 2.9 Sample surveys and censuses compared with register-based surveys. *Source*: Wallgren and Wallgren (2007, p. 101).

When we collect data by sending out questionnaires we know that all measurements in a record come from the same statistical unit. If we edit our data and find inconsistent records, then we know that there may be something wrong with the value of at least one of the variables. So, we edit to find errors in variables.

This is not the case when we integrate data from different sources to create a statistical register. In one record we have variables that come from different sources and if we edit and find inconsistent records, then we do not know if this is due to errors in variables or if the inconsistencies are due to errors in objects/units. In the same record we may have what we call a false hit, whereby some variables describe one unit and other variables describe another unit that has the same identity. This is common in business statistics as enterprises change over time, often without changing their identity numbers.

Table 2.5 illustrates two problems that always become methodological issues when we integrate sources. There will always be some mismatch, but how should it be treated? There will also be differences between variable values for similar or identical variables in different sources, but how should that issue be treated?

The wage sums in Table 2.5 are based on two administrative sources with high quality and after editing for errors in variables and errors in units. But still there are differences between the sources, so the question is how to estimate the yearly wage sum. If we first analyse the mismatch, then we must judge whether some of the 91 and 62 legal units that occur in only one source are the same units but with different identity numbers. We must also decide how to interpret the fact that in 2870 cases the wage sum reported in the yearly source is bigger than the monthly, and in 1518 cases we find the opposite circumstance.

This must be judged with subject-matter competence – and it is wise to discuss with persons at the tax authorities how data should be interpreted. When we work with administrative data we must always have good relations with the administrative authorities that give us the data.

In this case we have three possible estimators:

- Using only the monthly source, the estimated wage sum is 3954.3 SEK millions.
- Using only the yearly source, the estimated wage sum is 3989.3 SEK millions.

Table 2.5 Comparing similar or identical variables in two sources.

	Yearly wage sums, agricultural enterprises, SEK millions			
Sources:	Monthly source	Yearly source	Final estimate	Number of LeU
Only in yearly	–	2.3	2.3	91
In both sources, monthly $<$ yearly	1817.7	1864.6	1841.2	2870
In both sources, monthly $=$ yearly	1214.1	1214.1	1214.1	5736
In both sources, monthly $>$ yearly	920.5	908.4	914.5	1518
Only in monthly	2.0	–	2.0	62
Estimate:	3954.3	3989.3	3974.0	10277

- Using both sources in the best way, if we judge that the sources are equally reliable we can take the average of the two measurements we have for all legal units, so that the estimated wage sum is 3974.0 SEK millions. Instead of publishing two inconsistent estimates we will publish only one.

2.7 Conclusions

A fundamental principle in statistical science is that available data are used as efficiently as possible. If the statistical agencies want their published statistics to be regarded as trustworthy, estimates from different surveys must be consistent with each other. This means that we should use administrative data for statistical purposes by constructing a system of harmonized statistical registers. The populations of different registers must be decided after comparison with other registers. The statistical units in different registers must also be determined so that the entire system will function efficiently and produce statistics of good quality.

For agricultural statistics it is important that the Farm Register can be improved and made consistent with the Business Register. If the same data are used for both registers it will be possible to gradually harmonize them. This will also open up new possibilities for agricultural statistics as new sources can then be linked to holdings in the Farm Register.

The demand for new kinds of agricultural statistics describing ‘wide agriculture’ was mentioned in Section 2.3. Our main conclusion is that a system of statistical registers based on many administrative sources can meet these demands and that the statistical possibilities appear to be promising.

References

Selander, R. (2008) Comparisons between the Business Register and Farm Register in Sweden. Paper presented to the First International Workshop on Technology and Policy for Accessing Spectrum (TAPAS 2006).

Statistics Sweden (2007) Register-based economic statistics based on a standardised register population (in Swedish). *Background Facts, Economic Statistics*, 2007: 6.

Wallgren, A. and Wallgren, B. (2007) *Register-Based Statistics: Administrative Data for Statistical Purposes*. Chichester: John Wiley & Sons, Ltd.

Wallgren, A. and Wallgren, B. (2008) Correcting for coverage errors in enterprise surveys – a register-based approach. In *Proceedings of Q2008, European Conference on Quality in Official Statistics*.

3

Alternative sampling frames and administrative data. What is the best data source for agricultural statistics?

Elisabetta Carfagna and Andrea Carfagna
Department of Statistical Sciences, University of Bologna, Italy

3.1 Introduction

Many different data on agriculture are available in most countries in the world. Almost everywhere various kinds of data are collected for administrative purposes (administrative data). In some countries, a specific data collection is performed with the purpose of producing agricultural statistics, using complete enumeration or sample surveys based on list or area frames (a set of geographical areas) or both. Rationalization is felt as a strong need by many countries, since they deliver different and sometimes non-comparable data; moreover, maintaining different data acquisition systems is very expensive.

We compare sample surveys and administrative data and perform an analysis of the advantages, disadvantages, requirements and risks of direct tabulation of administrative data. Then we focus on other ways of taking advantage of administrative data in order to identify the best data source for the different agricultural systems. Finally, we delineate some kinds of combined use of different frames for producing accurate agricultural statistics.

Agricultural Survey Methods Edited by Roberto Benedetti, Marco Bee, Giuseppe Espa and Federica Piersimoni

3.2 Administrative data

In most countries in the world, administrative data on agriculture are available, based on various acquisition systems. Definitions, coverage and quality of administrative data depend on administrative requirements; thus they change as these requirements change. Their acquisition is often regulated by law; thus they have to be collected regardless of their cost, which is very difficult to calculate since most of the work involved is generally performed by public institutions. The main kinds of administrative data relevant to agricultural statistics are records concerning taxation, social insurance and subsidies. These data are traditionally used for updating a list created by a census. The result is a sampling frame for carrying out sample surveys in the period between two successive censuses (most often 4–10 years).

The extraordinary increase in the ability to manipulate large sets of data, the capacity of some administrative departments to collect data through the web (which allows rapid data acquisition in a standard form) and budget constraints have prompted the exploration of the possibility of using administrative data more extensively and even of producing statistics through direct tabulation of administrative data.

3.3 Administrative data versus sample surveys

A statistical system based on administrative data allows money to be saved and the response burden to be reduced. It also has advantages that are typical of complete enumeration, such as producing figures for very detailed domains (not only geographical) and estimating transition over time. In fact, statistical units in a panel sample tend to abandon the survey after some time and comparisons over time become difficult; whilst units are obliged to deliver administrative data, or at least interested in doing so.

Some countries in the world try to move from an agricultural statistical system based on sample surveys to a system based on administrative registers, where a register is a list of objects belonging to a defined object set and with identification variables that allow updating of the register itself.

When a sample survey is carried out, first the population is identified, then a decision is taken about the parameters to be estimated (for the variables of interest and for specific domains) and the levels of accuracy to be reached, taking into account budget constraints.

When statistics are produced on the basis of administrative registers, the procedure is completely different, since data have already been collected. Sometimes objects in the registers are partly the statistical units of the population for which statistics have to be produced and partly something else; thus evaluating undercoverage and overcoverage of registers is very difficult.

3.4 Direct tabulation of administrative data

Two interesting studies (Selander *et al.*, 1998; Wallgren and Wallgren, 1999), financed jointly by Statistics Sweden and Eurostat, explored the possibility of producing statistics on crops and livestock through the Integrated Administrative and Control System (IACS, created for the European agricultural subsidies) and other administrative data. After a comparison of the IACS data with an updated list of farms, the first study came to the following conclusion: 'The IACS register is generally not directly designed for statistical needs. The population that applies for subsidies does not correspond to the population of farms which should be investigated. Some farms are outside IACS and some farms are inside this system but do not provide complete information for statistical needs. Supplementary sample surveys must be performed, or the statistical system will produce biased results. To be able to use IACS data for statistical production the base should be a Farm Register with links to the IACS register.'

3.4.1 Disadvantages of direct tabulation of administrative data

When administrative data are used for statistical purposes, the first problem to be faced is that the information acquired is not exactly that which is needed, since questionnaires are designed for specific administrative purposes. Statistical and administrative purposes require different kinds of data to be collected and different acquisition methods (which strongly influence the quality of data). Strict interaction between statisticians and administrative departments is essential, although it does not guaranty that a good compromise solution can be found. Statistics Sweden began long ago to take advantage of administrative data, though as Wallgren and Wallgren observe in Chapter 2 of this book: 'The Business Register and the Farm Register at Statistics Sweden are not harmonized today.'

The list of codes adopted for administrative and for statistical purposes should be harmonized, but this is not an easy task. For example, the legend used in the IACS is much more detailed for some land use types than that adopted by the Italian Ministry of Policies for Agriculture, Food and Forest (MIPAAF) for producing crop statistics (AGRIT project; for a description of the AGRIT survey, see Chapters 13 and 22 of this book) and is less detailed for others, mainly due to the different aims of the data collection. Extracting the data from the IACS system in order to obtain the number of hectares of main crops is not an easy task. Moreover, the acquisition date of the IACS data does not allow information to be collected on yields; thus, an alternative data source is needed in order to estimate these important parameters.

Administrative data are not collected for purely statistical purposes, with the guarantee of confidentiality and of no use for other purposes (unless aggregated); they are collected for specific purposes which are very relevant for the respondent such as subsidies or taxation. On the one hand, this relevance should guarantee accurate answers and high quality of data; on the other, specific interests of respondents can generate biased answers. For example, the IACS declarations have a clear aim; thus the units that apply for an administrative procedure devote much attention to the records concerning crops with subsides based on area under cultivation, due to the checks that are carried out, and less attention to the areas of other crops. In Sweden (see Selander *et al.*, 1998; Wallgren and Wallgren, 1999), for crops with subsidies based on area and for other crops which are generally cultivated by the same farms, the bias is low, but for other crops the downward bias can be about 20%. Moreover, for very long and complicated questionnaires the risk of collecting poor-quality data is high.

Unclear dynamics can be generated by checks carried out on the IACS data, since some farmers may decide not to apply for subsides even if they are available, others may tend to underestimate the areas to avoid the risks and the consequences of the checks, and still others may inflate their declarations, hoping to escape the checks.

The most critical point, when using administrative data for statistical purposes, is that their characteristics can change from one year to the next depending on their aims and without regard to statistical issues. For example, a simplification of Swedish tax legislation made farm incomes virtually indistinguishable from other forms of income in tax returns; thus tax returns could no longer be used to produce statistics on farmers' income. Another very important example is given by the change in common rules for direct support schemes under the Common Agricultural Policy applied from 2005, which strongly simplified aid applications.[1]

A premium per hectare now applies only in isolated cases, such as the specific quality premium for durum wheat, the protein crop premium, the crop-specific payment for rice and the aid for energy crops. The law does not require information on other crops. Some Italian regions still request information on the area covered by each crop in the farm, but the farmers know that subsidies are not linked to this information and they tend to give less accurate answers; moreover, the response burden is very high. For these reasons, at present the IACS data cannot be the main source for crops statistics.

3.5 Errors in administrative registers

A pillar of sampling theory is that, when a sample survey is carried out, much care can be devoted to the collection procedure and to the data quality control, since a relatively small amount of data is collected; thus, non-sampling errors can be limited. At the same time, sampling errors can be reduced by adopting efficient sample designs. The result is that very accurate estimates can often be produced with a relatively small amount of data.

The approach of administrative registers is the opposite: a huge amount of data is collected for other purposes and sometimes a sample of those data is checked to apply sanctions and not for evaluating data quality or understanding if some questions can be misleading in the questionnaire.

3.5.1 Coverage of administrative registers

As already stated, evaluating the level of undercoverage and overcoverage of administrative registers is very difficult. The studies mentioned above made an analysis of record linkage results using the IACS records and a list of farms created by the census and updated. Telephone numbers enabled the identification of 64.1% of objects in the list of farms and 72.0% of objects in the IACS; much better results were achieved by also associating other identification variables, such as organization numbers (BIN) and personal identification numbers (PIN): 85.4% of objects in the list of farms and 95.5% of objects in the IACS. However, only 86.6% of objects in the IACS and 79% of objects in the list of farms have a one-to-one match; others have a one-to-many or many-to-many match and 4.5% of the IACS objects and 14.6% of objects in the list of farms have no match at all.

Another example is given by the number of farms by size (classes of utilized agricultural area (UAA)) in Italy according to the results of the survey concerning the structure of farms (FSS) in 2007 and the national authority for subsidies (AGEA): see Greco *et al.* (2009). Table 3.1 shows large differences, particularly for very small farms. For

[1] `http://eur-lex.europa.eu/LexUriServ/LexUriServ.do?uri=OJ:L:2003:270:0001:-0069:EN:PDF`

Table 3.1 Number of farms by size in Italy in 2007 according to the results of the farm structure survey and the national authority for subsidies.

Size (UAA)	Number of farms (AGEA)	Number of farms (FSS)	Difference	Difference (%)
Without UAA	1 684	1 674	10	0.59
Less than 1 ha	690 084	436 974	253 110	36.68
From 1 to 1.99 ha	279 164	394 930	−115 766	−41.47
From 2 to 4.99 ha	300 216	397 117	−96 901	−32.28
From 5 to 9.99 ha	175 695	202 560	−26 865	−15.29
From 10 to 19.99 ha	123 732	122 747	985	0.80
From 20 to 49.99 ha	94 077	83 423	10 654	11.32
From 50 ha	47 819	40 015	7 804	16.32
Total	1 712 471	1 679 440	33 031	1.93

some regions, the difference is even larger and can be only partially explained by the different farm definitions. Moreover, comparability over time is strongly influenced by the change in the level of coverage in the different years and can give misleading results.

3.6 Errors in administrative data

Estimation of parameters has a meaning only if the reference population is well defined; while, in most cases, administrative registers are constituted by a series of elements which cannot be considered as belonging to the same population from a statistical viewpoint. For instance, for the IACS the applicants are not necessarily the holders. Therefore, producing statistics about the population of farms requires a very good record linkage process for evaluating coverage problems (for a detailed analysis of record matching methods and problems, see Winkler, 1995; D'Orazio *et al.*, 2006).

Direct tabulation from a register is suggested for a specific variable if the sum of the values for that variable presented by all the objects in the register is an unbiased estimator of the total for this variable. Biased estimates can be produced when direct tabulation is applied to data affected by errors, since some objects can present inflated values, others can have the opposite problem and compensation is not guaranteed; furthermore, some objects that are in the administrative register should not be included and others which are not included should be in the register.

For example, let us consider the IACS declarations for a crop C. These data are affected by errors of commission (some parcels declared as covered by crop C are covered by another crop or their area is inflated) and omission (some parcels covered by crop C are not included in the IACS declarations or their area is less than the true value). If errors of commission and omission cancel out, the sum of declarations for crop C is an unbiased estimator of the area of this crop; but the quality control of the IACS data is not concerned with errors of omission.

3.6.1 Quality control of the IACS data

Errors of commission could be estimated through quality control of a probabilistic sample of the declarations. Quality control of the IACS data is carried out every year on a sample of declarations; however, its aim is not to produce an unbiased estimate of the error of commission, but to detect irregularities, thus the sample selection is not strictly probabilistic.

In 2003, at the Italian national level, for ten controlled crops (or groups of crops) the error was 48 591 ha, 3.9% of the total declared area of 1 270 639 ha. For an important crop like durum wheat (with a national declared area of 1 841 230 ha), 23 314 checks were carried out, corresponding to an area of 347 475 ha (19% of the declared area) and the error was 12 223 ha (3.5% of the area checked). The situation is very different for other crops, such as leguminous crops, for which the error was 1052 ha, 16% of the area checked (6568 ha).

If we consider specific geographical domains, for example the area of six provinces out of nine in Sicily, the error for durum wheat in 2000 was 16%. Even if we assume that no error of omission is committed, we cannot say that, at a national level, errors of commission for durum wheat amount to 3.5% and reduce the total area of this percentage in order to eliminate any upward bias, because the sample selection of the IACS quality control is purposive, since, as already stated, its aim is to detect irregularities and not to estimate the level of errors of commission; thus it probably tends to overestimate errors of commission.

It is evident that quality control is carried out for different purposes for statistical surveys and administrative registers and thus gives different results that should not be confused.

3.6.2 An estimate of errors of commission and omission in the IACS data

Errors of commission and omission in the IACS data were estimated in the study carried out by Consorzio ITA (AGRIT 2000) in the Italian regions of Puglia and Sicily for durum wheat in 2000. In both regions, an area frame sample survey based on segments with permanent physical boundaries was done. The ITA estimates of durum wheat cultivation were 435 487.3 ha in Puglia (with a coefficient of variation (CV) of 4.8%) and 374 658.6 ha in Sicily (CV 5.9%). Then, for each segment, the area of durum wheat deriving from the declarations was computed and the resulting estimates (IACS estimates) were compared with the ITA estimates.

The IACS estimates where smaller than the ITA estimates (6.9% less in Puglia and 16.0% in Sicily). Also the sum of the IACS declarations (based on IACS records) was smaller than the ITA estimates (10.4% less in Puglia and 12.2% in Sicily).

Parcels declared covered by durum wheat were identified on each sample segment. For some of these parcels, declared areas equalled areas detected on the ground by the area sample survey; for others, there was a more or less relevant difference. Finally, these differences were extrapolated to the population and the errors of commission were estimated: 7.8% of the sum of the declarations in Puglia and 8.4% in Sicily. A comparison with the ITA estimates suggests the presence of a relevant error of omission, which is about 13.9% of the ITA estimate in Puglia and 23.3% in Sicily. High levels of omission error are probably due partly to incorrect declarations and partly to farmers who did not apply. This study highlights the effect of assuming the IACS data unaffected by errors of omission.

It also shows that when quality control is carried out to uncover irregularities, the declarations can be influenced by complex dynamics, which are difficult to foresee and can generate biases. It should be borne in mind that durum wheat is one of the crops for which subsidies are paid based on area under cultivation, thus considered reliable by the Swedish studies mentioned.

The study described also suggests that statistics for small domains produced with administrative registers can be unreliable due to different dynamics in different domains.

3.7 Alternatives to direct tabulation

One approach to reducing the risk of bias due to undercoverage of administrative registers and, at the same time, avoiding double data acquisition is to sample farms from a complete and updated list and perform record linkage with the register in order to capture register data corresponding to farms selected from the list. If the register is considered unreliable for some variables, related data have to be collected through interviews as well as data not found in the register due to record linkage difficulties.

3.7.1 Matching different registers

Countries with a highly developed system of administrative registers can capture data from the different registers to make comparisons, to validate some data with some others and to integrate them. Of course, very good identification variables and a very sophisticated record linkage system are needed.

The main registers used are the annual income verifications in which all employers give information on wages paid to all persons employed, the register of standardized accounts (based on annual statements from all firms), the VAT register (based on VAT declarations from all firms) and the vehicle register (vehicles owned by firms and persons).

The combined use of these registers improves the coverage of the population and the quality of data through comparison of data in the different registers and allows the socio-economic situation of rural households to be described. However, it does not solve all the problems connected with undercoverage and incorrect declaration.

The statistical methodological work that needs to be done in order to be able to use multiple administrative sources is very heavy (see Wallgren and Wallgren, 1999, 2007):

- editing of data;
- handling of missing objects and missing values;
- linking and matching;
- creating derived objects and variables.

Furthermore, the work that needs to be done for quality assurance is also heavy:

- contacts with suppliers of data;
- checking of received data;
- analysis of causes and extent of missing objects and values;
- data imputation;
- checking of causes and extent of mismatch;

- evaluating objects and variables and reporting inconsistencies between registers;
- reporting deficiencies in metadata;
- carrying out register maintenance surveys.

Although technical and methodological capabilities for performing record matching have improved in recent years (statistical matching, block matching, etc. – for a review see D'Orazio *et al.*, 2006) all this is a considerable amount of work, since it has to be carried out on entire registers, but its cost is not generally evaluated; moreover, the effect of mismatch or imperfect match or statistical match on estimates is not easy to evaluate (see Moriarity and Scheuren, 2001; Scanu and Conti, 2006).

3.7.2 Integrating surveys and administrative data

Administrative data are often used to update the farm register and to improve the sample design for agricultural surveys. At Statistics Canada, a possible integration of administrative data and surveys is foreseen: 'Efforts are underway to explore the wealth of data available and to determine how they can be used to better understand our respondents and to improve sample selection to reduce burden' (Korporal, 2005).

A review of statistical methodologies for integration of surveys and administrative data is given in the ESSnet Statistical Methodology Project on Integration of Surveys and Administrative Data (ESSnet ISAD, 2008), which focuses mainly on probabilistic record linkage, statistical matching and micro integration processing. For a review of the methods for measuring the quality of estimates when combining survey and administrative data, see Lavallée (2005).

Problems connected with confidentiality issues are frequent when data have to be retrieved from different registers as well as when survey data are combined with register data. One example is given by the Italian experiment on the measurement of self-employment income within EU-SILC (European Union Statistics on Income and Living Conditions); see ESSnet ISAD (2008). Since 2004, the Italian team has carried out multi-source data collection, based on face-to-face interview and on linkage of administrative with survey data in order to improve data quality on income components and relative earners by means of imputation of item non-responses and reduction of measurement errors. Administrative and survey data are integrated at micro level by linking individuals through key variables. However, 'the Personal Tax Annual Register, including all the Italian tax codes, cannot be used directly by Istat (the Italian Statistical Institute). Therefore, record linkage has to be performed by the tax agency on Istat's behalf. The exact linkage performed by the Tax Agency produces 7.8% unmatched records that are partially retrieved by means of auxiliary information (1.5%).'

3.7.3 Taking advantage of administrative data for censuses

When a census or a sample survey of farms has to be carried out, administrative registers may be of considerable value in updating the list to be used. The IACS data are the most important source of information for this purpose, although in some cases the same farm is included several times for different subsidies and clusters based on auxiliary variables have to be created in order to identify the farm that corresponds to the different records.

Administrative data could also be used in order to reduce the respondent burden during the census data collection if administrative data can be considered reliable for some variables. A study was conducted by Istat (De Gaetano and Greco, 2009) in preparation for the next census of agriculture and was based on the comparison of the number of hectares covered by specific crops according to a sample survey of farms and the corresponding data in the IACS. It found, at a national level, differences of 9% for vineyards, 27% for olive trees and 58% for citrus trees; moreover, there are huge differences in some regions. This study suggests confining the use of the IACS data to updating the list of farms.

3.7.4 Updating area or point sampling frames with administrative data

Since refreshing the stratification of area or point sampling frames is generally very expensive, administrative data seem to be a cheap and valuable source of information that should be used for updating these kinds of stratification. The crucial point is being sure to update and not to corrupt the stratification.

In 2001, the AGRIT project started to move from an area frame to an unclustered point frame (Carfagna, 2007). The 1.2 million points of the sampling frame were stratified by the following classes: 'arable land', 'permanent crops', 'permanent grass', 'forest', 'isolated trees and rural buildings', 'other' (artificial areas, water, etc.). In March 2009, the MIPAAF carried out a test in order to assess the feasibility of annually updating the stratification of the 1.2 million points of the sampling frame with the ICAS declarations. The test was done on a sample of 84 444 points that had been surveyed by GPS in 2008. Almost 70% of these points (57 819) were linked to the IACS declarations. Agreement on land use was found for 22 919 points in the case of arable land (only 64.86%) and for 7379 points for permanent crops (69.63%). Consequently, the MIPAAF decided not to use the IACS data for updating the 1.2 million points of the sampling frame. Thus, before updating a stratification with administrative data we suggest carrying out similar tests.

3.8 Calibration and small-area estimators

Administrative registers can also be used at the estimator level as follows: the statistical system is based on a probabilistic sample survey with data collected for statistical purposes whose efficiency is improved by the use of register data as auxiliary variable in calibration estimators (Deville and Särndal, 1992; Särndal, 2007; see also Chapter 7 of this book).

The MIPAAF (AGRIT 2000) used the IACS data as auxiliary variable in a regression estimator. The CV of the estimates was reduced from 4.8% to 1.3% in Puglia and from 5.9% to 3.0% in Sicily. By way of comparison, the Landsat TM remote sensed data used as auxiliary variable allowed a reduction of the CVs in Puglia to 2.7% and in Sicily to 5.6% (for the cost efficiency of remote sensing data see Carfagna, 2001b).

As already stated, in recent years, the information included in the IACS data has become very poor and thus could be used as auxiliary variable only for very few crops.

When the available registers are highly correlated with the variables for which the parameters have to be estimated, the approach described has many advantages:

1. Register data are included in the estimation procedure, thus different data are conciliated in one datum.

2. A large reduction of the sample size and thus of survey costs and of respondent burden can be achieved.

3. If the sampling frame is complete and duplication has been eliminated there is no risk of undercoverage.

4. Data are collected for purely statistical purposes, so are not influenced and corrupted by administrative purposes.

The disadvantages are that the costs and the respondent burden are higher than when direct tabulation is performed. A detailed comparison should be made with the costs of a procedure using multiple administrative sources and sample surveys for maintaining registers. There is also the difficulty of producing reliable estimates for small domains, since this approach assumes a small sample size; thus, just a few sample units are allocated in small domains and corresponding estimates tend to be unreliable.

Administrative data could also be used as auxiliary variable in small-area estimation (Rao, 2003; see also Chapters 9 and 13 of this book) in order to achieve good precision of estimates in small domains where few sample units are selected. Small-area estimation methods use the link between the variable to be estimated and the covariable (auxiliary variable) outside the area and are model-dependent.

3.9 Combined use of different frames

When various incomplete registers are available and information included in their records cannot be directly used for producing statistics, a sample survey has to be designed.

Administrative data are most often used to create one single sampling frame, although on the basis of two or more lists. This approach should be used only if the different lists contribute essential information to complete the frame and the record matching gives extremely reliable results; otherwise, the frame will be still incomplete and have much duplication.

An alternative approach is to treat these registers as multiple incomplete lists from which separate samples can be selected for sample surveys. All the observations can be treated as though they had been sampled from a single frame, with modified weights for observations in the intersection of the lists (single-stage estimation). Of course, this procedure gives unbiased estimates only if the union of the different frames covers the whole population. Another possibility is to adopt an estimator that combines estimates calculated on non-overlapping sample units belonging to the different frames with estimates calculated on overlapping sample units (two-stage estimation). These two ways of treating different lists do not require record matching of listing units of the different frames (a process that is notoriously error-prone when large lists are used). Some estimators require the identification of identical units only in the overlap samples, and others have been developed for cases in which these units cannot be identified (see Hartley, 1962, 1974; Fuller and Burmeister, 1972).

Completeness has to be assumed: every unit in the population of interest should belong to at least one of the frames.

3.9.1 Estimation of a total

In a multiple-frame survey, probability samples are drawn independently from the frames $A_1, \ldots, A_Q$, $Q \geq 2$. The union of the Q frames is assumed to cover the finite population of interest, U. The frames may overlap, resulting in a possible $2^Q - 1$ non-overlapping domains. When $Q = 2$, the survey is called a dual-frame survey.

For simplicity, let us consider the case of two frames (A and B), both incomplete and with some duplication, which together cover the whole population. The frames A and B generate $2^2 - 1 = 3$ mutually exclusive domains: a (units in A alone), b (units in B alone), and ab (units in both A and B). N_A and N_B are the frame sizes, N_A, N_B and N_{AB} are the domain sizes.

The three domains cannot be sampled directly since we do not know which units belong to each domain and samples S_A and S_B, of sizes n_A and n_B, have to be selected from frames A and B. Thus n_a, n_{ab}^A, n_{ab}^B and n_b (the subsamples sizes of $S_A(n_A)$ and $S_B(n_B)$ respectively which fall into the domains a, ab and b) are random numbers and a post-stratified estimator has to be adopted for the population total.

For simple random sampling in both frames, when all the domain sizes are known, a post-stratified estimator of the population total is

$$\hat{Y} = N_a \bar{y}_a + N_{ab}(p\bar{y}_{ab}^A + q\bar{y}_{ab}^B) + N_b \bar{y}_b, \tag{3.1}$$

where p and q are non-negative numbers such that $p + q = 1$; $\bar{y}_a$ and $\bar{y}_b$ denote the respective sample means of domains a and b; and $\bar{y}_{ab}^A$ and $\bar{y}_{ab}^B$ are the sample means of domain ab, relative respectively to subsamples n_{ab}^A and n_{ab}^B. $N_a \bar{y}_a$ is an estimate of the incompleteness of frame B.

3.9.2 Accuracy of estimates

Hartley (1962) proposed to use the variance for proportional allocation in stratified sampling as an approximation of the variance of the post-stratified estimator of the population total with simple random sampling in the two frames (ignoring finite-population corrections):

$$\mathrm{var}(\hat{Y}) \approx \frac{N_A^2}{n_A}\left(\sigma_a^2(1-\alpha) + p^2\sigma_{ab}^2\alpha\right) + \frac{N_B^2}{n_B}\left(\sigma_b^2(1-\beta) + q^2\sigma_{ab}^2\beta\right),$$

where σ_a^2, σ_b^2 and σ_{ab}^2 are the population variances within the three domains, $\alpha = N_{ab}/N_A$ and $\beta = N_{ab}/N_B$. Under a linear cost function, the values for n_A/N_A, p and n_B/N_B minimizing the estimator variance can be determined (see Hartley, 1962).

Knowledge of the domain sizes is a very restrictive assumption that is seldom satisfied. Domain sizes are often known only approximately, due to the use of out-of-date information and lists, which makes difficult to determine whether a unit belongs to any other frame. In such a case, the estimator of the population total given in equation (3.1) is biased and the bias remains constant as the sample size increases. Many estimators that do not need domain sizes were proposed by Hartley (1962), Lund (1968) and Fuller and Burmeister (1972).

3.9.3 Complex sample designs

Complex designs are generally adopted in the different frames to improve the efficiency, and this affects the estimators. Hartley (1974) and Fuller and Burmeister (1972) considered the case in which at least one of the samples is selected by a complex design, such as stratified or multistage sampling.

Skinner and Rao (1996) proposed alternative estimators under complex designs where the same weights are used for all the variables. In particular, they modified the estimator suggested by Fuller and Burmeister for simple random sampling in the two frames, in order to achieve design consistency under complex designs, while retaining the linear combination of observations and simple form. For a review of multiple frame estimators, see Carfagna (2001a).

Lohr and Rao (2000) showed that the pseudo maximum likelihood estimator (PMLE) proposed by Skinner and Rao (1996) combined high efficiency with applicability to complex surveys from two frames, and Lohr and Rao (2006) proposed optimal estimators and PMLEs when two or more frames are used. They also gave some warnings on the use of multiple frames: 'If samples taken from the different frames use different questionnaires or modes of administration, then care must be taken that these do not create biases'. Moreover, the methods proposed for estimation with overlapping frames assume that domain membership can be determined for every sampled unit; thus estimates can be biased with misclassification if the domain means differ.

Ferraz and Coelho (2007) investigated the estimation of population totals incorporating available auxiliary information from one of the frames at the estimation stage, for the case of a stratified dual-frame survey. They assumed that samples are independently selected from each one of two overlapping frames, using a stratified sample design, not necessarily based on the same stratification variables, and proposed combined and separated versions of ratio type estimators to estimate a population total. They also derived approximate variance formulas using Taylor's linearization technique and showed through a Monte Carlo simulation study that incorporating auxiliary information at the estimation stage in stratified dual-frame surveys increases the precision of the estimate.

From a general viewpoint, whatever the sample design in the two frames, using the Horvitz–Thompson estimators of the totals of the different domains, the estimator of the population total is given by

$$\hat{Y} = \hat{Y}_a + p\hat{Y}^A_{ab} + q\hat{Y}^B_{ab} + \hat{Y}_b. \tag{3.2}$$

When the sample selection is independent in the two frames, we have

$$\mathrm{cov}(\hat{Y}_a, \hat{Y}_b) = \mathrm{cov}(\hat{Y}_a, \hat{Y}^B_{ab}) = \mathrm{cov}(\hat{Y}_a, \hat{Y}^A_{ab}) = \mathrm{cov}(\hat{Y}^A_{ab}, \hat{Y}^B_{ab}) = 0 \tag{3.3}$$

and the variance of the population total in equation (3.2) is

$$\begin{aligned}\mathrm{var}(\hat{Y}) =& \mathrm{var}(\hat{Y}_a) + p^2\mathrm{var}(\hat{Y}^A_{ab}) + (1-p^2)\mathrm{var}(\hat{Y}^B_{ab}) \\ &+ 2p\ \mathrm{cov}(\hat{Y}_a, \hat{Y}^A_{ab}) + 2(1-p)\mathrm{cov}(\hat{Y}_b, \hat{Y}^B_{ab}).\end{aligned} \tag{3.4}$$

Thus, the value of p that minimizes the variance in equation (3.4) is

$$p_{opt} = \frac{\text{var}\left(\hat{Y}^B_{ab}\right) + \text{cov}\left(\hat{Y}_b, \hat{Y}^B_{ab}\right) - \text{cov}\left(\hat{Y}_a, \hat{Y}^A_{ab}\right)}{\text{var}\left(\hat{Y}^A_{ab}\right) + \text{var}\left(\hat{Y}^B_{ab}\right)}. \tag{3.5}$$

The optimum value for p is directly related to the precision of $\hat{Y}^A_{ab}$, thus it can assume very different values for the different parameters to be estimated.

3.10 Area frames

When completeness is not guaranteed by the combined use of different registers, an area frame should be adopted in order to avoid the bias, since an area frame is always complete, and remains useful fora long time (Carfagna, 1998; see also Chapter 11 of this book).

The completeness of area frames suggests their use in many cases:

- if another complete frame is not available;
- if an existing list of sampling units changes very rapidly;
- if an existing frame is out of date;
- if an existing frame was obtained from a census with low coverage;
- if a multi-purpose frame is needed for estimating many different variables linked to the territory (agricultural, environmental, etc.).

Area frame sample designs also allow objective estimates of characteristics that can be observed on the ground, without the need to conduct interviews. Besides, the materials used for the survey and the information collected help to reduce non-sampling errors in interviews and are a good basis for data imputation for non-respondents. Finally, area sample survey materials are becoming cheaper and more accurate.

Area frame sample designs also have some disadvantages, such as the cost of implementing the survey programme, the necessity of many cartographic materials, the sensitivity to outliers and the instability of estimates. If the survey is conducted through interviews and respondents live far from the selected area unit, their identification may be difficult and expensive, and missing data will tend to be relevant.

3.10.1 Combining a list and an area frame

The most widespread way to avoid instability of estimates and to improve their precision is to adopt a multiple-frame sample survey design. For agricultural surveys, a list of very large operators and of operators that produce rare items is combined with the area frame. If this list is short, it is generally easy to construct and update. A crucial aspect of this approach is the identification of the area sample units included in the list frame. When units in the area frame and in the list sample are not detected, the estimators of the population totals have an upward bias.

Sometimes a large and reliable list is available. In such cases the final estimates are essentially based on the list sample. The role of the area frame component in the multiple-frame approach is essentially solving the problems connected with incompleteness of the list and estimating the incompleteness of the list itself. In these cases, updating the list and record matching to detect overlapping sample units in the two frames are difficult and expensive operations that can produce relevant non-sampling errors (Vogel, 1975; Kott and Vogel, 1995).

Combining a list and an area frame is a special case of multiple-frame sample surveys in which sample units belonging to the lists and not to the area frame do not exist (domain b is empty) and the size of domain ab equals N_B (frame B size: the list size, which is known).

This approach is very convenient when the list contains units with large (thus probably more variable) values of some variable of interest and the survey cost of units in the list is much lower than in the area frame (Kott and Vogel 1995; Carfagna, 2004). In fact, when A is an area frame, $\hat{Y}_b$ in equation (3.2) is zero as well as $\text{cov}(\hat{Y}_b, \hat{Y}_{ab}^B)$ in equations (3.3) and (3.4); thus, we have

$$\hat{Y} = \hat{Y}_a + p\hat{Y}_{ab}^A + q\hat{Y}_{ab}^B$$

with variance

$$\text{var}(\hat{Y}) = \text{var}(\hat{Y}_a) + p^2\text{var}(\hat{Y}_{ab}^A) + (1 - p^2)\text{var}(\hat{Y}_{ab}^B) + 2p\text{cov}(\hat{Y}_a, \hat{Y}_{ab}^A), \tag{3.6}$$

and the value of p that minimizes the variance in equation (3.6) is

$$p_{opt} = \frac{\text{var}\left(\hat{Y}_{ab}^B\right) + \text{cov}\left(\hat{Y}_a, \hat{Y}_{ab}^A\right)}{\text{var}\left(\hat{Y}_{ab}^A\right) + \text{var}\left(\hat{Y}_{ab}^B\right)}. \tag{3.7}$$

The optimum value of p in equation (3.7) depends on the item; it can take very different values for the different variables and is directly related to the precision of $\hat{Y}_{ab}^A$. If the precision of the estimate of the total for the overlapping sampling units is low, ($\text{var}(\hat{Y}_{ab}^A)$ is high) its contribution to the final estimate is low; thus, in some applications, the value of p is chosen equal to zero and the resulting estimator is called a screening estimator, since it requires the screening and elimination from the area frame of all the area sampling units included in the list sample:

$$\hat{Y} = \hat{Y}_a + \hat{Y}_{ab}^B,$$

with variance

$$\text{var}(\hat{Y}) = \text{var}(\hat{Y}_a) + \text{var}(\hat{Y}_{ab}^B).$$

3.11 Conclusions

The extraordinary increase in the ability to manipulate large sets of data makes it feasible to explore the more extensive use of administrative data and the creation of statistical systems based on administrative data. This is a way to save money, reducing response burden, producing figures for very detailed domains and allowing estimation of transition over time.

However, definitions, coverage and quality of administrative data depend on administrative requirements; thus they change as these requirements change – in the European Union a relevant example is given by the IACS data. Then information acquired is not exactly that needed for statistical purposes and sometimes objects in the administrative registers are partly the statistical units of the population for which the statistics have to be produced and partly something else; thus evaluating undercoverage and overcoverage is difficult.

Administrative departments collect data for specific purposes and carry out quality control in order to detect irregularities rather than to evaluate data quality. Such quality control can be misleading if used to estimate errors of commission and will have an effect on the declarations that can be influenced by complex dynamics, which are difficult to foresee and can produce a bias. Moreover, the comparability over time is strongly influenced by the change in the level of coverage from year to year.

The combined use of administrative registers improves the coverage of the population and the quality of data through comparison of data in the different registers and allows the socio-economic situation of rural households to be described, but very good identification variables and a very sophisticated record linkage system are needed and a great deal of statistical methodological work has to be done. The effect of imperfect matching on the estimates of parameters should be evaluated.

An approach to reducing the risk of bias due to undercoverage of administrative registers and, at the same time, avoiding double data acquisition is to sample farms from a complete and updated list and carry out record linkage with the administrative registers in order to capture register data corresponding to farms selected from the list. If the register is considered unreliable for some variables, related data have to be collected through interviews as well as data not found in the register due to record linkage difficulties.

A completely different way of taking advantage of administrative registers is to improve the efficiency of estimates based on a probabilistic sample survey by the use of register data as auxiliary variable at the design level (e.g. stratification) as well as at the estimator level in calibration estimators. The improvement in efficiency allows the same precision to be achieved, reducing sample size, survey costs and response burden.

When various incomplete registers are available but information included in their records cannot be directly used and a sample survey has to be designed, these registers can be treated as multiple incomplete lists from which separate samples can be selected. This way of treating different lists does not require record matching of listing units of the different lists.

When completeness is not guaranteed by the different registers, an area frame should be adopted in order to avoiding bias, since an area frame is always complete, and remains useful for a long time.

The most widespread way to avoid instability of estimates based on an area frame and to improve their precision is to adopt a multiple-frame sample survey design that combines the area frame with a list of very large operators and of operators that produce rare items. This approach is very convenient when the list contains units with large (thus probably more variable) values of some variable of interest and the survey cost of units in the list is much lower than in the area frame.

Acknowledgements

The authors are most grateful to many people in the following organizations for their support: FAO, Eurostat, World Bank, United States Department of Agriculture, Italian Ministry of Policies for Agriculture, Food and Forest, Istat, Consorzio ITA, AGEA, Statistics Sweden, UNECE and University of Pernambuco.

References

Carfagna, E. (1998) Area frame sample designs: a comparison with the MARS project. In T.E. Holland and M.P.R. Van den Broecke (eds) *Proceedings of Agricultural Statistics 2000*, pp. 261–277. Voorburg, Netherlands: International Statistical Institute.

Carfagna, E. (2001a) Multiple frame sample surveys: advantages, disadvantages and requirements. In International Statistical Institute, Proceedings, Invited papers, International Association of Survey Statisticians (IASS) Topics, Seoul, August 22-29, 2001, 253–270.

Carfagna, E. (2001b) Cost-effectiveness of remote sensing in agricultural and environmental statistics. In *Proceedings of the Conference on Agricultural and Environmental Statistical Applications in Rome* (CAESAR), 5–7 June, Vol. 3, pp. 618–627. http://www.ec-gis.org/.

Carfagna, E. (2004) List frames, area frames and administrative data: are they complementary or in competition? Invited paper presented to the Third International Conference on Agricultural Statistics, Cancún (Mexico), 2–4 November 2004. http://www.nass.usda.gov/mexsai/.

Carfagna, E. (2007) A comparison of area frame sample designs for agricultural statistics. *Bulletin of the International Statistical Institute*, 56th Session (Lisbon), Proceedings, Special Topics Contributed Paper Meetings (STCPM11), Agricultural and Rural Statistics, 1470, 22–29 August 22-29, pp. 1–4.

Consorzio ITA (2000) AGRIT 2000 Innovazione tecnologica. Studio per l'integrazione dati ADRIT-PAC, Ministero delle Politiche Agricole e Forestali.

De Gaetano, L. and Greco, M. (2009) Analisi di qualit dei dati amministrativi: confronto Clag e Agea. Comitato consultivo per la preparazione a livello regionale del sesto Censimento generale dell'agricoltura.

Deville, J.C. and Särndal, C.E. (1992) Calibration estimators in survey sampling, *Journal of the American Statistical Association*, 85, 376–382.

D'Orazio, M., Di Zio, M. and Scanu, M. (2006) *Statistical Matching: Theory and Practice*. Chichester: John Wiley & Sons, Ltd.

ESSnet ISAD (2008) Report of WP1. State of the art on statistical methodologies for integration of surveys and administrative data.

Ferraz, C. and Coelho, H.F.C. (2007) Ratio type estimators for stratified dual frame surveys. In Proceedings of the 56th Session of the ISI.

Fuller, W.A. and Burmeister, L.F. (1972) Estimators for samples from two overlapping frames. *Proceedings of the Social Statistics Section*, American Statistical Association, pp. 245–249.

Greco, M., Bianchi, G., Chialastri, L., Fusco, D. and Reale, A., (2009) Definizione di azienda agricola. Comitato consultivo per la preparazione a livello regionale del sesto Censimento generale dell'agricoltura.

Hartley, H.O. (1962) Multiple-frame surveys. *Proceedings of the Social Statistics Section*, American Statistical Association, pp. 203–206.

Hartley, H.O. (1974) Multiple Frame Methodology and Selected Applications, *Sankhyā, Series C*, 36, 99–118.

Korporal, K.D. (2005, Respondent relations challenges for obtaining agricultural data. *Proceedings of Statistics Canada Symposium 2005: Methodological Challenges for Future Information Needs*. http://www.statcan.gc.ca/pub/11-522-x/11-522-x2005001-eng.htm.

Kott, P.S. and Vogel, F.A. (1995) Multiple-frame business surveys. In B.G. Cox, D.A. Binder, B.N. Chinnappa, A. Christianson, M.J. Colledge and P.S. Kott (eds), *Business Survey Methods*, pp. 185–201. New York: John Wiley & Sons, Inc.

Lavallée, P. (2005) Quality indicators when combining survey data and administrative data. In *Proceedings of Statistics Canada Symposium 2005: Methodological Challenges for Future Information Needs*. http://www.statcan.gc.ca/pub/11-522-x/11-522-x2005001-eng.htm.

Lohr, S.L. and Rao, J.N.K. (2000) Inference from dual frame surveys. *Journal of the American Statistical Association*, 95, 271–280.

Lohr, S.L. and Rao, J.N.K. (2006) Estimation in multiple-frame surveys. *Journal of the American Statistical Association*, 101, 1019–1030.

Lund, R.E. (1968) Estimators in multiple frame surveys. *Proceedings of the Social Statistics Section*, American Statistical Association, pp. 282–288.

Moriarity, C. and Scheuren, F. (2001) Statistical matching: a paradigm for assessing the uncertainty in the procedure. *Journal of Official Statistics*, 17, 407–422.

Rao, J.N.K. (2003) *Small Area Estimation*. Hoboken, NJ: John Wiley & Sons, Inc.

Särndal C.E. (2007) The calibration approach in survey theory and practice. *Survey Methodology*, 33, 99–119.

Scanu, M. and Conti, P.L. (2006) Matching noise: formalization of the problem and some examples. *Rivista di Statistica Ufficiale*, 1, 43–55.

Selander, R., Svensson, J., Wallgren, A. and Wallgren, B. (1998) How should we use IACS data? Statistics Sweden.

Skinner, C.J. and Rao, J.N.K. (1996) Estimation in dual frame surveys with complex designs. *Journal of the American Statistical Association*, 91, 349–356.

Vogel, F.A. (1975) Surveys with overlapping frames – problems in applications. *Proceedings of the Social Statistics Section*, American Statistical Association, pp. 694–699.

Wallgren, A. and Wallgren, B. (1999) How can we use multiple administrative sources? Statistics Sweden.

Wallgren, A. and Wallgren, B. (2007) *Register-Based Statistics: Administrative Data for Statistical Purposes*. Chichester: John Wiley & Sons, Ltd.

Winkler, W.E. (1995) Matching and record linkage. In B.G. Cox, D.A. Binder, B.N. Chinnappa, A. Christianson, M.J. Colledge and P.S. Kott (eds), *Business Survey Methods*, pp. 355–384. New York: John Wiley & Sons, Inc.

4

Statistical aspects of a census

Carol C. House
US Department of Agriculture, National Agricultural Statistics Service, USA

4.1 Introduction

This chapter provides a basic overview of the statistical aspects of planning, conducting and publishing data from a census. The intention is to demonstrate that most (if not all) of the statistical issues that are important in conducting a survey are equally germane to conducting a census.

In order to establish the scope for this chapter, we begin by reviewing some basic definitions. *Webster's New Collegiate Dictionary* defines a 'census' to be 'a count of the population' and 'a property evaluation in early Rome' (Webster's, 1977). However, we will want to utilize a broader definition. The International Statistical Institute (ISI) in its *Dictionary of Statistical Terms* defines a census to be 'the complete enumeration of a population or group at a point in time with respect to well-defined characteristics' (International Statistical Institute, 1990). This definition is of more use. We now look at the term 'statistics' to further focus the chapter. Again from the ISI we find that statistics is the 'numerical data relating to an aggregate of individuals; the science of collecting, analysing and interpreting such data'. Together these definitions render a focus for this chapter – those issues germane to the science and/or methodology of collecting, analysing and interpreting data through what is intended to be a complete enumeration of a population at a point in time with respect to well-defined characteristics. Further, this chapter will direct its discussion to agricultural censuses. Important issues include the (sampling) frame, sampling methodology, non-sampling error, processing, weighting, modelling, disclosure avoidance, and data dissemination. This chapter touches on each of these issues as appropriate to the chapter's focus on censuses of agriculture.

Agricultural Survey Methods Edited by Roberto Benedetti, Marco Bee, Giuseppe Espa and Federica Piersimoni

4.2 Frame

Whether conducting a sample survey or a census, a core component of the methodology is the sampling frame. The frame usually consists of a listing of population units, but alternatively it might be a structure from which clusters of units can be delineated. For agricultural censuses, the frame is likely to be a business register or a farm register. Alternatively it might be a listing of villages from which individual farm units can be delineated during data collection. The use of an area frame is a third common alternative. Often more than a single frame is used for a census. Papers presented at the Agricultural Statistics 2000 conference highlight the diversity of sampling frames used for agricultural censuses (Sward *et al.*, 1998; Kiregyera, 1998; David, 1998).

There are three basic statistical concerns associated with sampling frames: coverage, classification and duplication. These concerns are equally relevant whether the frame will be used for a census or sampled for a survey.

4.2.1 Coverage

Coverage deals with how well the frame fully delineates all population units. The statistician's goal should be to maximize coverage of the frame and to provide measures of undercoverage. For agricultural censuses, coverage often differs by size of farming operation. Larger farms are covered more completely, and smaller farms less so. Complete coverage of smaller farms is highly problematic, and statistical organizations have used different strategies to deal with this coverage problem.

The Australian Bureau of Statistics (Sward *et al.*, 1998) intentionally excludes smaller farms from its business register and census of agriculture. It focuses instead on production agriculture, and maintains that its business register has good coverage for that target population. Statistics Canada (Lim *et al.*, 2000) has dropped the use of an area frame as part of its census of agriculture, and is conducting research on using various sources of administrative data to improve coverage of its farm register. Kiregyera (1998) reports that a typical agriculture census in Africa will completely enumerate larger operations (identified on some listing), but does not attempt to enumerate completely the smaller operations because of the resources required to do so. Instead it will select a sample from a frame of villages or land areas, and delineate small farms within the sampled areas for enumeration. In the United States, the farm register used for the 1997 Census of Agriculture covered 86.3% of all farms, but 96.4% of farms with gross value of sales over $10 000 and 99.5% of the total value of agricultural products. The USA uses a separate area sampling frame to measure undercoverage of its farm register, and has published global measures of coverage. They are investigating methodology to model undercoverage as part of the 2002 census and potentially publish more detailed measures of that coverage.

4.2.2 Classification

A second basic concern with a sampling frame is whether frame units are accurately classified. The primary classification is whether the unit is, in fact, a member of the target population, and thus should be represented on the frame. For example, in the

USA there is an official definition of a farm: an operation that sold \$1000 or more of agricultural products during the target year, or would normally sell that much. The first part of the definition is fairly straightforward, but the second causes considerable difficulty with classification.

Classification is further complicated when a population unit is linked with, or owned by, another business entity. This is an ongoing problem for all business registers. The statistician's goal is to employ reasonable, standardized classification algorithms that are consistent with potential uses of the census data. For example, a large farming operation may be a part of a larger, vertically integrated enterprise which may have holdings under semi-autonomous management in several dispersed geographic areas. Should each geographically dispersed establishment be considered a farm, or should the enterprise be considered a single farm and placed only once on the sampling frame? Another example is when large conglomerates contract with small, independent farmers to raise livestock. The larger firm (contractor) places immature animals with the contractee who raises the animals. The contractor maintains ownership of the livestock, supplies feed and other input expenses, then removes and markets the mature animals. Which is the farm: the contractor, the contractee, or both?

4.2.3 Duplication

A third basic concern with a sampling frame is duplication. There needs to be a one-to-one correspondence between population units and frame units. Duplication occurs when a population unit is represented by more than one frame unit. Similar to misclassification, duplication is an ongoing concern with all business registers. Software is available to match a list against itself to search for potential duplication. This process may eliminate much of the duplication prior to data collection. Often it is important in a census or survey to add questions to the data collection instrument that will assist in a post-collection evaluation of duplication. In conjunction with its 1997 Census of Agriculture, the USA conducted a separate 'classification error study'. For this study, a sample of census respondents was re-contacted to examine potential misclassification and duplication, and to estimate levels of both.

4.3 Sampling

When one initially thinks of a census or complete enumeration, statistical sampling may not seem relevant. However, in the implementation of agricultural censuses throughout the world, a substantial amount of sampling has been employed. David (1998) presents a strong rationale for extensive use of sampling for agricultural censuses, citing specifically those conducted in Nepal and the Philippines. The reader is encouraged to review his paper for more details. This chapter does not attempt an intensive discussion of different sampling techniques, but identifies some of the major areas where sampling has (or can be) employed.

Reducing costs is a major reason why statistical organizations have employed sampling in their census processes. We have already discussed how agricultural censuses in Africa, Nepal and the Philippines have used sampling extensively for smaller farms.

Sampling may also be used in quality control and assessment procedures. Examples include: conducting a sample survey of census non-respondents to assist in non-response adjustment; or conducting a specialized follow-up survey of census respondents to more carefully examine potential duplication and classification errors. The USA uses a sample survey based on an area frame to conduct a coverage evaluation of its farm register and census. It may be advantageous in a large collection of data to subdivide the population and use somewhat different questionnaires or collection methodologies on each group. Here again is a role for sampling. For example, in order to reduce overall respondent burden some organizations prepare both aggregated and detailed versions of a census questionnaire and use statistical sampling to assign questionnaire versions to the frame units. Alternatively sampling may facilitate efforts to evaluate the effect of incentives, to use pre-census letters as response inducements, or to examine response rates by different modes of data collection.

4.4 Non-sampling error

Collection of data generates sampling and non-sampling errors. We have already discussed situations in which sampling, and thus sampling error, may be relevant in census data collection. Non-sampling errors are always present, and generally can be expected to increase as the number of contacts and the complexity of questions increases. Since censuses generally have many contacts and fairly involved data collection instruments, one can expect them to generate a fairly high level of non-sampling error. In fact, David (1998) uses expected higher levels of non-sampling error in his rationale for avoiding complete enumeration in censuses of agriculture: a census produces 'higher non-sampling error which is not necessarily less than the total error in a sample enumeration. What is not said often enough is that, on account of their sizes, complete enumeration [censuses of agriculture] use different, less expensive and less accurate data collection methods than those employed in the intercensal surveys'.

Two categories of non-sampling error are response error and error due to non-response.

4.4.1 Response error

The literature (Groves, 1989; Lyberg *et al.*, 1997) is fairly rich in discussions of various components of this type of error. Self-enumeration methods can be more susceptible to certain kinds of response errors, which could be mitigated if interviewer collection were employed. Censuses, because of their large size, are often carried out through self-enumeration procedures. The Office for National Statistics in Britain (Eldridge *et al.*, 2000) has begun to employ cognitive interviewing techniques for establishment surveys much the same as they have traditionally employed for household surveys. They conclude that the 'use of focus groups and in-depth interviews to explore the meaning of terms and to gain insight into the backgrounds and perspectives of potential respondents can be very valuable'. They further conclude regarding self-administered collection that 'layout, graphics, instructions, definitions, routing etc. need testing'. Kiregyera (1998) additionally focuses readers' attention on particular difficulties that are encountered when collecting information from farmers in developing countries. These include the 'failure of holders to provide accurate estimates of crop area and production ... attributed to many causes including lack of knowledge about the size of fields and standard measurement units, or unwillingness to report correctly for a number of reasons (e.g. taboos, fear of taxation, etc.)'.

The statistician's role is fourfold: to understand the 'total error' profile of the census, to develop data collection instruments and procedures that minimize total error, to identify and correct errors during post collection processing, and to provide, as far as reasonable, measures of the important components of error.

4.4.2 Non-response

The statistician's role in addressing non-response is very similar to his/her role in addressing response error: to understand the reasons for non-response, to develop data collection procedures that will maximize response, to provide measures of non-response error, and to impute or otherwise adjust for those errors.

Organizations employ a variety of strategies to maximize response. These include publicity, pre-collection contacts, and incentives. Some switch data collection modes between waves of collection to achieve higher response rates. Others are developing procedures that allow them to target non-response follow-up to those establishments which are most likely to significantly impact the estimates (McKenzie, 2000).

A simple method for adjusting for unit non-response in sample surveys is to modify the sampling weights so that respondent weights are increased to account for non-respondents. The assumption in this process is that the respondents and non-respondents have similar characteristics. Most often, the re-weighting is done within strata to strengthen the basis for this assumption. A parallel process can be used for censuses. Weight groups can be developed so that population units within groups are expected to be similar in relationship to important data items. All respondents in a weight group may be given a positive weight, or donor respondents may be identified to receive a positive weight. Weight adjustment for item non-response, although possible, quickly becomes complex as it creates a different weight for each item.

Imputation is widely used to address missing data, particularly that due to item non-response. Entire record imputation is also an appropriate method of addressing unit non-response. Manual imputation of missing data is a fairly widespread practice in data collection activities. Many survey organizations have been moving towards more automated imputation methods because of concerns about consistency and costs associated with manual imputation, and to improve the ability to measure the impact of imputation. Automating processes such as imputation is particularly important for censuses because of the volume of records that must be processed.

Yost *et al.* (2000) identify five categories of automated imputations: (i) deterministic imputation, where only one correct value exists (such as the missing sum at the bottom of a column of numbers); (ii) model-based imputation, that is, use of averages, medians, ratios, regression estimates, etc. to impute a value; (iii) deck imputation, where a donor questionnaire is used to supply the missing value; (iv) mixed imputation, where more than one method used; and (v) the use of expert systems. Many systems make imputations based on a specified hierarchy of methods. Each item on the questionnaire is resolved according to its own hierarchy of approaches, the next being automatically tried when the previous method has failed. A nearest-neighbour approach based on spatial 'nearness' may make more sense for a census, where there is a greater density of responses, than it would in a more sparsely distributed sample survey.

4.5 Post-collection processing

Post-collection processing involves a variety of different activities, several of which (imputation, weighting, etc.) are discussed in other sections of this chapter. Here we will briefly address editing and analysis of data. Because of the volume of information associated with a census data collection, it becomes very important to automate as many of these edit and analysis processes as possible. Atkinson and House (2001) address this issue and provide several guiding principles that the National Agricultural Statistics Service is using in building an edit and analysis system for use on the 2002 Census of Agriculture: (a) automate as much as possible, minimizing required manual intervention; (b) adopt a 'less is more' philosophy to editing, creating a leaner edit that focuses on critical data problems; and (c) identify problems as early as possible.

Editing and analysis must include the ability to examine individual records for consistency and completeness. This is often referred to as 'micro' editing or 'input' editing. Consistent with the guiding principles discussed above, the Australian Bureau of Statistics has implemented the use of significance criteria in input editing of agricultural data (Farwell and Raine, 2000). They contend that 'obtaining a corrected value through clerical action is expensive (particularly if respondent re-contact is involved) and the effort is wasted if the resulting actions have only a minor effect on estimates'. They have developed a theoretical framework for this approach.

Editing and analysis must also include the ability to perform macro-level analysis or output editing. These processes examine trends for important subpopulations, compare geographical regions, look at data distributions and search for outliers. Desjardins and Winkler (2000) discuss the importance of using graphical techniques to explore data and conduct outlier and inlier analysis. Atkinson and House concur with these conclusions and further discuss the importance of having the macro-analysis tool integrated effectively with tools for user-defined ad hoc queries.

4.6 Weighting

When one initially thinks of a census, one thinks of tallying up numbers from a complete enumeration, and publishing that information in a variety of cross tabulations that add to the total. This chapter has already discussed a variety of situations in which weighting may be a part of a census process. In this section we focus on the interaction between weighting and the rounding of data values.

Many of the important data items collected in an agricultural census are intrinsically 'integral' numbers, making sense only in whole increments (i.e. the number of farms, number of farmers, number of hogs, etc.). For these data, desirable characteristics of the census tabulation is to have integer values at all published levels of disaggregation, and to have those cells sum appropriately to aggregated totals.

The existence of non-integer weights creates non-integer weighted data items. Rounding each of the multiple cell totals creates the situation that they may not add to rounded aggregate totals. This issue can be addressed in one of several ways. In the USA, the census of agriculture has traditionally employed the technique of rounding weights to integers, and then using these integerized weights. An alternative would be to retain the non-integer weights and round the weighted data to integers. A recent evaluation of census data in the USA (Scholetsky, 2000) showed that totals produced using the rounded weighted data values were more precise than the total produced using the integerized weights except for the demographic characteristics, number of farms, and ratio per farm estimates. A drawback to using rounded weighted data values is the complexity these procedures add to storing and processing information.

4.7 Modelling

Modelling can be effective within a census process by improving estimates of small geographical areas and rare subpopulations. Small-area statistics are perhaps one of the most important products from a census. However, a number of factors may impact the census's ability to produce high-quality statistics at fairly disaggregate levels. The highly skewed distribution of data, which is intrinsic to the structure of modern farming, creates estimation difficulties. For example, many larger operations have production units which cross the political or geographical boundaries used in publication. If data are collected for the large operation and published as if the 'whole' farm is contained within a single geographical area, this result will be an overestimate of agricultural production within that area and a corresponding underestimate within surrounding areas. Mathematical models may be used effectively to prorate the operation totals to appropriate geographical areas.

Census processes for measuring and adjusting non-response, misclassification and coverage may produce acceptable aggregate estimates while being inadequate for use at the more disaggregate publication levels. Statistical modelling and smoothing methodology may be used to smooth the measures so that they produce more reasonable disaggregate measures. For example, for the 1997 Census of Agriculture the USA provided measures of frame coverage at the state level for farm counts for major subpopulations. They are evaluating several smoothing techniques that, if successful, may allow the 2002 census release to include coverage estimates at the county level instead of just state level, and for production data as well as farm counts.

Although a census may be designed to collect all information from all population units, there are many cases in which circumstances and efficiencies require that census data not stand alone. We have already discussed methodologies in which a separate survey may be used to adjust census numbers for non-response, misclassification and/or coverage. Sometimes sources of administrative data are mixed with census data to reduce respondent burden or data collection costs. Most often the administrative data must be modelled to make them more applicable to the census data elements. Alternatively, some census collection procedures utilize a 'long' and 'short' version of the questionnaire so that all respondents are not asked every question. To combine the data from these questionnaire versions may also require some kind of modelling.

4.8 Disclosure avoidance

The use of disclosure avoidance methodology is critically important in preparing census and survey data for publication. Disclosure avoidance can be very complex for agricultural census publications because of the scope, complexity and size of these undertakings. Disclosure avoidance is made more difficult by the highly skewed nature of the farm population. Data from large, or highly specialized, farming operations are hard to disguise, especially when publishing totals disaggregated to small geographical areas.

Disclosure avoidance is typically accomplished through the suppression of data cells at publication. A primary suppression occurs when a cell in a publication table requires suppressing because the data for the cell violate some rule or rules defined by the statistical agency. Typical rules include:

(a) *threshold rule*: the total number of respondents is less than some specified number, i.e. the cell may be suppressed if it had fewer than 20 positive responses.

(b) (n, k) *rule*: a small number of respondents constitute a large percentage of the cell's value, for example a (2,60) rule would say to suppress if 2 or fewer responses made up 60% or more of the cell's value.

(c) *p per cent rule*: if a reported value for any respondent can be estimated within some specified percentage.

Secondary suppression occurs when a cell becomes a disclosure risk from actions taken during the primary suppression routines. These additional cells must be chosen in such a way as to provide adequate protection to the primary cell and at the same time make the value of the cell mathematically underivable.

Zayatz *et al.* (2000) have discussed alternatives to cell suppression. They propose a methodology that adds 'noise' to record level data. The approach does not attempt to add noise to each publication cell, but uses a random assignment of multipliers to control the effect of the noise on different types of cells. This results in the noise having the greatest impact on sensitive cells, with little impact on cells that do not require suppression.

4.9 Dissemination

Data products from a census are typically extensive volumes of interconnected tables. The Internet, CD-ROMs and other technical tools now provide statistical agencies with exciting options for dissemination of dense pools of information. This chapter will discuss several opportunities to provide high-quality data products.

The first component of a quality dissemination system is metadata, or data about the data. Dippo (2000) expounds on the importance of providing metadata to users of statistical products and on the components of quality metadata.

> Powerful tools like databases and the Internet have vastly increased communication and sharing of data among rapidly growing circles of users of many different categories. This development has highlighted the importance of metadata, since easily available data without appropriate metadata could sometimes be more harmful than beneficial.
>
> Metadata descriptions go beyond the pure form and contents of data. Metadata are also used to describe administrative facts about data, like who created them, and when. Such metadata may facilitate efficient searching and locating of data. Other types of metadata describe the processes behind the data, how the data were collected and processed, before they were communicated or stored in a database. An operational description of the data collection process behind the data (including e.g. questions asked to respondents) is often more useful than an abstract definition of the 'ideal' concept behind the data.

The Internet has become a focal point for the spread of information. Web users expect: to have sufficient guidance on use; to be able to find information quickly, even if they do not know precisely what they are looking for; to understand the database organization and naming conventions; and to be able to easily retrieve information once it is found. This implies the need, at a minimum, for high-quality web design, searchable databases, and easy-to-use print and download mechanisms. The next step is to provide tools such as interactive graphical analysis with drill-down capabilities and fully functional interactive query systems. Graphs, charts and tables would be linked, and users could switch between these different representations of information. Finally, there would be links between the census information and databases and websites containing information on agriculture, rural development, and economics.

4.10 Conclusions

Conducting a census involves a number of highly complex statistical processes. One must begin with a quality sampling frame, in which errors due to undercoverage, misclassification and duplication are minimized. There may be opportunities in which statistical sampling will help bring efficiency to the data collection or facilitate quality control measurements. Non-sampling errors will be present, and the design must deal effectively with both response and non-response errors. Post-collection processing should allow both micro- and macro-analysis. Census processing will probably involve weighting and some type of modelling. The dissemination processes should prevent disclosure of respondent data while providing useful access by data users.

References

Atkinson, D. and House, C. (2001) A generalized edit and analysis system for agricultural data. In *Proceedings of the Conference on Agricultural and Environmental Statistical Application in Rome*. International Statistical Institute.

David, I. (1998) Sampling strategy for agricultural censuses and surveys in developing countries. In T.E. Holland and M.P.R. Van den Broecke (eds) *Proceedings of Agricultural Statistics 2000*, pp. 83–95. Voorburg, Netherlands: International Statistical Institute.

Desjardins, D. and Winkler, W. (2000) Design of inlier and outlier edits for business surveys. In *Proceedings of the Second International Conference on Establishment Surveys*. Alexandria, VA: American Statistical Association.

Dippo, C. (2000) The role of metadata in statistics. In *Proceedings of the Second International Conference on Establishment Surveys*. Alexandria, VA: American Statistical Association.

Eldridge, J., Martin, J. and White, A. (2000) The use of cognitive methods to improve establishment surveys in Britain. In *Proceedings of the Second International Conference on Establishment Surveys*. Alexandria, VA: American Statistical Association.

Farwell, K. and Raine, M. (2000) Some Current Approaches to Editing in the ABS. In *Proceedings of the Second International Conference on Establishment Surveys*, pp. 529–538. Alexandria, VA: American Statistical Association.

Groves, R. (1989) *Survey Errors and Survey Costs*. New York: John Wiley & Sons, Inc.

International Statistical Institute (1990) *A Dictionary of Statistical Terms*. Harlow: Longman Scientific & Technical for the International Statistical Institute.

Kiregyera, B. (1998) Experiences with Census of Agriculture in Africa. In T.E. Holland and M.P.R. Van den Broecke (eds) *Proceedings of Agricultural Statistics 2000*, pp. 71–82. Voorburg, Netherlands: International Statistical Institute.

Lim, A., Miller, M., Morabito, J. (2000) Research into improving frame coverage for agricultural surveys at Statistics Canada. In *Proceedings of the Second International Conference on Establishment Surveys*. Alexandria, VA: American Statistical Association.

Lyberg, L., Biemer, P., Collins, M., De Leeuw, E., Dippo, C., Schwarz, N. and Trewin, D. (eds) (1997) *Survey Measurement and Process Quality*. New York: John Wiley & Sons, Inc.

McKenzie, R. (2000) A framework for priority contact of non respondents. In *Proceedings of the Second International Conference on Establishment Surveys*. Alexandria, VA: American Statistical Association.

Scholetzky, W. (2000) Evaluation of integer weighting for the 1997 Census of Agriculture. RD Research Report No. RD-00-01, National Agricultural Statistics Service, US Department of Agriculture, Washington, DC.

Sward, G., Hefferman, G. and Mackay, A. (1998) Experience with annual censuses of agriculture. In T.E. Holland and M.P.R. Van den Broecke (eds) *Proceedings of Agricultural Statistics 2000*, pp. 59–70. Voorburg, Netherlands: International Statistical Institute.

Webster's (1977) *Webster's New Collegiate Dictionary*. Springfield, MA: G. & C. Merriam.

Yost, M., Atkinson, D., Miller, J., Parsons, J., Pense, R. and Swaim, N. (2000) Developing a state of the art editing, imputation and analysis system for the 2002 Agricultural Census and beyond. Unpublished staff report, National Agricultural Statistics Service, US Department of Agriculture, Washington, DC.

Zayatz, L., Evans, T. and Slanta J. (2000) Using noise for disclosure limitation of establishment tabular data. In *Proceedings of the Second International Conference on Establishment Surveys*. Alexandria, VA: American Statistical Association.

5

Using administrative data for census coverage

Claude Julien
Business Survey Methods Division, Statistics Canada, Canada

5.1 Introduction

The Canadian Census of Agriculture will undergo a significant change in 2011: it will be conducted using only a mail-out/mail-back collection methodology. This is quite a change given that, until 2001, the Census relied mostly on a drop-off collection methodology with field enumerators canvassing the whole country. In order to ensure the transition from field enumeration to mail-out collection without affecting the coverage of farms, Statistics Canada has made increasing use of administrative data, notably its farm register, as well as significant improvements to the quality of this register over time. This chapter provides a chronological review of the last three censuses of agriculture. It highlights the progressive use of Statistics Canada's farm register and other administrative sources to assure the coverage of the census, peaking with the use of the register as a mail-out list in the upcoming 2011 Census.

The chapter begins with an overview of the main components of Statistics Canada's agriculture statistics programme that have traditionally or recently contributed to census coverage assurance. Each census conducted since 1996 is then described with particular attention to the increased use of administrative data to assure and measure census coverage.

Agricultural Survey Methods Edited by Roberto Benedetti, Marco Bee, Giuseppe Espa and Federica Piersimoni

5.2 Statistics Canada's agriculture statistics programme

Statistic Canada's agriculture statistics programme consists of a central farm register; numerous annual, sub-annual and occasional surveys; the use of administrative data, including tax data; and a census conducted every 5 years. Each of these components is described in this section.

Farm register The farm register is a database of farm operations and operators. Its creation dates back to the late 1970s, while its present design and technology were developed in the mid-1990s. It contains key identifiers such as farm name and operator name, address, telephone number, sex and date of birth. An operator can be associated with more than one farm and up to three operators can be associated with a single farm. It is updated after each census, as well as with regular survey feedback. The register contains many more records than there are farms or operators in the population. Usually, a farm operation or operator record is removed from the register only after there have been two censuses in which it did not respond.

Agriculture surveys Every year, Statistics Canada conducts between 40 and 45 agriculture surveys on crop production, livestock, farm finances, farm practices, and so on. Every survey is run from the farm register. Most surveys are conducted from regional offices using computer-assisted telephone interview technology. Survey feedback on business status or changes in operators or address/phone information is recorded during the interview and put on the farm register within a few weeks.

Tax data In order to reduce response burden and improve the quality of farm income and expenses data, Statistics Canada's agriculture statistics programme has used corporate and individual tax data since the mid-1980s. The tax universe comprises individual tax returns reporting farm income and corporate tax returns coded to the farm sector (North American Industrial Classification System) or reporting farm income or expenses. Basic information on names, addresses and farm income is available electronically for the whole population. Samples of tax records are selected and processed every year to collect more detailed financial information for statistical purposes. Furthermore, the whole tax population is matched to the farm register as a means to combine tax data with survey data for statistical analysis. The matching is done using a series of direct matches on key identifiers. Since 2000, this matching operation has become a key element for census coverage. It is described in step 1 of Section 5.4.

Census of Agriculture The Census of Agriculture is conducted every 5 years, in years ending in 1 or 6. A unique feature is that it is conducted concurrently with the Census of Population. Up to the 2001 Census, this feature was the main element of the census coverage strategy. It also enables Statistics Canada to link both censuses and create an agriculture–population database combining farm data with the demographic, social and economic data of its operators and their families. The next three sections of this chapter describe how the use of administrative data has evolved in assuring the coverage of the last three censuses.

5.3 1996 Census

In the 1996 Census the coverage assurance procedures were mostly conducted during field operations (Statistics Canada, 1997). Enumerators canvassed the country and dropped off a Census of Population questionnaire at each occupied dwelling. An agriculture questionnaire was also given when (a) a respondent declared at drop-off that someone in the household operated a farm or (b) there was an indication that someone living in the dwelling operated a farm. The Census of Population questionnaire also asked whether a member of the household operates a farm (henceforth referred to as the Step D question). The completed population and agriculture questionnaires were mailed back to the enumerators, who proceeded to reconcile the households that reported 'Yes' to the Step D question but had not mailed back an agriculture questionnaire (this operation is called the Step D field follow-up).

The use of administrative sources for coverage assurance during collection was limited to producing small lists of farms that required special collection arrangements, for example institutional farms or community pastures, and farms located in remote areas or urban-rural fringes. Other lists were mostly produced from the farm register or the previous census. Lists of farms from other sources, such as lists of producers from trade associations, were also used to validate certain census counts.

When processing and analysing the census data, it became apparent that they were lacking in coverage. To correct this situation, it was decided to make more use of the farm register to identify and contact farms that may not have been counted. Only a portion of the register was up to date, that being the farm operations that had recently received survey feedback. Consequently, some external lists were also used independently. This unplanned use of administrative data succeeded in raising the census coverage to historical levels. However, it was fairly costly and inefficient due to the short time used to set up automated and manual processes and the duplication of information from the various sources. Finally, there was no measurement of coverage.

Based on this experience, improvements were made to assure the coverage of farms in a more efficient manner. First, it was decided that the farm register would be the only administrative source to be used directly to assure census coverage. All other sources would be matched and reconciled to the farm register rather than being used independently. This was done to avoid duplication and contacting farms needlessly. Second, it was decided to use tax data to improve the coverage of the farm register. Although tax records (farm taxfilers) are conceptually different from farm register records (farm operators or contacts), they have the clear advantage over any other administrative source of covering all types of farms across the whole country (i.e. across all provinces and territories) at roughly the same level.

5.4 Strategy to add farms to the farm register

The following describes the process used to improve the coverage of a main source (M) with an external source (E), in this case the farm register and the tax data, respectively. This basic process is used in other components of the overall coverage assurance strategy described later in this chapter.

5.4.1 Step 1: Match data from E to M

Every year, tax returns are matched to the farm register. Individuals' tax records are matched to operator records using key identifiers (name, sex, date of birth and address), while corporate tax records are matched to operations using other key identifiers (farm name and address). The matches are done using direct matching techniques. Very strong matches are accepted automatically; weaker matches are verified; and the weakest matches are automatically rejected. The thresholds between what is 'very strong', 'weaker' and 'weakest' have been determined by prior studies and results observed over time. Every year, approximately 80% of all farm register operations and 70% of all tax records are matched together. Indirect or probabilistic matching techniques were also considered, but rejected due to marginal gains for the additional costs.

5.4.2 Step 2: Identify potential farm operations among the unmatched records from E

As indicated above, approximately 30% of all tax records do not match with the farm register. This is mostly due to conceptual differences, reporting errors and processing errors. Most of the unmatched tax records report very small farm income and do not involve much farming activity, if any. As a result, a cut-off sample of the farms reporting the largest farm income is selected for further processing. The size of the sample depends on the level of perceived coverage of the register, the timely need to improve its coverage and costs. For example, the sample would be much larger just before the census when the coverage of the register is likely at its lowest and the need for this coverage is at its highest.

The particular challenge with tax records is to avoid duplication caused by individuals reporting income from the same farm operations (i.e. partnerships). To identify such individuals, the reported farm income was used as an additional identifier. Individuals reporting the same 'unique looking' farm income located within the same geography were considered as being likely to operate the same farm. For example, the following individuals were considering to be operating the same farm:

- individuals located in the same municipality and reporting a very high farm income such as $145 656;
- individuals located at the same address and reporting a medium farm income such as $45 545.

As a counterexample, individuals who reported different farm income or a rounded farm income, such as $125 000, were not grouped together.

5.4.3 Step 3: Search for the potential farms from E on M

The matching step described in step 1 is based on the principle that a tax record is not matched until proven otherwise. For the purpose of improving the coverage of the farm register, this principle is reversed. In order to bring the tax record to the next step (collection), it must be determined with confidence that it is not on the register. The tax record must be searched by clerical staff using queries or printouts to find tax records that were not found previously due to incomplete information, nicknames, spelling or recording errors. Although this is a costly and time-consuming operation, it avoids contacting farm operations needlessly; also it is an operation that needs to be done later in the process anyway.

5.4.4 Step 4: Collect information on the potential farms

The potential farms are contacted by telephone to determine whether they are actual farms, as well as to obtain more up-to-date and accurate information on operators, key identifiers and basic production data (farm area, head of cattle, etc.).

5.4.5 Step 5: Search for the potential farms with the updated key identifiers

This final step combines some of the matches and searches conducted in steps 1 and 3 to determine whether a farm is on the farm register or should be added to it.

Julien (2000, 2001) and Miller *et al.* (2000) provide more information on the development and implementation of this process.

5.5 2001 Census

The collection methodology used in the 2001 Census was basically the same as in the 1996 Census (Statistics Canada, 2002). There were just a few adjustments, none of which had a significant impact on the coverage assurance operations conducted in the field. However, the coverage assurance strategy changed significantly once the census questionnaires were sent to head office and captured. In particular, for the first time since the 1981 Census, an assessment of coverage error was conducted.

5.5.1 2001 Farm Coverage Follow-up

A snapshot of the farm register was taken on Census Day, 14 May 2001. The snapshot included only farm operations that were coded as in business at the time. Based on previous census data and recent survey data, the majority of the farms in the snapshot records were flagged as 'must-get' and in scope for the Farm Coverage Follow-up (FCFU) operation. Shortly after the collection and capture of census questionnaires, a process similar to the one described in Section 5.4 was in progress between the census records as the main source (M) and the snapshot records as the external source (E). The must-get snapshot records that were not found among the census records were sent for collection. The follow-up collection period lasted 3–5 weeks, depending on the province. The FCFU farms with their updated data from collection were matched quickly to the census records once more; those that were confirmed as active and not enumerated by the field collection operation (i.e. not duplicates) were added to the census database in time for release.

5.5.2 2001 Coverage Evaluation Study

The snapshot records that were neither flagged as must-gets nor found among the census records composed the survey frame for a Coverage Evaluation Study (CES). A sample of these farms was selected, prepared, collected, matched and searched much as in the FCFU, but at a later time. The farms that were confirmed as active and not counted

Table 5.1 Census of Agriculture coverage statistics, 2001 and 2006

	2001	2006
Field collection	243 464	203 510
Farm register snapshot	299 767	303 653
Follow-up collection		
Population	9 543	70 229
Added to Census database	3 459	25 863
Final Census count	24 6923	229 373
Coverage Evaluation Study		
Frame	93 656	34 686
Sample	10 085	9 720
Farms missed in sample	1 668	2 839
Farms missed in population	14 617	7 839
Coverage error rate		
Farms	5.6%	3.4%
Farm area	N/A	1.3%
Farm sales	N/A	0.9%

in the census were weighted to produce an estimate of coverage error. It was the first measurement of coverage error produced since the 1981 Census, and it was produced quickly enough to be released at the same time as the census counts, one year after Census Day (Statistics Canada, 2002). Summary numbers on the coverage of the 2001 Census of Agriculture are provided in Table 5.1. The percentage of farms missed was estimated at 5.6%. Over half of the farms missed were very small; they reported less than $10 000 in farm sales. Consequently, the undercoverage of most farm commodity and financial data was much likely less than the undercoverage of farm counts.

The more extensive use of the farm register to assure census coverage proved to be successful. It also highlighted the importance of having good-quality information on key identifiers used for matching. During the processing of FCFU and CES data, a number of duplicate records had to be reconciled. Further research on improving the quality of the key identifiers also showed that some of the farms coded as missed in the CES had, in fact, been enumerated.

5.6 2006 Census

The collection methodology changed significantly in 2006 (Statistics Canada, 2006). First, the Census of Population used a mail-out/mail-back methodology to enumerate 70% of the population, while the rest of the population was canvassed using field enumerators as in previous censuses. In addition, all questionnaires were mailed back to a central location for data capture and editing rather than to field enumerators. The former change was concentrated in larger urban areas and affected a little less than 6% of the farm population. However, the latter modification greatly reduced the amount of coverage assurance conducted in the field, as it eliminated the Census of Population Step D follow-up described in Section 5.3. These changes and the experiences of the previous census required further improvements to the quality of the farm register. Additional efforts were made to increase the coverage of farms as well as to improve the completeness and accuracy of key identifiers.

A large editing and clean-up operation was carried out to improve the quality of the key identifiers (Lessard, 2005). The simple inversion of day, month or year of birth and the identification of invalid numbers led to over 7000 corrections to the date of birth. The matches on names and addresses to tax records, the links to the Census of Population, i.e. the agriculture–population links mentioned in Section 5.2, and matches to other administrative sources provided nearly 11 000 birthdates for the operator records on the register. Addresses and telephone numbers were also edited and standardized to confirm their completeness and accuracy.

The improvements to the key identifiers revealed a number of potential duplicate farms on the register. Prior to the census, several operators associated with more than one farm were contacted to determine the number of farms they were operating. As a result, over 5000 duplicate farms were removed from the farm register.

Tax records were not only used to add farms to the register, but also to change the status of some farm records from out of business to in business. Through negotiations with provincial departments and other organizations (e.g. marketing boards, organic farmers associations, crop insurance), a number of administrative lists of farms became available and were matched and reconciled to the farm register. Such an operation, which required quite a lot of time and resources a few years before, was run fairly smoothly, even with large lists, with the more up-to-date register.

In 2006, the Census of Population captured names, addresses and telephone numbers for the first time. This information was needed to conduct failed edit and non-response follow-up by telephone from the central processing offices. This information was also used to compensate for the loss of the Step D field follow-up by setting up a three-way matching operation between the census of agriculture, the farm register and the census of population. The results of this matching operation were used to conduct a Missing Farms Follow-up (MFFU) operation and a CES (Lachance, 2005).

5.6.1 2006 Missing Farms Follow-up

The 2006 MFFU was similar to the 2001 FCFU. A snapshot of the farm register was produced on Census Day. With additional knowledge and experience gained in the previous census, some farms coded as out of business were also included on the snapshot. It provided the addresses of farm operators or farm operations located in the Census of Population mail-out/mail-back area. The majority of snapshot records were flagged as must-gets for the MFFU. The main difference compared to 2001 was that households that had responded 'Yes' to Step D in the Census of Population (Step D households) were used to (1) automatically code some snapshot farms as active on Census Day and (2) to identify farms that were neither covered by the Census of Agriculture nor present on the farm register.

More specifically, snapshot records that (a) were not must-get farms, (b) had not matched to the Census of Agriculture records, (c) had reported farm activity in a survey in the 6 months prior to the MFFU collection period, and (d) had matched to a Step D household were automatically added to the census database. These small farms were assumed to be active on Census Day and not contacted to avoid needless response burden. In a similar manner, MFFU farms that had (a) not matched to census records, (b) had not responded in MFFU collection (i.e. not contacted or refused), and had either (c.1) reported farm activity in a survey in the 6 months prior to the MFFU collection period or (c.2) matched to a Step D household were also automatically added to the census database. All farms that were added automatically to the census database had their data imputed using census data from other farms and auxiliary information from the previous census, the register and recent surveys.

To compensate for the loss of the Step D field follow-up, the Step D households that had matched to neither the Census of Agriculture nor the farm register were considered for the MFFU. Prior censuses have shown that many false positive responses occur among Step D households – that is, when probed with more specific questions, these households report that they do not operate a farm. In spite of efforts to improve the wording of the question, many households reported operating a farm in the Census of Population, including over 36 000 that did not match to other sources. The Census of Agriculture had set up a strategy for using other geographic, demographic and financial characteristics of the households to identify ones that were more likely to operate a farm. The other Step D households would be considered for the Coverage Evaluation Survey. As a result, over 18 000 Step D households were selected for the MFFU, 7000 of which confirmed operating a farm that had not been enumerated. Like other MFFU records, these farms were added to the census database prior to data release.

5.6.2 2006 Coverage Evaluation Study

The CES was conducted much like the 2001 CES. The sampling frame included the snapshot records that (a) had been excluded from the MFFU, (b) had not matched to a Census of Agriculture record and (c) were coded as in business on the snapshot. The frame also included all other Step D households that (a) had matched to neither the snapshot nor the census and (b) had been excluded from the MFFU.

The proportion of farms missed was estimated at 3.4% (see Table 5.1). This time, commodity and financial data were asked for and the undercoverage of total farm area and farm sales was estimated at 1.3% and 0.9%, respectively. Furthermore, the CES records were processed quickly enough to produce estimates of undercoverage that were timely enough to be used in the validation and certification of census counts. Once again, the coverage error estimates were released at the same time as the farm counts (Statistics Canada, 2007).

The 2006 Census proved that the farm register, combined with other administrative sources, had become essential to assure the coverage in an efficient and timely manner. With the changes in field collection methodology, the register contributed to 11% of the final farm count compared to 1% in 2001. It also contributed to reducing the undercoverage of farms from 5.6% to 3.4%. Finally, the timely processing of agriculture, farm register and population records brought critical quality indicators to the analysis and validation of decreasing farm counts prior to their release.

However, the 2006 Census did not prove that the farm register and administrative data can assure census coverage by themselves. Data collected concurrently by the Census of Population contributed a little over 7000 farms (3.1% of the final count) and identified another 1700 farms that were missed. These results, both positive and negative, are critical given that the Census of Population will be taking further steps towards a full mail-out census in 2011 when approximately 80% of the population will be mailed out.

The Census of Population will also introduce other changes to encourage respondents to fill their questionnaires on-line. These changes, the experiences gained in 2006 and further improvements to the farm register have led Census of Agriculture to decide to go 'full mail-out' in 2011 with the farm register as its only source for names, addresses and telephone numbers.

5.7 Towards the 2011 Census

In preparing the farm register for a full mail-out census, other operations were added to those introduced since the 1996 Census. Firstly, more effort and resources are being invested in the process which identifies and negotiates access to external lists. Priority is placed on larger lists with broad coverage (e.g. a registration list of all farms in a given province) followed by specific lists for more typically undercovered or specialty operations (e.g. beekeeper lists, organic producers).

Secondly, a Continuous Farm Update (CFU) is now conducted on a quarterly basis. At every occasion, approximately 1000 potential duplicate farms are contacted by telephone, while 3000 potential new farms identified from various sources are contacted by mail. The mail-out is done at the start of the quarter and telephone follow-ups with non-respondents are done toward the end of the quarter. So far, the mail return rate is approximately 35%. The CFU is used, among other things, to assess the accuracy and usefulness of external lists on a sample basis before deciding to use them more extensively.

Thirdly, efforts to further improve the quality of the farm register will culminate with a much larger sample for the CFU in 2010. This sample will also include operations on the farm register that will not have been contacted by a survey since the 2006 Census, operations with incomplete addresses (contacted by telephone) and out-of-business farms that have signals indicating that they may be back in business or coded inaccurately.

Finally, the mail-out methodology is being tested in the 2009 National Census Test. The primary objective for the Census of Agriculture will be to test the full mail-out collection methodology, including all systems and processes that support it. The test will be conducted on nearly 2800 farms located in the Census of Population test areas and 4500 other farms located throughout Canada. The test will evaluate the accuracy of the mailing addresses extracted from the farm register. For example, post office returns (i.e. questionnaires that could not be delivered) will be quantified and analysed to identify potential problems and find possible solutions.

5.8 Conclusions

In 20 years, Statistics Canada has made immense progress in the use of administrative data to assure the coverage in its census of agriculture. From the use of small lists of farms to assist field collection operations before the 2001 Census, administrative data maintained on a central farm register will be used in a full mail-out census in 2011. This evolution can be summarized in the following lessons learned.

- The use of administrative data is a good means of assuring coverage of farms, especially the large and medium sized ones. To assure good coverage of all farms, it is best to combine administrative data with another, much broader source, such as a concurrent or recent census of population in order to identify all farm operators.

- Gather all sources of administrative information into a single source, such as a central farm register database. Use only that source to assure and measure the coverage of the census.
- Make this database the central element from which all or most intercensal survey activity is carried out. In one direction, the farm register is used to create survey frames and select samples; in the other direction, survey feedback provides updates on the farms' status and on the persons operating them.
- Match all possible sources to the register. Set up a matching and reconciliation protocol and apply it as systematically as possible to all sources. Contact the farms that do not match to the register to get accurate identifiers before deciding to add them to the register.
- Invest time and resources in keeping key identifiers as complete and accurate as possible, including farm names and operation name, sex, date of birth, address and telephone number. Data available from the external sources, commercially available address list or telephone directories, as well as address verification software should be used to update key identifiers.
- Use all information on farms from other sources that match to the farm register to complete or correct key identifiers on the register; that is, use other sources not only to add records to the register, but also to update existing records.
- Administrative sources vary greatly in terms of coverage and accuracy of information. Prioritize those with broader and timely coverage. Investigate the usefulness of an administrative source on a sample or pilot-test basis first before using it on an extensive basis.
- The farm register usually evolves on a near-daily basis through survey feedback. Take a snapshot of it as a reference population for the census; in return, update the register with data collected by the census.
- Use previous census data, recent survey data or any other key characteristics to prioritize records on the reference population; this prioritization should be carried out before the census collection operation. Not all records can be contacted or reconciled to the census. The portion of the reference population with the lowest priority (mostly smaller farms) should be processed on a sample basis to produce estimates of undercoverage.
- Correctly enumerating farms operated by the same operator is important and very difficult. Contact multi-farm operators before the census to ensure the accuracy of the register.
- Update the register on a regular basis, rather than in one large operation just before the census. The updates can reduce undercoverage in surveys conducted between censuses. Such surveys are, in return, a good means to further confirm the existence of new farms.

In spite of all the efforts put into creating and maintaining a farm register, the experiences in the recent censuses of agriculture at Statistics Canada have shown some limitations on its ability to assure full coverage of farms, in particular smaller farms. In 2006, the concurrent census of population still contributed, by itself, 3.1% of all farms in the final count and identified another 0.7% that had been missed. It is essential that a census conducted mainly from a farm register be supplemented by another collection operation, such as one based on an area frame or, as is the case in Canada, by a concurrent census of population.

The National Census Test conducted in May 2009 will provide a good assessment of a full mail-out census off the farm register. If all goes according to plan, the farm register will be the main element of the coverage assurance strategy for the 2011 Census. Until then, the farm register will be updated regularly with survey feedback, tax data and other administrative lists.

Afterwards, the need to rely on administrative information to update the farm register will only grow further, as Statistics Canada has recently decided to assess the feasibility of integrating the farm register into its central business register. As a result, census coverage will rely even more on administrative data of all sorts, tax data in particular. The methods and systems recently developed to use sub-annual tax data in business surveys will likely be very useful. From a more long-term perspective, the results of the 2011 Coverage Evaluation Study and the potential impact of running agriculture surveys and censuses from the business register may require looking at means of integrating census undercoverage estimates into the final census counts in the 2016 Census.

Acknowledgements

The author wishes to thank Steven Danford, Lynda Kemp and Dave MacNeil for their valuable comments and suggestions during the writing of this chapter, as well as Martin Lachance and Martin Lessard for pursuing the research and development in this area.

References

Julien, C. (2000) Recent developments in maintaining survey frames for agriculture surveys at Statistics Canada. In *Proceedings of the Second International Conference on Establishment Surveys*. Alexandria, VA: American Statistical Association.

Julien, C. (2001) Using administrative data for census coverage. *Conference on Agricultural and Environmental Statistics in Rome (CAESAR)*, Essays No. 12-2002, Volume II, Istat, pp. 303–312.

Lachance, M. (2005) La couverture du recensement de l'agriculture 2006. *Actes du 5-ème colloque francophone sur les sondages*. http://www.mat.ulaval.ca/la_recherche/sondages_2005/actes_du_col-loque_de_quebec/index.html.

Lessard, M. (2005) Évaluation, amÉlioration de la qualitÉ du registre des fermes et des bÉnÉfices retirÉs. *Actes du 5-ème colloque francophone sur les sondages*. www.mat.ulaval.ca/la_recherche/sondages_2005/actes_du_colloque_de_que-bec/index.html.

Miller, M., Lim, A. and Morabito, J. (2000) Research into improving frame coverage for agriculture surveys at Statistics Canada. In *Proceedings of the Second International Conference on Establishment Surveys*. Alexandria, VA: American Statistical Association.

Statistics Canada (1997) *1996 Census Handbook*. Catalogue No. 92-352-XPE, 20-32.

Statistics Canada (2002) *2001 Census of Agriculture – Concepts, Methodology and Data Quality*. http://www.statcan.gc.ca/pub/95f0301x/4064744-eng.htm.

Statistics Canada (2006) *Overview of the Census of Population: Taking a Census of Population – Data Collection*. http://www12.statcan.gc.ca/english/census06/reference/dictionary/ovpop2b.-cfm#10.

Statistics Canada (2007) *2006 Census of Agriculture – Concepts, Methodology and Data Quality*. http://www.statcan.gc.ca/pub/95-629-x/2007000/4123850-eng.htm.

Part II

SAMPLE DESIGN, WEIGHTING AND ESTIMATION

6

Area sampling for small-scale economic units

Vijay Verma, Giulio Ghellini and Gianni Betti
Department of Quantitative Methods, University of Siena, Italy

6.1 Introduction

Much discussion of sampling methods, including in textbooks on the subject, is confined to the design of population-based surveys, that is, surveys in which households (or sometimes individual persons) form the ultimate units of selection, collection and analysis. The theory and practice of large-scale population-based sample surveys is reasonably well established and understood.

In this chapter we discuss some special considerations which arise in the design of samples for what may be termed economic surveys, as distinct from population-based household surveys. By *economic surveys* we mean surveys concerned with the study of characteristics of economic units, such as agricultural holdings, household enterprises, own-account businesses and other types of establishments in different sectors of the economy. We address some issues concerning sampling in surveys of small-scale economic units in developing countries, units which, like households, are small, numerous and widely dispersed in the population, but differ from households in being much more heterogeneous and unevenly distributed. Units which are medium to large in size, few in number or are not widely dispersed in the population normally require different approaches, often based on list frames. There is a vast body of literature concerning sampling economic and other units from lists (e.g. Cox *et al.*, 1995). We are concerned here with a different type of problem.

In specific terms, the sampling issue addressed is the following. The population of interest comprises small-scale economic units ('establishments') of different types

Agricultural Survey Methods Edited by Roberto Benedetti, Marco Bee, Giuseppe Espa and Federica Piersimoni

('sectors'). A single integrated sample of establishments covering different sectors is required. However, there is a target sample size to be met separately for each type or sector of units. There is no list frame for the direct selection of establishments covering the whole population. For this and/or other reasons, the sample has to be selected in multiple stages: one or more area stages, followed by listing and selection of establishments within each (ultimate) area unit selected. For the selection of establishments within each area it is highly desirable, for practical reasons, to apply a uniform procedure and the same sampling rate for different types (sectors) of units. The central requirement of such a design is to determine how the overall selection probability of establishments may be varied across area units such that the sample size requirements by economic sector are met. This has to be achieved under the constraint that the procedure for the final selection of establishments within any sample area is identical for different types of establishments. Below we propose a practical procedure for achieving this. We develop a useful technique which involves defining 'strata of concentration' classifying area units according to the predominant type(s) of economic units contained in the area, and use this structure to vary the selection probabilities in order to achieve the required sample sizes by type (sector) of units. We develop various strategies for achieving this, and evaluate them in terms of the efficiency of the design. This type of design has a wide variety of practical applications, such as surveys of small agricultural holdings engaged in different types of activities, small-scale economic units selling or producing different goods, or – as an example from the social field – surveys involving the selection of schools while controlling the allocation of the sample according to ethnic group of the student population, under the restriction that no reference to the ethnic group is allowed in the procedure for selecting students within any given school.

6.2 Similarities and differences from household survey design

The type of sample designs used in 'typical' household surveys provides a point of departure in this discussion of multi-stage sampling of small-scale economic units. Indeed, there may often be a one-to-one correspondence between such economic units and households, and households rather than the economic units themselves may directly serve as the ultimate sampling units. Nevertheless, despite much common ground with sampling for population-based household surveys, sampling small-scale economic units involves a number of different and additional considerations.

It is useful to begin by noting some similarities between the two situations, and then move on to identify and develop special features of sampling for surveys of small-scale economic units.

6.2.1 Probability proportional to size selection of area units

National or otherwise large-scale household surveys are typically based on multi-stage sampling designs. Firstly, a sample of area units is selected in one or more stages, and at the last stage a sample of ultimate units (dwellings, households, persons, etc.) is selected within each sample area. Increasingly – especially in developing countries – a more or less standard two-stage design is becoming common. In this design the first stage consists of the selection of area units with probability proportional to some measure of size (N_k), such as the estimated number of households or persons in area k from some external source providing such information for all areas in the sampling frame. At the second stage, ultimate units are selected within each sample area with probability inversely proportional to a (possibly alternative) measure of size N'_k . The overall probability of selection of a unit in area k is

$$f_k = \frac{aN_k}{N} \frac{b}{N'_k} = \frac{N_k}{N'_k} f. \tag{6.1}$$

Here a, b, N and f are constants, a being the number of areas selected and N the total of N_k values in the population; if N'_k refers to the actual (current) size of the area then b is the constant number of ultimate units selected per sample area, giving $ab = n$ as the sample size; the constant f stands for an overall sampling rate

$$f = \frac{ab}{N} = \frac{n}{N}.$$

The denominator in (6.1) may be the same as N_k (the measure of size used at the first stage), in which case we get a 'self-weighting' sample with $f_k = f =$ constant. Alternatively, it may be the actual size of the area, in which case we get a 'constant take' design, that is, with a constant number b of ultimate units selected from each sample area irrespective of the size of the area. It is also possible to have N'_k as some alternative measure of size, for instance representing a compromise between the above two designs. In any case, N_k and N'_k are usually closely related and are meant to approximate the actual size of the area.[1]

It is common in national household surveys to aim at self-weighting or approximately self-weighting designs. This often applies at least within major geographical domains such as urban–rural divisions or regions of the country.

The selection of ultimate units within each sample area requires lists of these units. Existing lists often have to be updated or new lists prepared for the purpose to capture the current situation. No such lists are required for areas not selected at the first stage. The absence of up-to-date lists of ultimate units for the whole population is a major reason for using area-based multi-stage designs.

Now let us consider a survey of small-scale economic units such as agricultural holdings or other types of household enterprises in similar circumstances. Just like households, such units tend to be numerous and dispersed in the population. Indeed, households themselves may form the ultimate sampling units in such surveys, the economic units of interest coming into the sample through their association with households. Similar to the situation with household surveys, typically no up-to-date lists of small-scale economic units are available for the entire population. This requires resorting to an area-based multi-stage design, such as that implied by (6.1) above.

Despite the above-noted similarities, there are certain major differences in the design requirements of population-based household surveys and surveys of small-scale establishments (even if often household-based). These arise from differences in the type and distribution of the units involved and in the reporting requirements.

[1] Throughout this chapter, 'size' refers to the number of ultimate units in the area, not to its physical size.

6.2.2 Heterogeneity

Household surveys are generally designed to cover the entire population uniformly. Different subgroups (households by size and type, age and sex groups in the population, social classes, etc.) are often important analysis and reporting categories, but (except possibly for geographical subdivisions) are rarely distinct design domains. By contrast, economic units are characterized by their heterogeneity and by much more uneven spatial distribution. The population comprises multiple economic sectors – often with great differences in the number, distribution, size and other characteristics of the units in different sectors – representing types of economic activities to be captured in the survey, possibly using different questionnaires and even different data collection methodologies. Separate and detailed reporting by sector tends to be a much more fundamental requirement than it is in the case of different population subgroups in household surveys. The economic sectors can, and often do, differ greatly in size (number of establishments in the population) and in sample size (precision) requirements, and therefore in the required sampling rates. Hence an important difference from household surveys is that economic surveys covering different types of establishments often require major departures from a self-weighting design.

6.2.3 Uneven distribution

These aspects are accentuated by uneven geographical distribution of establishments of different types. Normally, different sectors to be covered in the same survey are distributed very differently across the population, varying from (1) some sectors concentrated in a few areas, to (2) some sectors widely dispersed throughout, but with (3) many unevenly distributed sectors, concentrated or dispersed to varying degrees. These patterns of geographical distribution have to be captured in the sampling design. True, population subgroups of interest in household surveys can also differ in their distribution (such as that implied in the typology of 'geographical', 'cross' and 'mixed' subclasses proposed by Kish *et al.* (1976), but normally type (2) rather than type (3) predominates in household surveys. By contrast, often situation (3) predominates in economic surveys. The sampling design must take into account the pattern of distribution in the population of different types of units.

6.2.4 Integrated versus separate sectoral surveys

There are a number of other factors which make the design of economic surveys more complex than that of household surveys. Complexity arises from the possibility that the ultimate units used in sample selection may not be of the same type as the units involved in data collection and analysis. The two types of units may lack one-to-one correspondence. For instance, the ultimate sampling units may be (often are) households, each of which may represent no, one, or more than one establishment of interest. For instance, the same household may undertake different types of agricultural activities. Hence, seen in terms of the ultimate sampling units (e.g. households), different sectors (domains) of establishment are not disjoint but are overlapping. This gives rise to two possible design strategies: (1) an integrated design, based on a common sample of units such as households, in which all sectors of activity engaged in by a selected unit would be covered simultaneously; and

(2) separate sectoral designs in which, in terms of the sampling units such as households, the samples for distinct sector populations may overlap. In each sectoral survey, activity of the selected households pertaining only to the sector concerned would be enumerated.

Separate sectoral surveys tend to be more costly and difficult to implement. The overlap between the sectoral samples may be removed by characterizing the sampling units (households) in terms of their predominating sector. This helps to make the sampling process more manageable. However, this precludes separate sectoral surveys: if the sample for any particular sector is restricted only to the households in which that sector predominates over all other sectors, the coverage of the sector remains incomplete. An integrated design, covering different sectors simultaneously, is often preferable.

6.2.5 Sampling different types of units in an integrated design

An integrated multi-stage design implies the selection of a common sample of areas to cover all types of units of interest in a single survey. The final sampling stage involves the selection of ultimate units (e.g. establishments) within each selected area. In such a survey, in practice it is often costly, difficult and error-prone to identify and separate out the establishments into different sectors and apply different sampling procedures or rates by sector within each sample area. Hence it is desirable, as far as possible, to absorb any differences in the sampling requirements by sector at preceding area stage(s) of sampling, so as to avoid having to treat different types of units differently at the ultimate stage of sampling. This means that, for instance, any differences in the required sampling rates for different sectors have to be achieved in the preceding stages, by distinguishing between areas according to their composition in terms of different types of establishments.

The cost of such (very desirable) operational simplification is, of course, the increased complexity of the design this may involve.

6.3 Description of the basic design

Let us now consider some basic features of an area-based multi-stage sampling design for an integrated survey covering small-scale economic units of different types. The population of units comprises a number of sectors, such as different types of establishments. We assume that, on the basis of some criterion, each establishment can be assigned to one particular sector. Sample size requirements in terms of number of establishments $n_{\cdot i}$ have been specified for each sector i. The available sampling frame consists of area units (which form the primary sampling units in a two-stage design, or the ultimate area units in a design with multiple area stages). Information is available on the expected number of establishments N_{ki} in each area unit k by sector i, and hence on their total $N_{k\cdot}$ for the area and total $N_{\cdot i}$ for each sector.[2] It is assumed that the above measures are scaled such that $N_{\cdot i}$ approximates the actual number of establishments belonging to sector i.

[2] Note that this information requirement by sector is more elaborate than a single measure of population size normally required for probability proportional to-size sampling in a household survey. It is important that, especially in the context of developing countries with limited administrative sources, potential sources such as population, agricultural and economic censuses are designed to yield such information, required for efficient design of surveys of small-scale economic units during the post-census period.

The required average overall probability for selecting establishments varies by sector:

$$f_i = \frac{n_{\cdot i}}{N_{\cdot i}}. \tag{6.2}$$

However, the design requirement is that within each area, different types of establishments are selected at the same rate, say g_k, giving the *expected* number of units of sector i contributed to the sample by area k as

$$n_{ki} = g_k N_{ki}.$$

Note that the reference here is to the expected value of the contribution of the area to the sample. In a multi-stage design, the actual number of units contributed by any area will be zero if the area is not selected at the preceding stage(s), and generally much larger if the area has been selected.

The sample size constraints by sector require that the g_k are determined such that the relationship

$$\sum_k g_k . N_{ki} = n_{\cdot i} \tag{6.3}$$

is satisfied simultaneously for all sectors i in the most efficient way. The criterion of efficiency also needs to be defined (see next section).

The sum in (6.3) is over all areas k in the population (and not merely the sample). As noted, the above formulation assumes that at the ultimate sampling stage, establishments within a sample area are selected at a uniform rate, g_k, irrespective of the sector. As noted earlier, this is a very desirable feature of the design in practice. It is often costly, difficult and error-prone to identify and separate out the ultimate survey units into different sectors and apply different sampling procedures or rates by sector. The preceding sampling stages are assumed to absorb any difference in the required sampling rates by sector by incorporating those in the area selection probabilities in some appropriate form.

The most convenient (but also the most unlikely) situation in the application of (6.3) is when units of different types (sectors) are geographically completely segregated, that is, when each area unit contains establishments belonging to only one particular sector. Denoting areas in the set containing establishments of sector i only (and of no other sector) by $k(i)$, it can be seen that (6.3) is simply (6.2):

$$g_{k(i)} = f_i = \frac{n_{\cdot i}}{N_{\cdot i}}. \tag{6.4}$$

In reality the situation is more complex because area units generally contain a mixture of establishments of different types (sectors), and a simple equation like (6.4) cannot be applied. Clearly, we should inflate the selection probabilities for areas containing proportionately more establishments from sectors which need to be over-sampled, and vice versa. These considerations need to be quantified more precisely. We know of no exact or theoretical solutions to equation (6.3), and have to seek empirical (numerical) solutions determining g_k for different areas depending on their composition in terms of different types of establishments, solutions which involve trial and error and defy strict optimization.

6.4 Evaluation criterion: the effect of weights on sampling precision

Equation (6.2), if it can be applied, gives an equal probability or self-weighting sample for each sector i meeting the sample size requirements in terms of number of establishments $n_{.i}$ to be included.

The constraint is that within each area different types of establishments are selected at the same rate g_k determined by (6.3). This means that the establishment probabilities of selection have to vary within the same sector depending on the area units from which they come. This design is generally less efficient than a self-weighting sample within each sector.

6.4.1 The effect of 'random' weights

The design effect, which measures the efficiency of a sample design compared to a simple random sample of the same size, can be decomposed under certain assumptions into two factors:

- the effect of sample weights;
- the effect of other aspects of the sample design, such as clustering and stratification.

We are concerned here with the first component – the effect of sample weights on precision. This effect is generally to inflate variances and reduce the overall efficiency of the design. The increase in variance depends on the variability in the selection probabilities $g_{k.}$ or the resulting design weights. (The design weight to be applied at the estimation stage is inversely proportional to the overall selection probability, i.e. proportional to $1/g_{k.}$). We use the following equations to compare different choices of the g_k values in terms of their effect on efficiency of the resulting sample in an empirical search for the best, or at least a 'good', solution.

It is reasonable to assume that in the scenario under discussion weights are 'external' or 'arbitrary', arising only from sample allocation requirements, and are therefore essentially uncorrelated with population variances. It has been established theoretically and empirically that the effect of such weighting tends to persist uniformly across estimates for diverse variables and population subclasses, including estimates of differentials and trends, and is well approximated by the expression (Kish, 1992)

$$D^2 = \left(1 + cv^2(w_u)\right),$$

where $cv(w_u)$ is the coefficient of variation of the weights of the ultimate units in the sample. The expression approximates the factor by which sampling variances are inflated, that is, the effective sample size is reduced.

With weights w_u for individual units u in the sample of size n, the above expression can be written as

$$D^2 = \frac{\frac{1}{n}\sum w_u^2}{\left(\frac{1}{n}\sum w_u\right)^2}$$

the sum being over units in the sample, or, for sets of n_k units with the same uniform weight w_k,

$$D^2 = \frac{\sum n_{k.} w_k^2 / \sum n_k}{\left(\sum n_{k.} w_k / \sum n_k\right)^2}.$$

6.4.2 Computation of D^2 from the frame

It is useful to write the above equations in terms of weights of units in the population, so that different design strategies can be evaluated without actually having to draw different samples:

$$D^2 = \frac{\sum 1/w_u}{N} \frac{\sum w_u}{N},$$

or, for sets of N_{ki} units with the same uniform weight w_k,

$$D_i^2 = \frac{\sum N_{ki}/w_k}{\sum N_{ki}} \frac{\sum N_{ki} w_k}{\sum N_{ki}}, \qquad w_k = \frac{1}{g_k}. \tag{6.5}$$

6.4.3 Meeting sample size requirements

The above equations can be applied to the total population or, as done in (6.5), separately to each sector i. As noted, it is assumed that subsampling within any area k is identical for all sectors i, implying uniform weights $w_k = 1/f_k$ for all types of units in the area. The average of D_i^2 values over I sectors,

$$\bar{D}^2 = \sum_i \frac{D_i^2}{I}, \tag{6.6}$$

may be taken as an overall indicator of the inflation in variance due to weighting. The objective is to minimize this indicator by appropriately choosing the g_k values satisfying the required sample size constraints (6.3).

The loss $\bar{D}^2$ due to variation in weights has to be balanced against how closely we are able to obtain the target sample sizes by sector. We may measure the latter by, for instance, the mean absolute deviation

$$\bar{E} = \frac{\sum_i |\Delta n_i|}{\sum_i n_i}, \tag{6.7}$$

where n_i is the target sample size for sector i, and Δn_i is the discrepancy between this and the sample size achieved with the selection procedure being considered.

The loss due to weighting also has an effect on the effective sample size actually achieved in each sector. The effective sample size $n'_{i.}$ in the presence of arbitrary weights is smaller than actual sample size $n_{i.}$ by the factor $1/D_i^2$. The sectoral sample size constraints should be seen as applying to $n'_{i.}$ rather than to $n_{i.}$. Usually, as we try to satisfy the sample size constraints more precisely, the loss in precision due to weighting tends to increase, which may be more marked in certain sectors than others. Often sample size constraints have to be relaxed in order to limit the loss in efficiency due to non-uniform weights.

6.5 Constructing and using 'strata of concentration'

6.5.1 Concept and notation

In order to apply variable sampling rates to achieve the required sample size by sector, it is useful to begin by classifying areas into groups on the basis of which particular sector predominates in the area. The basic idea is as follows.

For each sector, the corresponding 'stratum of concentration' is defined to consist of the set of area units in which that sector 'predominates' in the sense defined below. One such stratum corresponds to each sector. The objective of constructing such strata is to separate out areas according to an important aspect of their composition in terms of economic sectors. In order to distinguish these from 'ordinary' strata used for sample selection, we will henceforth refer to them as 'StrCon'.

Denote by N_{ki} the number of ultimate units (establishments) of sector i, $i = 1, \ldots, I$, in area k; by $N_{k\cdot}$ the total number of establishments in area k (all sectors combined); and by $N_{\cdot i}$ the total number of establishments of sector i in the population. Let A_i be the number of areas containing an establishment of sector i (areas with $N_{ki} > 0$). Let $B_i = N_{\cdot i}/A_i$ be the average number of establishments of sector i per area (counting only areas containing at least one such unit). Write $R_{ki} = N_{ki}/B_i$ for the 'index of relative concentration' of sector i in area k. Let j he index identifying StrCon, that is, the group of area units k in which sector $i = j$ has the largest R_{ki} value. Where necessary, $k(j)$ is used to refer to any area k in a given StrCon j. Hence, for instance, summing over areas in the frame for StrCon j: $N_{ji} = \sum_{k(j)} N_{ki}$, $N_{j\cdot} = \sum_{k(j)} N_{k\cdot}$.

The sector (i) which has the largest value of R_{ki} in the area unit concerned (k) is the StrCon for the area. In this way each area unit can be assigned to exactly one StrCon (apart from any ties, which may be resolved arbitrarily), the number of such strata being exactly the same as the number of sectors involved.

Note that the 'index of relative concentration' R_{ki} has been defined in relative terms: the number of units of a particular sector i in area k in relation to the average number of such units per area. Defining this simply in terms of the actual number of units would result in automatic over-domination by the largest sectors. However, for a sector not large in overall size but concentrated within a small proportion of the areas, the average per area (including zeros) would tend to be small, making the sector likely to be the dominant one in too many areas. Hence we find it appropriate to exclude areas with $N_{ki} = 0$ in computing the average B_i.

6.5.2 Data by StrCon and sector (aggregated over areas)

Once the StrCon has been defined for each area unit, the basic information aggregated over areas on the numbers of units classified by sector and StrCon can be represented as in Table 6.1. By definition, there is a one-to-one correspondence between the sectors and the StrCon. Subscript i (rows 1 to I) refers to the sector, and j (columns 1 to I) to StrCon. Generally, any sector is distributed over various StrCon; any StrCon contains establishments from various sectors, in fact it contains all the establishments in area units included in it. The diagonal elements ($i = j$) predominate to the extent that sectors are geographically segregated, and each tends to be concentrated within its 'own' StrCon.

Table 6.1 Classification of units by sector and the 'strata of concentration'.

StrCon	$j = 1$	...	j	...	...	$j = I$	Total	Target sample size	Sample size achieved
Sector									
$i = 1$	$i = j = 1$							$n_{\cdot 1}$	
...		$i = j$							
...			$i = j$						
i			N_{ji}	$i = j$			$N_{\cdot i}$	$n_{\cdot i}$	$\sum_j g_j N_{ji}$
$i = I$						$i = j = I$		$n_{\cdot I}$	
Total			$N_{j\cdot}$				$N_{\cdot\cdot}$	$n_{\cdot\cdot}$	$\sum_j g_j N_{ji}$
Sampling rate applied	g_1		g_j			g_I			
Sample size achieved			$n_{j\cdot} = g_j N_{j\cdot}$				$\sum_j g_j N_{j\cdot}$		

6.5.3 Using StrCon for determining the sampling rates: a basic model

The use of StrCon as defined above is the fundamental aspect of the proposed strategy for determining the selection probabilities to be applied to area units for the purpose of achieving the required sample size by sector in terms of the number of establishments, under the constraint that the procedure for sampling establishments within a selected area units is the same for different types of establishments.

We will first consider a simple model using a uniform sampling rate within each StrCon, but varied across StrCon with the objective of obtaining the required allocation by sector. This basic model will be illustrated numerically in the next section and some possible refinements noted.

The basic model is to apply a constant overall sampling probability g_j to all establishments in all areas $k(j)$ belonging to StrCon j. The expected sample sizes obtained with this procedure are

$$n_{ji} = g_j N_{ji}, \qquad n_{j\cdot} = \sum_i n_{ji} = g_j N_{j\cdot\cdot}$$

The g_j values have to be determined such that the obtained sample sizes by sector i,

$$\sum_j g_j N_{ji} = \sum_j n_{ji} \approx n_{\cdot i}, \tag{6.8}$$

agree with the required sizes $n_{\cdot i}$ simultaneously for all sectors, at least approximately.

The solution to (6.8) is trivial if all establishments are concentrated along the diagonal of Table 6.1, that is, when each area contains establishments from only one sector: as noted in Equation (6.4), the solution is simply $g_j = f_i$ for StrCon $j = i$.

At the other extreme, when establishments of all types are uniformly dispersed across the population areas, no solution of the above type is possible: we cannot select establishments of different types at different rates simply by varying the selection probabilities at the area level, without distinguishing between different types of establishments in the same area.

In fact the simple procedure (6.8) gradually ceases to give useful results as establishments of different types become increasingly uniformly dispersed across the population. This is because satisfying it requires increasingly large differences among StrCon sampling rates, thus reducing the efficiency of the resulting sample.

Nevertheless, the basic procedure is found to be quite useful in practice. It is common in real situations for establishments of different types to be fairly concentrated or at least unevenly distributed in the population.

Following some numerical illustrations of the basic model, some more flexible empirical refinements are suggested below for situations where that model appears insufficient.

6.6 Numerical illustrations and more flexible models

6.6.1 Numerical illustrations

Table 6.2 shows three simulated populations of establishments. The distribution of the establishments by sector is identical in the three populations – varying linearly from

Table 6.2 Number of establishments, by economic sector and 'stratum of concentration' (three simulated populations).

Stratum of concentration: population 1							
Sector	StrCon1	StrCon2	StrCon3	StrCon4	StrCon5	Total	% diagonal
1	**5000**	0	0	0	0	5000	100%
2	0	**4000**	0	0	0	4000	100%
3	0	0	**3000**	0	0	3000	100%
4	0	0	0	**2000**	0	2000	100%
5	0	0	0	0	**1000**	1000	100%
All	5000	4000	3000	2000	1000		
% diagonal	100%	100%	100%	100%	100%		**100%**
Stratum of concentration: population 2							
Sector	StrCon1	StrCon2	StrCon3	StrCon4	StrCon5	Total	% diagonal
1	**3443**	509	431	394	223	5000	69%
2	512	**2400**	437	420	231	4000	60%
3	439	478	**1460**	393	230	3000	49%
4	354	373	326	**785**	162	2000	39%
5	202	215	196	158	**229**	1000	23%
All	4950	3975	2850	2150	1075		
% diagonal	70%	60%	51%	37%	21%		**55%**
Stratum of concentration: population 3							
Sector	StrCon1	StrCon2	StrCon3	StrCon4	StrCon5	Total	% diagonal
1	**1964**	835	806	790	605	5000	39%
2	703	**1542**	656	609	490	4000	39%
3	559	554	**1071**	454	362	3000	36%
4	387	365	309	**680**	259	2000	34%
5	187	179	158	167	**309**	1000	31%
All	3800	3475	3000	2700	2025		
% diagonal	52%	44%	36%	25%	15%		**37%**

5000 in sector 1 to 1000 in sector 5. The populations differ in the manner in which the establishments are assumed to cluster geographically. In population 1 different sectors are geographically completely separated – each area unit containing establishments from only one sector. We can construct five StrCon corresponding to the five sectors as described in the previous sections. In population 1 these completely coincide with the sectors – StrCon1, for instance, containing all sector 1 establishments and none from any other sector. All the cases in the sector-by-StrCon cross tabulation lie on the diagonal. The last row of the panel shows the percentage of establishments in each StrCon which are in the diagonal cell, that is, are from the sector corresponding to the StrCon; the last column shows the percentage of establishments in each sector which are in the diagonal cell, that is, are from the StrCon corresponding to the sector. In the special case of population 1 all these figures are 100%, of course.

Table 6.3 Required and achieved sample sizes: population 1.

SAMPLE	Required	Achieved n by StrCon							
Sector	n	StrCon1	StrCon2	StrCon3	StrCon4	StrCon5	Total	D_i^2	E_i
1	267	267	0	0	0	0	267	1.00	0
2	239	0	239	0	0	0	239	1.00	0
3	207	0	0	207	0	0	207	1.00	0
4	169	0	0	0	169	0	169	1.00	0
5	119	0	0	0	0	119	119	1.00	0
All	1000	267	239	207	169	119	1000	1.00	0
Sampling rate		5.3%	6.0%	6.9%	8.4%	11.9%			
Weight		1.25	1.12	0.97	0.79	0.56			

The proportion of cases in the diagonal are lower for the other two populations: 55% in population 2 and 37% in population 3 in our simulation. Apart from the overall level, the pattern across StrCon of proportions diagonal happens to be very similar in the two populations, as can be seen from the last row in each panel. However, the pattern across sectors happens to differ markedly between the two populations, as can be seen from the last column.

Table 6.3 shows the selection of a sample ($n = 1000$) with given target sample sizes by sector from population 1. This is a very simple case: it involves no more than applying the required overall sampling rate for each sector to its corresponding StrCon, since the two in fact are identical in this case. The sample weights shown in the last row are inversely proportional to the probability of selection of establishments in the StrCon or the corresponding sector. The weights have been scaled to average 1.0 per establishment in the sample.

Table 6.4 shows the results of sample selection for population 2, with the same required sample allocation as in the previous illustration. Its second panel shows D_i^2, the design effect (factor by which effective sample size is reduced) for each sector as a result of variation in sample weights introduced to meet the target sample sizes in population 2. The third panel shows the difference between the achieved and target sample sizes by sector; the sum of their absolute values as a percentage of the total sample size appears in the last row of the panel. Columns show the results obtained by sequential application of a simple iterative procedure aimed at making the achieved sample sizes closer to the target sizes by sector. The mean absolute deviation $\bar{E}$ declined from over 12% initially to under 2% after five iterations. On the other hand, the mean design effect $\bar{D}^2$ increases from nearly 1.0 initially to around 1.5 after five iterations. This is the price to be paid for meeting the target sample sizes more closely. These variations are shown in Figure 6.1.

The first panel of Table 6.4 shows the sample weights for each StrCon, and how these change by iteration. The contrast among the sectors in these weights becomes more pronounced with each iteration.

The corresponding results for population 3 are shown in Table 6.5 and Figure 6.2. In this population, establishments in each sector are more widely dispersed across the population, and consequently the basic procedure proposed here gives less satisfactory results than the previous illustration. For example, the mean absolute deviation, which is high (16%) to begin with, remains quite high (8%) even after five iterations. At the same

Table 6.4 Sample weights, design effect due to weights, and deviation from target sample size: results with five iterations, population 2.

		Iteration					
		0	1	2	3	4	5
		Sample weight (average = 1.00)					
Stratum of concentration							
StrCon1		1.26	1.48	1.65	1.79	1.87	1.93
StrCon2		1.13	1.28	1.44	1.57	1.68	1.76
StrCon3		0.97	1.04	1.13	1.23	1.32	1.40
StrCon4		0.80	0.73	0.71	0.71	0.71	0.72
StrCon5		0.56	0.38	0.29	0.25	0.22	0.21
		D_i^2 (design effect due to weights)					
Sector							
1		1.04	1.12	1.20	1.27	1.32	1.36
2		1.04	1.12	1.21	1.29	1.35	1.40
3		1.04	1.13	1.22	1.31	1.39	1.44
4		1.05	1.16	1.27	1.37	1.44	1.50
5		1.09	1.29	1.52	1.73	1.90	2.03
All		**1.05**	**1.16**	**1.28**	**1.39**	**1.48**	**1.55**
	Required n	E_i (mean deviation)					
Sector							
1	267	35	18	9	4	1	0
2	239	23	17	11	8	5	4
3	207	5	8	9	8	7	6
4	169	−21	−12	−7	−4	−3	−2
5	119	−42	−32	−22	−15	−10	−7
All	1000	0	0	0	0	0	0
% mean absolute deviation		**12.5**	**8.7**	**5.9**	**4.0**	**2.7**	**1.9**

time, the loss due to weighting increases more and more rapidly with iteration, reaching 2.5 after five iterations.

Note also how the difference between StrCon sampling rates and hence the associated weights becomes more marked with iterations, the StrCom5-to-StrCon1 ratio becoming nearly 20:1 after five iterations from the initial value close to 2:1. This points to the need to seek solutions more flexible than the basic procedure developed above. Some useful possibilities are outlined below.

6.6.2 More flexible models: an empirical approach

As is clear from the above illustrations, depending on the numbers and distribution of units of different types and on the extent to which the required sampling rates by sector differ, a basic model like (6.8) may be too inflexible, and may result in large variations in design weights and hence in large losses in efficiency of the design. It may even prove impossible to satisfy the sample allocation requirements in a reasonable way.

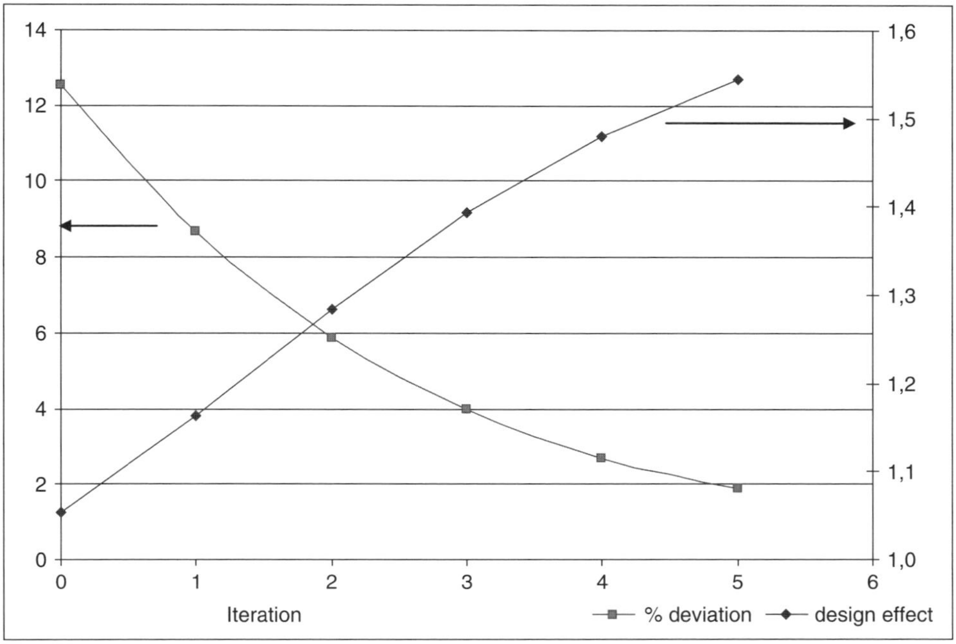

Figure 6.1 Design effect due to weights versus percentage mean absolute deviation from target sample sizes by sector: population 2.

Lacking a general theoretical solution, we have tried more flexible empirical approaches for defining the sampling rates to be applied to area units in order to meet the required sample allocation, and for achieving this more efficiently.

Basically, the approach has involved supplementing (6.8) by further modifying the area selection probabilities in a more targeted fashion within individual StrCon. Instead of taking the overall sampling rate g_j as a constant for all area units $k(j)$ in StrCon j, we may introduce different rates $g_{k(j)}$ which may vary among groups of area units within j or, in principle, even by individual area. The models we have tried and found empirically useful are all of the form

$$g_{k(j)} = g_j((1 - P_{kj}) + (1 + c_j h(P_{kj}))P_{kj}), \tag{6.9}$$

where P_{kj} is the proportion, for a given area k, of establishments belonging to the sector $(i = j)$ which corresponds to StrCon j of the area. Parameters g_j and c_j are constants for StrCon j, and h is a function of P_{kj}. Note that in the basic model (6.8), we take $c_j = 0$, $h = 1$, so that $g_{k(j)} = g_j$, the same for all area units in the StrCon. Form (6.9) allows the sampling rate to be varied according to the extent to which establishments in the sector corresponding to the area's StrCon (i.e. in the sector having the largest R_{ki} value in the area, as defined in Section 6.5) predominate in the area concerned. This form allows greater flexibility in achieving the target sample sizes by sector, with a better control on the increase in variance resulting from variations in sample weights (factor D_i^2 described in Section 6.4). Empirical choice of parameters of a function of the type (6.9) is made on the basis of balance between (1) meeting the target sample sizes by sector as well as possible, and (2) limiting the loss due to variations in sample weights.

Table 6.5 Sample weights, design effect due to weights, and deviation from target sample size: results with five iterations, population 3.

		Iteration					
		0	1	2	3	4	5
		Sample weight (average = 1.00)					
Stratum of concentration							
StrCon1		1.36	1.84	2.46	3.20	4.08	5.10
StrCon2		1.21	1.50	1.84	2.22	2.61	3.01
StrCon3		1.05	1.15	1.27	1.38	1.47	1.55
StrCon4		0.86	0.80	0.78	0.77	0.76	0.75
StrCon5		0.61	0.44	0.35	0.31	0.28	0.27
		D_i^2 (design effect due to weights)					
Sector							
1		1.07	1.26	1.52	1.83	2.20	2.61
2		1.06	1.22	1.44	1.69	1.98	2.28
3		1.06	1.21	1.40	1.63	1.89	2.16
4		1.07	1.23	1.45	1.72	2.01	2.35
5		1.10	1.36	1.71	2.11	2.57	3.06
All		**1.07**	**1.25**	**1.50**	**1.80**	**2.13**	**2.49**
	Required n	E_i (mean deviation)					
Sector							
1	267	55	46	41	37	35	33
2	239	23	19	15	13	11	9
3	207	−6	−7	−8	−9	−10	−10
4	169	−29	−24	−22	−20	−18	−17
5	119	−43	−34	−27	−21	−17	−14
All	1000	0	0	0	0	0	0
% mean absolute deviation		**15.6**	**13.1**	**11.3**	**10.0**	**9.1**	**8.4**

Our empirical approach has involved the following steps:

- Choosing a form for $h(P_{kj})$ in (6.9).
- Choosing parameter c_j.
- Using the above in (6.9) to iteratively determine g_j values by StrCon which meet the sample allocation requirements n_i by sector.
- Computing the implied losses in efficiency due to weighting from (6.5) for each sector, and their average over sectors $\bar{D}^2$ from (6.6).
- Computing the mean absolute deviation between the achieved and target sample sizes, $\bar{E}$ from (6.7).
- Making a choice of the combination $(\bar{D}^2, \bar{E})$ from the range of possible combinations obtained with the given model and parameters.

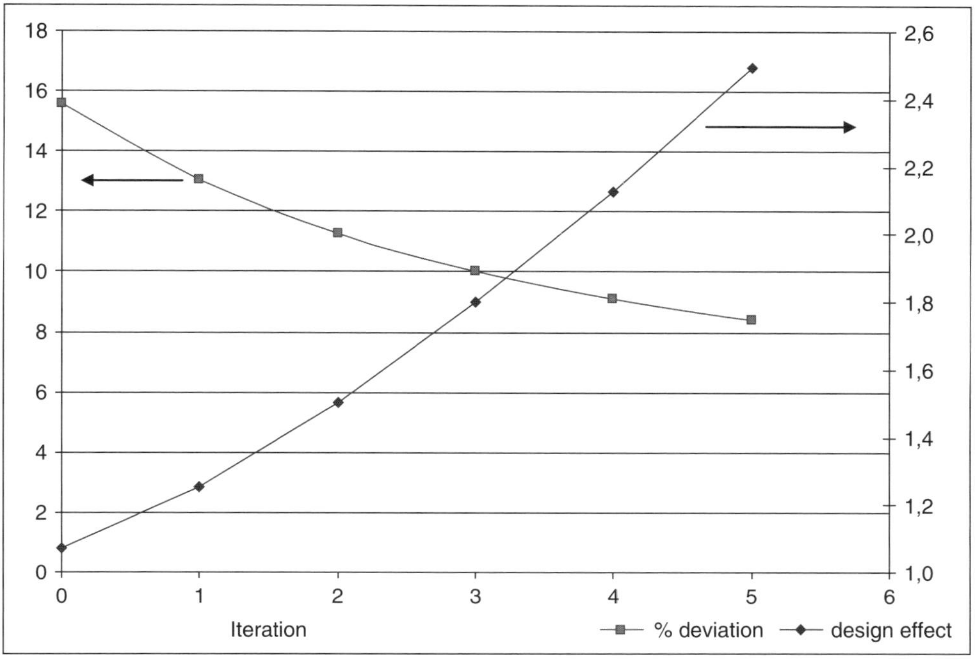

Figure 6.2 Design effect due to weights versus percentage mean absolute deviation from target sample sizes by sector: population 3.

- Finally, comparing this outcome against a range of others outcomes obtained with different models and parameters (h, c), and choosing the most reasonable solution from among those computed.

The objective is to identify and choose the 'best' model, at least among those empirically evaluated. Comparing large numbers of numerical trials points at least to the direction in which we should be moving in the choice of the design parameters.

Here are examples of some of the forms which we have investigated and compared. Concerning function h, in practice we have hitherto taken its functional form to be the same for all StrCon in a given trial of the procedure, that is, independent of particular j. These included the following:

Basic: $c_j = h = 1$, simply reducing (6.9) to its basic form (6.8): $g_{k(j)} \equiv g_j$.

Constant: The overall selection probability of units in the sector corresponding to the area's StrCon j adjusted by a constant factor c_j with $h = 1$: $g_{k(j)} = g_j((1 - P_{kj}) + (1 + c_j)P_{kj})$.

Linear: The above measures adjusted by taking $h = P_{kj}$, the proportion of units in the area belonging to the sector corresponding to the area's StrCon: $g_{k(j)} = g_j((1 - P_{kj}) + (1 + c_j P_{kj})P_{kj})$.

S-shaped: The above with more elaborate variation, for instance with $h(P_{kj}) = P_{kj}^2(3 - 2P_{kj})$, and so on. The last mentioned form appeared a good one at least for one survey tried.

As for parameter c_j in the above equations, a simple choice we have tried is to take it as a function of the overall sampling rate required in sector $(i = j)$ corresponding to StrCon j, such as

$$c_j = \left(\left(\frac{f_i}{f}\right)^{\alpha} - 1\right), \qquad \alpha \geq 0, \quad i = j.$$

6.7 Conclusions

The specific examples given above must be taken as merely illustrative of the type of solutions we have looked into to the basic design problem in multi-stage sampling of heterogeneous and unevenly distributed small-scale economic units. The solution depends on the nature of the population at hand, and we have used the approach sketched above in designing a number of samples, covering diverse types of surveys in different situations (countries). These have included surveys of agricultural and non-agricultural small-scale units, and even a survey of schools where the objective was to control the sample allocation by ethnic group.

In this last example, it was not ethically permissible to identify and sample differentially students of different ethnic groups: all variations in the required overall sampling rates by ethnic group had to be achieved by appropriately adjusting the school selection probabilities according to prior and approximate information on the schools' ethnic composition.

In conclusion, one important aspect of the sampling procedures described above should be noted. In describing these procedures, we have not explicitly considered any aspects of the structure of the required sample except for the following two: (1) that the selection of establishments involves a multi-stage design, with the final selection of establishments preceded by one or more area stages; and (2) that certain target sample sizes in terms of the number of establishments in different sectors are to be met. We have mainly addressed issues concerned with meeting the sample allocation requirements, which of course is a fundamental aspect of the design for surveys of the kind under discussion. Any real sample may, however, also involve other complexities and requirements, such as stratification, variation in sampling rates and even different selection procedures by geographical location or type of place, selection of units using systematic sampling with variable probabilities, etc. These determine the actual sample selection procedures to be used.

As far as possible, the issue of sample allocation – meeting the externally determined target sample size requirements by sector – should be isolated from the structure and process of actual sample selection. One technique for meeting the sample allocation requirements automatically, without complicating the sample selection procedures determined by other aspects of the design, is to incorporate the sample allocation requirements into the size measures of units used for probability proportional to size (PPS) selection (Verma, 2008).

Consider the two-stage PPS selection equation of the type (6.1). Ignoring the difference between the expected and actual unit sizes (N_k, N_k') not relevant here, that equation gives a constant overall selection probability f for an establishment. Suppose that for an area k, the required overall probability is to be a variable quantity g_k. A selection equation of the form

$$\frac{ag_k^{(1)}N_k}{\sum_k g_k^{(1)}N_k}\frac{b}{N_k/g_k^{(2)}} = \frac{ab}{\sum_k g_k^{(1)}N_k}g_k^{(1)}g_k^{(2)} = g_k^{(1)}g_k^{(2)}g = g_k \tag{6.10}$$

gives the required variation by area in the establishment selection probabilities. The terms in (6.10) are as follows. The required g_k may be factorized into three parts: $g_k^{(1)}$, the relative selection probability of the area; $g_k^{(2)}$, the relative selection probability of an establishment within the area; and a scaling constant g, being the overall selection probability of establishments in areas with $g_k^{(1)}g_k^{(2)} = 1$.

At the second stage of selection, the expected sample size from the area unit is $bg_k^{(2)}$ establishments; it is common in practice to keep this size constant throughout, that is, take $g_k^{(2)} = 1$.

The first stage of selection is equivalent to ordinary PPS selection of areas using modified size measures $M_k = N_k g_k^{(1)}$, or $N_k g_k/g$ if $g_k^{(2)} = 1$, as is common.[3]

With size measures adjusted as above, we can apply a uniform selection procedure (such as systematic PPS sampling with a constant selection interval) to all the areas, irrespective of differences among areas in the required overall selection probabilities g_k.

To summarize, once the size measures are so adjusted, the required differences in the sampling rates by sector are automatically ensured and it is no longer necessary to apply the sample selection operation separately by StrCon or even by sector. Once assigned, the units always 'carry with themselves' their size measure – incorporating the relative selection probabilities required for them, irrespective of details of the selection process – and the required allocation is automatically ensured, at least in the statistically expected sense.

The use of adjusting size measures to simplify the selection operation is a useful and convenient device, applicable widely in sampling practice. The advantage of such separation of allocation and selection aspects is that the structure and process of selection (stratification, multiple sampling stages, subsampling, etc.) can be determined flexibly, purely by considerations of sampling efficiency, and need not be constrained by the essentially 'external' requirements of sample allocation.

When the two aspects overlap in practice – for instance, when a whole design domain is to be over-sampled – that is a coincidental rather than an essential aspect of the design.

Acknowledgements

This is a revised and enlarged version of a paper originally presented at Conference on Agricultural and Environmental Statistical Applications in Rome (CAESAR), 4–8 June 2001, and subsequently published in Verma (2001).

References

Cox, B.G., Binder, D.A., Chinnappa, B.N., Christianson, A., Colledge, M.J. and Knott, P.S. (1995) *Business Survey Methods*. New York: John Wiley & Sons, Inc.

[3] Similarly, the second stage is ordinary inverse-PPS selection with size measures deflated as $S_k = N_k/g_k^{(2)}$; note that $M_k/S_k = g_k^{(1)}g_k^{(2)} = g_k/g$.

Kish, L. (1992) Weighting for unequal Pi. *Journal of Official Statistics*, 8, 183–200.

Kish, L., Groves, R. and Krotky, K. (1976) Sampling errors for fertility surveys. Occasional Paper No. 17, World Fertility Survey, International Statistical Institute.

Verma, V. (2001) Sample designs for national surveys: surveying small-scale economic units. *Statistics in Transition*, 5, 367–382.

Verma, V. (2008) *Sampling for Household-based Surveys of Child Labour*. Geneva: International Labour Organization.

7

On the use of auxiliary variables in agricultural survey design

Marco Bee[1], Roberto Benedetti[2], Giuseppe Espa[1] and Federica Piersimoni[3]
[1]*Department of Economics, University of Trento, Italy*
[2]*Department of Business, Statistical, Technological and Environmental Sciences, University 'G. d'Annunzio' of Chieti-Pescara, Italy*
[3]*Istat, National Institute of Statistics, Rome, Italy*

7.1 Introduction

The national statistical institutes put a considerable effort into the design of their surveys, in order to fully exploit the auxiliary information, whether univariate or multivariate, useful for producing precise and reliable estimates.

This way of proceeding mainly involves techniques applied after the sample selection, as shown by the solid-line boxes in Figure 7.1. The most common survey set-up is typically based on a standard design, mostly formed by a stratification of the archive and by simple random sampling in the strata.

After the data-recording step, a second phase takes place, where the auxiliary information is used. The institutes focus their attention on this issue, developing and employing sophisticated estimators that can provide efficiency gains.

Most surveys conducted in the primary sector are based on this strategy. Among them we recall the Istat survey on red meat slaughtering, which will be used as an illustration in the following.

This procedure can be summarized as 'stratification/constrained calibration' and has spread quickly in practical applications. At the same time, the agricultural survey

Agricultural Survey Methods Edited by Roberto Benedetti, Marco Bee, Giuseppe Espa and Federica Piersimoni

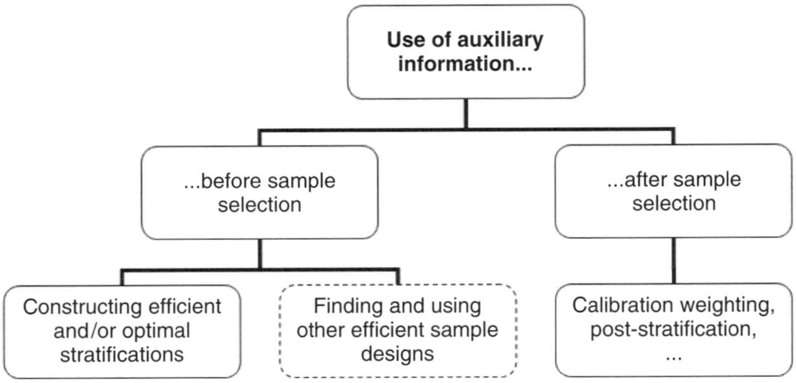

Figure 7.1 Use of auxiliary information.

framework has also changed in recent decades, becoming more and more specialized with respect to other surveys, in particular when the units are agricultural holdings. In this case, there are two specific issues (see also the introduction to this book). First, the target population and frame population are very different, and it is often necessary to employ census frames, because the registers are not accurately updated. Second, unlike business surveys, a rich set of auxiliary variables (not just dimensional variables) is available: consider, for example, the bulk of information provided by airplane or satellite remote sensing. See Section 3.8 and Chapters 9 and 13 of this book.

It is then natural to ask whether there are other sample designs (different from the 'stratification/constrained calibration' scheme) that allow all this auxiliary information to be exploited. The answer is in the affirmative: researchers have long focused their attention on the development of sample designs that allow frames containing a considerable amount of auxiliary information to be handled. These procedures take place mostly before sample selection (see the dashed box in Figure 7.1).

A further question is whether the combination of *ex ante* and *ex post* auxiliary information leads to further gains, in terms of efficiency of the estimators. This chapter tries to answer this question. To this end, we performed several controlled experiments for testing the efficiency of some sample selection criteria. The same criteria have also been applied to the Istat monthly slaughtering survey for the production of direct estimates and for adjusting the estimates after the selection of the sample.

Before examining the experiments, it is necessary to give a critical survey of the techniques upon which they are based. However, this survey is not merely relevant to the presentation of the experiments. It is, actually, completely in line with some of the goals of this book, in particular with the purpose of rethinking the usefulness, for people who perform daily agricultural surveys, of some new sophisticated methodologies. Our aim is that the discussion presented here should also shed some light on the framework and the way of implementing these methodologies.

In line with the sequence shown in Figure 7.1, this chapter begins with a review of the problem of stratifying an archive (Section 7.2). The main issues considered here are the goals, the necessary inputs and the problems that usually arise in practical applications. We try to treat separately the single- and multi-purpose case (corresponding respectively to one or more surveyed variables) with uni- or multivariate auxiliary information. We summarize the main results obtained in the literature and, in particular, we touch on some of the themes that are currently at the centre of the scientific debate:

- the stratification of strongly asymmetric archives (such as the agricultural holdings ones), with some attention to a formalization of the so-called cut-off sampling;
- the problem of strata formation in a 'global' approach where it is possible to consider simultaneously the choice of the number of stratifying variables, the number of class intervals for each variable and of the optimal allocation to strata.

In Section 7.3 we analyse the traditional sampling scheme with probabilities proportional to size and provide some remarks in favour of a selection based on non-constant probabilities. Section 7.4, which presents balanced sampling, concludes the discussion concerning the *ex ante* use of auxiliary information. This issue has quite intuitive features and, although proposed back in the 1940s, is of great importance because of some recent developments. Section 7.5 reviews the most general approach to the use of *ex post* auxiliary information: the so-called calibration weighting. Given the importance of this technique for the production of official statistics, we say a few words about the idea of an *a posteriori* correction of the expansion weights of the original design. Then we consider the use of this class of estimators for the correction of total non-responses and coverage errors and conclude by examining some operational and algorithmic issues. After giving a brief description of the Istat survey and of the features of the archive employed, Section 7.6 is devoted to the illustration of the main results of the experiments performed. The analysis also satisfies the need to come up with proposals for the redefinition of samples in surveys concerning agricultural holdings. Section 7.7 concludes the chapter with a brief discussion.

7.2 Stratification

Stratification is one of the most widely used techniques in finite population sampling. Strata are disjoint subdivisions of a population U, and the union of the strata coincides with the universe:

$$U = \cup_{h=1}^{H} U_i, \quad U_h \cap U_i = \emptyset, \quad h \neq i \in \{1, \ldots, H\}.$$

Each group contains a portion of the sample. Many business surveys employ stratified sampling procedures where simple random sampling without replacement is performed within each stratum; see, for example Sigman and Monsour (1995) and, for farm surveys, Vogel (1995). The essential objective consists in choosing a stratification strategy that allows for efficient estimation.

Stratified sampling designs differ with respect to (Horgan, 2006):

(i) the number of strata used;

(ii) how the sample is allocated to strata (allocation is the survey designer's specification of stratum sample size);

(iii) the construction of stratum boundaries.

Concerning (ii), a review of the allocation techniques for stratified samples has been given by Sigman and Monsour (1995, Section 8.2). It is, however, always necessary to treat separately the single-purpose and univariate set-up and the multipurpose and multivariate case. For the former (allocation for a survey that collects data for only one item and estimates a single population-level total) we mention:

- proportional allocation. The size of the sample in each stratum is proportional to the stratum's population size.
- optimal allocation. This criterion minimizes the product VC with either C or V held fixed, where V is the variable component of the variance of the estimator and C is the survey's variable cost, which changes along with the stratum sample size.
- Neyman's rule. Neyman (1934) derived a method of distributing the n sample elements among the H strata such that the variance of the stratified estimator is minimized.

In practice it is often the case that the actual sample sizes do not coincide with the optimal ones. For example, the allocation n_h to stratum h must be an integer, but this is not guaranteed in the optimal allocation. Kish (1976) and Cochran (1977, pp. 115–117) analyse the problem of inequality of n_h and n_h^* and assess its effects on variance. Moreover, the computation of optimal sample sizes requires the variances within the strata to be known. Typically, this is not true in applications. What is known to the survey designer is the variability within the strata of an auxiliary variable (usually a measure of size, on which the classical stratification by size is based) strongly correlated with the surveyed variable. The knowledge of this variability redefines the set-up as a single-purpose and univariate one (one auxiliary variable at hand). From a point of view recently defined as 'model-assisted', Dayal (1985) finds a sub-optimal allocation called 'x-optimal allocation'. Dayal (1985) assumes a linear model for the relation between the unknown values of the variable of interest and the values of an auxiliary variable. The Dayal allocation is computed by minimizing the expected value of the variance with respect to the model. An interesting technique based on the Dayal one is considered by Godfrey *et al.* (1984).

A multi-purpose survey collects data and produces estimates for two or more variables. Thus, each variable has a different variance V_i. Sigman and Monsour (1995, Section 8.2.3) distinguish two possible approaches to multi-purpose allocation: see Kish (1976) and Chromy (1987). One of these, called 'convex programming', involves attaching to each variance V_i a precision constraint and finding, among the allocations satisfying the constraints, the one with the minimal costs. As concerns this issue, it is worth citing the solution proposed by Bethel (1985, 1989) and its most important competitor, introduced by Chromy (1987). These two iterative algorithms are extremely useful for survey designers. In practical applications the Chromy algorithm is often preferred, in particular when the 'size' is large (many strata and many constraints), because the costs are lower and the convergence is quicker. However, in spite of some numerical results concerning convergence shown by Chromy (1987), a formal proof does not exist.

Next we consider item (iii), namely the construction of the strata, mentioning methodological and applied issues studied since the 1950s. The problem of optimal univariate stratification can be formulated in the following terms: given the variable of interest y and the number of strata H, we have to find the bounds $y_1, \ldots, y_{H-1}$ of the strata conditionally on $y_0 < y_1 < \ldots < y_H$, with $y_h \in [y_0, y_H]$ $(h = 1, \ldots, H-1)$ such that the resulting stratification has minimum variance. Therefore the optimality criterion implies a 'minimum variance stratification' (MVS).

Dalenius (1950) found an exact solution, revisited, among others, by Dalenius and Hodges (1959). He assumes a continuous distribution for the variable of interest with density $f(w)$, where $f(w) > 0$ for $w \in [y_0, y_H]$. The MVS solution to the optimization problem was solved by Dalenius (1957) under the assumption that y follows standard distributions such as the exponential. In practical applications, unfortunately, we rarely work with such densities, so that finding the minimum becomes a complicated problem, in particular from the computational point of view. Therefore, in general, Dalenius's (1950) approach is seldom applied. Much has been written on ways of overcoming this difficulty. The best-known approximating rule is the so-called 'cum$\sqrt{f}$ rule' (Dalenius and Hodges, 1959). They assume the stratification variable to be approximately uniform between two adjacent points y_h and y_{h+1}. However, several populations, including agricultural holdings, show a strong asymmetry (see below), with the consequence that this assumption is not satisfied.

Cochran (1961) proposed a performance-based comparison of four approximations, including the cum$\sqrt{f}$ rule, applied to real data consisting of discrete and asymmetric populations. The results show that the Dalenius and Hodges (1959) and Ekman (1959) approximations 'work well'.

The Dalenius and Hodges (1959) optimal approximation suffers of some drawbacks that limit its use in practical applications:

1. The variable y is usually unknown; thus, it is impossible to produce a stratification of its distribution.

2. The population is often finite, as in the application proposed in this chapter. It is therefore difficult to justify the use of a density function.

3. The algorithm can only find a univariate stratification, but most surveys, including those concerning the agricultural sector, aim to produce estimates for a set $\boldsymbol{y}$ of variables (multipurpose) in presence of a set $\boldsymbol{x}$ of auxiliary variables (multivariate).

As for the first issue, if x is a known auxiliary variable strongly correlated with y, it is possible to construct an optimal stratification by finding the bounds of the strata $x_1, \ldots, x_h, \ldots, x_{H-1}$. To do that, the procedures above are applied to x instead of y (Särndal *et al.*, 1992, Section 12.6). Turning to the second issue, we again use the auxiliary variable x; it is indeed possible to draw a histogram based on the N values of the auxiliary variable known at the population level. This histogram, built with C ($C \gg H$) bins of equal width, is the starting point. By means of this histogram it is possible to approximate the cum$\sqrt{f}$ rule. When a positive auxiliary variable x is available, there are many ways to combine stratification, allocation and estimation. However, they are beyond the scope of this chapter; for details of two interesting strategies see Särndal *et al.* (1992, Section 12.6).

Concerning the third drawback above, Kish and Anderson (1978) analyse the issue of finding the boundaries of the strata with two auxiliary variables x_1 and x_2 and two target variables y_1 and y_2. They apply the cum$\sqrt{f}$ rule separately to x_1 and x_2 obtaining respectively H_1 and H_2 strata. Finally, the sample is allocated proportionally to the $H = H_1 H_2$ strata.

An extension of the Dalenius and Hodges (1959) algorithm to the case of K auxiliary variables $x_1, \ldots, x_K$ is given by Jarque (1981), who developed for the case $K > 2$ the results obtained by Ghosh (1963) and Sadasivan and Aggarwal (1978).

An alternative approach involves employing techniques borrowed from multivariate statistics for the determination of the boundaries of the strata in multivariate surveys: see, among others, Hagood and Bernert (1945) for the use of principal components and Green *et al.* (1967), Golder and Yeomans (1973), Heeler and Day (1975), Yeomans and Golder (1975), Mulvey (1983) and Julien and Maranda (1990) for the use of cluster analysis. These examples are linked by the common aim of finding homogeneous groups (with minimum variance) according to predefined distance measures. In some works in this field, unfortunately, it is not clear how this is related to the goals of the surveys to be performed. This drawback does not seem to affect, in our opinion, the method proposed by Benedetti *et al.* (2008), who define as 'atomized stratification' the formation of strata by combining all possible classes from any of the auxiliary variables in use.

Atomized stratification can be interpreted as an extreme solution to the problem of stratum formation, since, between the cases of no stratification and the use of atomized stratification, there exists a full range of opportunities to select a stratification whose subpopulations can be obtained as unions of atoms.

Benedetti *et al.* (2008) propose carrying out this selection through the definition of a tree-based stratified design (Breiman *et al.*, 1984; Bloch and Segal, 1989) in a multivariate and multipurpose framework. They form strata by means of a hierarchical divisive algorithm that selects finer and finer partitions by minimizing, at each step, the sample allocation required to achieve the precision levels set for each surveyed variable. The procedure is sequential, and determines a path from the null stratification, namely the one whose single stratum matches the population, to the atomized one.

The aim of this approach is to give the possibility of combining stratification and sample allocation. This means that Benedetti *et al.* (2008) are concerned with the choice of the number of stratifying variables, the number of class intervals for each variable and the optimal Bethel allocation to strata. Concerning this choice, the stratification tree methodology cannot be reduced to the standard solution of the multivariate stratification problem, that is, the use of multivariate techniques such as cluster analysis and principal components. As a matter of fact, this branch of the literature does not use (or uses only indirectly) the variables of interest, but only the auxiliary variables, and the allocation issue is neglected.

A different approach is necessary when the population is markedly asymmetric. Usually, asymmetry is positive, as few units are very large and most are small. This is the typical situation of business surveys. This is often true in the agricultural sector as well, where small family-owned holdings coexist with large industrial companies.

Stratifying populations with such features without using ad hoc criteria such as the ones presented below may result in overestimation of the population parameters (Hidiroglou and Srinath, 1981).

As usual in business surveys, we assume that the population of interest is positively skewed, because of the presence of few 'large' units and many 'small' units. If one is interested in estimating the total of the population, a considerable percentage of the observations gives a negligible contribution to the total. On the other hand, the inclusion in the sample of the largest observations is essentially mandatory.

In such situations, practitioners often use partitions of the population into three sets (so-called 'cut-off' sampling): a take-all stratum whose units are surveyed entirely (U_C), a take-some stratum from which a simple random sampling is drawn (U_S) and a take-nothing stratum whose units are discarded (U_E). In other words, survey practitioners decide *a priori* to exclude from the analysis part of the population (e.g. holdings with less than five employees). However, this choice is often motivated by the desire to match administrative rules (in this case, the partition of holdings into small, medium and large). This strategy is employed so commonly in business surveys that its use is 'implicit' and 'uncritical', so that the inferential consequences of the restrictions caused to the archive by this procedure are mostly ignored.

The problem of stratifying into two strata (take-all and take-some) and finding the census threshold was first treated by Dalenius (1952) and Glasser (1962). Dalenius determined the census threshold as a function of the mean, the sampling weights and the variance of the population. Glasser derived the value of the threshold under the hypothesis of sampling without replacement a sample of size n from a population of N units. Hidiroglou (1986) reconsidered this problem and provided both exact and approximate solutions under a more realistic hypothesis. He found the census threshold when a level of precision concerning the mean squared error of the total was set *a priori*, without assuming a predefined sample size n.

In this context, the algorithm proposed by Lavallée and Hidiroglou (1988) is often used to determine the stratum boundaries and the stratum sample sizes. For the set-up where the survey variable and the stratification variable differ, Rivest (2002) proposed a generalization of the Lavallée–Hidiroglou algorithm. The Rivest algorithm includes a model that takes into account the differences between the survey and the stratification variable and allows the optimal sample size and the optimal stratum boundaries for a take-all/take-some design to be found.

In his thorough review, Horgan (2006) lists some interesting works that identify numerical problems arising when using the Lavallée–Hidiroglou algorithm. In particular, it is worth mentioning Gunning and Horgan (2004) who derive, under certain hypotheses, an approximation to the optimal solution that is more convenient because it does not require iterative algorithms. However, all these authors limit their attention to a single-purpose and univariate set-up. In a multi-purpose and multivariate set-up, Benedetti *et al.* (2010) propose a framework to justify cut-off sampling and to determine the census and cut-off thresholds. They use an estimation model that assumes the weight of the discarded units with respect to each variable to be known, and compute the variance of the estimator of the total and its bias, determined by violations of the aforementioned hypothesis. Benedetti *et al.* (2010) then implement a simulated annealing algorithm that minimizes the mean squared error as a function of multivariate auxiliary information at the population level.

7.3 Probability proportional to size sampling

Consider a set-up where the study variable y and a positive auxiliary variable x are strongly correlated. Intuitively, in such a framework it should be convenient to select the elements to be included in the sample with probability proportional to x. Probability proportional to size (PPS) sampling designs can be applied in two different set-ups: fixed-size designs without replacement (πps) and fixed-size designs with replacement (pps). Only πps will be considered here; an excellent reference about pps is Särndal *et al.* (1992, pp. 97–100). The πps sampling technique has become reasonably well known in business surveys (Sigman and Monsour, 1995).

Consider the π estimator of the total $t_y = \sum_U y_k$:

$$\hat{t}_{y\pi} = \sum_s \frac{y_k}{\pi_k}, \tag{7.1}$$

where $\pi_k > 0$ denotes the probability of inclusion of the kth unit ($k = 1, \ldots, N$) in the sample. Formula (7.1) is the well-known Horvitz–Thompson (HT) estimator. Suppose now that it is possible to implement a fixed-size without-replacement sampling design such that

$$\frac{y_k}{\pi_k} = c, \quad k = 1, \ldots, N, \tag{7.2}$$

where $c \in \mathbb{R}$. In this case, for every sample s, $\hat{t}_{y\pi} = nc$ would hold, where n is the fixed-size of s. Notice that $\hat{t}_{y\pi}$ is constant, so that its variance is equal to zero.

Although this case is reported only for illustrative purposes (there is no design satisfying (7.2), because it implies knowledge of all the y_k), it is often true that the (known) auxiliary variable x is approximately proportional to y. It follows that, if we assign to π_k a numerical value proportional to x_k, the ratio y_k/π_k turns out to be approximately constant, so that the variance of the estimator is small.

Formally, let $U = (1, \ldots, k, \ldots, N)$ be the surveyed population. The outcome s of a sampling design is denoted by an indicator variable defined as follows:

$$S_k = \begin{cases} 1 & k \in s, \\ 0 & \text{otherwise.} \end{cases}$$

By means of this variable it is easy to define the first- and second-order inclusion probabilities: $\pi_k = P(S_k = 1)$ and $\pi_{kl} = P(S_k = S_l = 1)$, respectively.

We consider simple random sampling without replacement with n fixed. Thus, every sample has the same probability of being selected. It is easy to see that this probability is given by

$$p(s) = \frac{1}{\binom{N}{n}},$$

so that $\pi_k = n/N = f$ $(k = 1, \ldots, N)$, the sampling fraction. Therefore $\sum_U \pi_k = n$.

Let us now go back to the main goal, namely estimating the total $t_y = \sum_U y_k$ of a finite population. We consider the linear statistic

$$L(\boldsymbol{y}, \boldsymbol{w}) = \sum_{k=1}^{N} w_k y_k S_k, \tag{7.3}$$

where the weights $\boldsymbol{w}$ are assumed to be known for all the units in U, whereas the variable of interest y is known only for the observed units. Notice that (7.3) reduces to (7.2) if $w_k = 1/\pi_k$. The estimator of the variance of (7.3), called Sen–Yates–Grundy estimator, is given by

$$\hat{V}(L(\boldsymbol{y}, \boldsymbol{w})) = -\frac{1}{2} \sum_{k=1}^{N} \sum_{l=1}^{N} (w_k y_k - w_l y_l)^2 \Delta_{kl} S_k S_l, \tag{7.4}$$

with $\Delta_{kl} = (1 - \pi_k\pi_l/\pi_{kl})$ and $\pi_{kl} > 0$ $(k, l = 1, \ldots, N)$. When $w_k = 1/\pi_k$, (7.3) is the HT estimator, which is unbiased for t_y.

Now let $\{\lambda_1, \ldots, \lambda_k, \ldots, \lambda_N\}$ be the set of inclusion probabilities for the sampling design, with $\lambda_k \in [0, 1)$ and $\sum_U \lambda_k = n$. Roughly speaking, the πps approach consists in determining a sampling design without replacement with inclusion probabilities $\pi_k \approx \lambda_k$ $(k = 1, \ldots, N)$. The problem is slightly different if we consider the dimensional variable x such that $x_k > 0$ $(k = 1, \ldots, N)$: in this case we have to find a sampling design such that $\pi_k \propto x_k$ $(k = 1, \ldots, N)$.

This problem is actually equivalent to the preceding one with $\lambda_k = nx_k/\sum_U x_k$, where the proportionality factor $n/\sum_U x_k$ is such that $\sum_U \pi_k = n$. Obviously, the condition $\pi_k \leq 1$ must be satisfied. When $n = 1$, this is true for any $k \in \{1, \ldots, n\}$. In contrast, when $n > 1$, some x_k can be very large, to the extent that $nx_k/\sum_U x_k > 1$ and therefore $\pi_k > 1$.

Fortunately, this difficulty can be solved by completely enumerating the stratum containing the largest units (Hidiroglou, 1986). Specifically, one proceeds as follows. Let

$$\pi_k \begin{cases} = 1 & \text{if } k : nk > \sum_U x_k, \\ \propto x_k & \text{otherwise.} \end{cases}$$

In other words, given n, π_k is equal to $(n - n_A)x_k/\sum_{U\setminus A} x_k$, where A is the set containing the units x_k such that $nx_k > \sum_U x_k$.

In general, a πps sampling design is expected to satisfy the following properties, which are also important because they allow feasible designs to be ranked:

1. The sample selection mechanism should be easy to implement.
2. The first-order inclusion probabilities have to be proportional to x_k.
3. The condition $\pi_{kl} > 0$ $(k \neq l)$ must hold for the second-order inclusion probabilities. Borrowing the terminology used in Särndal *et al.* (1992), this condition implies a *measurable* sampling design and is a necessary and sufficient condition for the existence of a consistent estimator of the variance of $\hat{t}_{y\pi}$.

In the latter set-up (i.e. for measurable designs) the following properties are typically required as well:

4. The second-order probabilities π_{kl} must be exactly computable, and the burden of computation should be low.
5. Δ_{kl} $(k \neq l)$ should be negative, in order to guarantee the non-negativity of the estimator of the variance (7.4).

In applications, the most common πps sampling design is systematic πps. Unfortunately, this approach cannot guarantee that $\pi_{kl} > 0$ $(k \neq l)$. This issue, known as the problem of estimating the variance of the estimators, is thoroughly discussed in Wolter (1985). Rosén (1997) introduces a new sampling design, called ordered sampling design, that constitutes an important contribution to the solution of the problems related to πps.

To conclude this review of πps, we would like to consider the algorithms used for the selection of a sample according to the πps methodology. Typically, three different cases are treated separately.

$n = 1$ This set-up is mostly unrealistic and only mentioned in the literature for didactic purposes.

$n = 2$ Many πps sampling designs have been proposed in the literature with $n = 2$. See Brewer and Hanif (1983) for a review and some comparisons. Here we confine ourselves to citing the solution proposed by Brewer (1963, 1979), involving a sequential selection where the probabilities, in any draw, are adjusted so that we obtain the inclusion probabilities $\pi_i = 2x_k / \sum_U x_i$ $(i = 1, \ldots, N)$.

$n > 2$ Hanif and Brewer (1980) and Brewer and Hanif (1983) list approximately 50 criteria for selecting a πps sample. These methods, with the exception of systematic selection, are quite complicated for $n > 2$. Because of this difficulty, some approximately πps procedures have been developed. Among these, it is worth recalling the one proposed by Sunter (1986) and Ohlsson (1998); see below. In particular, Sunter (1977a, 1977b) has proposed a simple sequential solution that does not require the proportionality of π_i and x_i.

A simple PPS procedure without replacement is the so-called Poisson sampling (PS): with each population unit is associated a random number uniformly distributed between 0 and 1: $U_i \overset{iid}{\sim} U[0, 1]$ $(i = 1, \ldots, N)$. The ith unit is included in the sample if $U_i \leq nx_i / \sum_U x_i = \lambda_i$. The main difficulty related to the use of PS when the sample size n is moderately large is that the actual sample size n_{PS} is random. More precisely, $n_{PS} \sim \text{Pois}(n)$, so that both its expected value and its variance are equal to n. This implies that the realized values of n_{PS} may be very different from n. Because of this problem, the original PS sampling procedure has been modified in order to get the predetermined sample size n. The technique is known as sequential Poisson sampling (SPS) and was introduced by Ohlsson (1998).

If we assume $nx_i / \sum_U x_i \leq 1$ for any $i \in \{1, \ldots, N\}$ (in practice this is equivalent to putting $\lambda_i = 1$ for all the units such that $nx_i / \sum_U x_i > 1$), then from the random numbers U_i we get the transformed random numbers

$$\psi_i = \frac{U_i}{x_i / \sum_U x_i}.$$

Using the numerical values ψ_i, a sample is obtained by means of SPS if it contains the n units of the population corresponding to the n smallest ψ_i. This modification allows a sample of size n to be obtained, but unfortunately SPS is not πps. However, some simulation results obtained by Ohlsson (1998) suggest that SPS is approximately πps, so that it seems reasonable to use a classical estimator of the total t_y.

7.4 Balanced sampling

The fundamental property of a balanced sample is that the HT estimators of the totals of a set of auxiliary variables are equal to the totals we wish to estimate. This idea dates back to the pioneering work by Neyman (1934) and Yates (1946). More recently, balanced sampling designs have been advocated by Royall and Herson (1973) and Scott *et al.* (1978), who pointed out that such sampling designs ensure the robustness of the estimators of totals, where 'robustness' essentially means 'protection of inference against a misspecified model'.

More formally, a sampling design $p(\mathbf{s})$ is said to be balanced with respect to the J auxiliary variables $x_1, \ldots, x_J$ if and only if it satisfies the so-called *balancing equations*:

$$\sum_s \frac{x_{kj}}{\pi_k} = \sum_U x_{kj}, \qquad j = 1, \ldots, J, \tag{7.5}$$

for all samples s such that $p(s) > 0$.

An important remark concerning the preceding definition is that in most cases a sampling design satisfying condition (7.5) does not exist. Thus, in practice, the aim is to find a design such that (7.5) is satisfied *approximately* instead of *exactly*. Even using this more general definition, implementing balanced sampling is a challenging task from the computational point of view; see Tillé (2006, Section 8.5) and the references therein for some details about various algorithms for obtaining a balanced sampling design. A major step forward was the introduction of the so-called *cube method* (Deville and Tillé, 2004; Chauvet and Tillé, 2006; see also Tillé, 2006, Chapter 8), which we now describe.

The name of the algorithm comes from the geometric idea upon which it is based: every sample can be described by the coordinates of a vertex of the hypercube $C = [0, 1]^N$ in $\mathbb{R}^N$.

We first need to introduce some notation. If we define the $(J \times N)$ matrix

$$\boldsymbol{A} = (\boldsymbol{a}_1 \cdots \boldsymbol{a}_k \cdots \boldsymbol{a}_N) = \begin{pmatrix} x_{11}/\pi_1 & \cdots & x_{k1}/\pi_k & \cdots & x_{N1}/\pi_N \\ \vdots & \ddots & \vdots & \ddots & \vdots \\ x_{1j}/\pi_1 & \cdots & x_{kj}/\pi_k & \cdots & x_{Nj}/\pi_N \\ \vdots & \ddots & \vdots & \ddots & \vdots \\ x_{1J}/\pi_1 & \cdots & x_{kJ}/\pi_k & \cdots & x_{NJ}/\pi_N \end{pmatrix},$$

then it is easy to see that $\boldsymbol{A\pi} = \boldsymbol{t_x}$, where $\boldsymbol{\pi}$ is the vector of inclusion probabilities and $\boldsymbol{t_x} = \sum_{k \in U} \boldsymbol{x}_k$ is the vector of the totals of the auxiliary variables. Moreover, if $\boldsymbol{S} = (S_1, \ldots, S_N)'$ is the random sample, we also have $\boldsymbol{AS} = \hat{\boldsymbol{t}}_{\boldsymbol{x}}$, where $\hat{\boldsymbol{t}}_{\boldsymbol{x}}$ is the vector of estimated totals of the auxiliary variables. It immediately follows that a sample is balanced if

$$\boldsymbol{AS} = \boldsymbol{A\pi}. \tag{7.6}$$

Geometrically, the system (7.6) defines the subspace $K = \boldsymbol{\pi} + \mathrm{Ker}(\boldsymbol{a})$ in $\mathbb{R}^N$. The basic idea consists in selecting randomly a vertex of $K \cap C$; to do this, Chauvet and Tillé (2006) propose to construct a sequence of random displacements in $K \cap C$.

The algorithm consists of two phases called the flight phase and the landing phase. In the flight phase the constraints are always exactly satisfied. The objective is to round off randomly to 0 or 1 almost all the inclusion probabilities. The landing phase aims at taking care as well as possible of the fact that (7.6) cannot always be exactly satisfied.

The flight phase can be detailed as follows. The starting value is $\boldsymbol{\pi}(0) = \boldsymbol{\pi}$. At each iteration $t = 0, \ldots, T$ the following steps three hare carried out:

1. Generate any vector $\boldsymbol{u}(t) = \{u_k(t)\} \neq 0$, not necessarily random, such that $\boldsymbol{u}(t)$ belongs to the kernel of $\boldsymbol{A}$, and $u_k(t) = 0$ if $\pi_k(t)$ is an integer.

2. Compute the largest values of $\lambda_1(t)$ and $\lambda_2(t)$ such that $0 \le \boldsymbol{\pi}(t) + \lambda_1(t)\boldsymbol{u}(t) \le 1$ and $0 \le \boldsymbol{\pi}(t) - \lambda_2(t)\boldsymbol{u}(t) \le 1$. Call these two values $\lambda_1^*(t)$ and $\lambda_2^*(t)$. Obviously $\lambda_1(t) > 0$ and $\lambda_2(t) > 0$.

3. Compute the next value of $\boldsymbol{\pi}$ as:

$$\boldsymbol{\pi}(t+1) = \begin{cases} \boldsymbol{\pi}(t) + \lambda_1^*(t)\boldsymbol{u}(t) & \text{with prob. } q(t); \\ \boldsymbol{\pi}(t) - \lambda_2^*(t)\boldsymbol{u}(t) & \text{with prob. } 1 - q(t), \end{cases}$$

where $q(t) = \lambda_2^*(t)/(\lambda_1^*(t) + \lambda_2^*(t))$.

The three steps are iterated until it becomes impossible to perform step 1. In this phase, finding a vector in the kernel of $\boldsymbol{A}$ can be computationally quite expensive. To overcome this difficulty, Chauvet and Tillé (2005a, 2005b, 2006) developed a fast algorithm for implementing the three steps above. The idea consists in 'replacing' $\boldsymbol{A}$ with a smaller matrix $\boldsymbol{B}$; this is a submatrix of $\boldsymbol{A}$ containing only $J+1$ columns of $\boldsymbol{A}$. From a technical point of view, it turns out that a vector $\boldsymbol{v}$ of Ker($\boldsymbol{B}$) is the basic ingredient for getting a vector $\boldsymbol{u}$ of Ker($\boldsymbol{A}$), because it is enough to insert zeros in $\boldsymbol{v}$ for each column of $\boldsymbol{B}$ that is not in $\boldsymbol{A}$. All the computations can be done using matrix $\boldsymbol{B}$, and this speeds up the algorithm dramatically.

This fast version of the flight phase is detailed, for example, in Tillé (2006, algorithm 8.4); the algorithm was implemented in SAS (Chauvet and Tillé 2005a, b) and in the R package `sampling`.

It can be shown that at convergence ($t = T$) the following three properties hold:

- $\mathrm{E}(\boldsymbol{\pi}(T)) = \boldsymbol{\pi}$.
- $\boldsymbol{A}\boldsymbol{\pi}(T) = \boldsymbol{A}\boldsymbol{\pi}$.
- The number of non-integer elements of $\boldsymbol{\pi}(T)$ is at most equal to the number of auxiliary variables.

At the end of the flight phase, if $\boldsymbol{\pi}^* = \boldsymbol{\pi}(T)$ contains no non-integer elements, the algorithm is completed. Otherwise, some constraints cannot be exactly satisfied. In the latter instance, the landing phase should be performed. A possible way of carrying out the landing phase is the use of an enumerative algorithm. In this case the problem involves solving a linear program that does not depend on the population size, but only on the number of balancing variables, so that the computational burden is acceptable; see Tillé (2006, Section 8.6.3).

7.5 Calibration weighting

One of the most relevant problems encountered in large-scale business (e.g. agricultural) surveys is finding estimators that are (i) efficient and (ii) derived in accordance with criteria of internal and external consistency (see below).

The methodology presented in this section satisfies these requirements. The class of *calibration estimators* (Deville and Särndal, 1992) is an instance of a very general approach to the use of auxiliary information in the estimation procedures in finite-population sampling. It is indeed possible to prove that the class contains all the estimators commonly used in sampling surveys.

The process of sampling estimation involves associating a weight with each unit in the sample. Roughly speaking, the methodology proposed by Deville and Särndal (1992) finds the weights (associated with a certain amount of auxiliary information) by means of a distance measure and a system of calibration equations. The weights are determined as a by-product of the classical approach used in surveys based on non-trivial sampling designs. This procedure can be summarized as follows:

1. Compute the initial weights $d_k = 1/\pi_k$ where π_k is, as usual, the probability of including the kth unit in the sample s.
2. Compute the quantities γ_k that correct the initial weights for total non-responses and external consistency requirements.
3. Compute the final weights $w_k = d_k \gamma_k$.

Let $U = \{1, \ldots, N\}$ be the finite surveyed population. Furthermore, let U_L be the set of units present in the list from which a sample $s^*(s^* \subseteq U_L)$ is drawn by means of design $p(\cdot)$. Thus $p(s^*)$ is the probability of selecting the sample s^*. Let n^* be the cardinality of s^*. Finally, let $s \subseteq s^*$ be the set of the n responding units.

For the kth element of the population, let y_k be the value taken by the surveyed variable y. For any y_k there exists a vector $\boldsymbol{x}_k = (x_{k1}, \ldots, x_{kj}, \ldots, x_{kJ})'$ containing the values taken by J auxiliary variables. Therefore, for each element k included in the sample $(k \in s)$ we observe the pair $(y_k, \boldsymbol{x}_k)$.

The totals of the J auxiliary variables for the population are the elements of the vector

$$\boldsymbol{t}_x = (t_{x1}, \ldots, t_{xj}, \ldots, t_{xJ})' = \left(\sum_U x_{k1}, \ldots, \sum_U x_{kj}, \ldots, \sum_U x_{kJ} \right)'$$

and are assumed to be known.

The goal of the survey is the estimation of the total t_y of the variable y. The estimator used is required to satisfy the following properties:

1. It should be approximately unbiased.
2. It should produce estimates of the totals $\boldsymbol{t}_x$ equal to the known values of these totals (*external consistency of the estimators*).
3. It should mitigate the biasing effect of the total non-responses $(n < n^*)$.
4. It should mitigate the effect caused by the so-called list undercoverage. This happens when the list from which a sample is drawn is such that it does not contain existing units $(U_L \subset U)$.

If there are no total non-responses (i.e. if $s^* = s$) and the list from which the sample is drawn is not undercovered $(U \subseteq U_L)$, then the direct estimator of the total t_y satisfies condition 1. This estimator is given by (7.1) and can be rewritten as $\hat{t}_{y\pi} = \sum_s d_k y_k$. The terms $d_k = 1/\pi_k$ are the direct weights of each unit $k \in s$, that is, the original weights produced by the sampling design.

Unfortunately, in most large-scale business surveys, (7.1) does not satisfy the conditions above. There are two reasons for this:

- For the direct estimate, $\hat{\boldsymbol{t}}_{x\pi} = \sum_s d_k x_k \neq \boldsymbol{t}_x$. In other words, the sample sum of the weighted auxiliary variables is not equal to the aggregate known total value. From an operational point of view, it is very important that property 2 be satisfied, because if the auxiliary variables $\boldsymbol{x}$ and the variable of interest y are strongly correlated, then the weights that give good results for $\boldsymbol{x}$ must give good results for y too.
- If $n^* > n$ and/or $U_L \subset U$, the direct estimate $\hat{\boldsymbol{t}}_{y\pi}$ is an underestimate of the total t_y, because $\mathrm{E}_p(\hat{\boldsymbol{t}}_{y\pi}) < t_y$, where $\mathrm{E}_p(\cdot)$ is the expected value with respect to the sampling design.

A class of estimators of the total that satisfy conditions 1–4 above under quite general assumptions is the class of calibration estimators (Deville and Särndal, 1992). Here we are concerned with the use of calibration estimators in absence of non-sampling errors. However, calibration weighting can be used to adjust for unit non-response and/or coverage errors under appropriate models. Kott (2006) extends the work of Folsom and Singh (2000). They showed that calibration weighting can adjust for known coverage errors and/or total non-response under quasi-randomization models. An earlier, linear version of calibration weighting for unit non-response adjustment can be found in Fuller *et al.* (1994). See also Lundström and Särndal (1999) and Särndal and Lundström (2005), which is the definitive reference on these issues.

Skinner (1999) explores the properties of calibration estimation in the presence of both non-response and measurement errors. In a set-up where the quantity of interest is the total of the variable y, Skinner (1999) assumes that the data comes from two sources. The values y_k^s and x_k^s are computed from the set of respondents. A second source gives the vector of population totals of J auxiliary variables $x_1^a, \ldots, x_J^a$. If y is measured without errors, we have that $y_k^s = y_k$, where y_k is the true value. Otherwise, $y_k^s - y_k$ is the measurement error. Skinner (1999) treats separately the two following cases:

(i) $x_k^s = x_k^a$, that is, both data sources lead to identical measurements;

(ii) $x_k^s \neq x_k^a$ for some responding k, that is, there is some variation between the measurements obtained in the sources.

Finally, Skinner (1999) shows that, in case (i), measurement error in the auxiliary variables does not tend to introduce bias in the calibration estimator but only causes a loss of precision. On the other hand, in case (ii), the use of calibration estimation may introduce severe bias. Calibration weighting should not, however, be dismissed immediately in this case. Skinner (1999) gives some suggestions for reducing the bias if y_k is also subject to error.

An estimator belonging to the class of calibration estimators can be denoted by $\hat{t}_{yw} = \sum_s w_k y_k$, where the weights w_k should be as close as possible to the weights π_k^{-1} of the original sampling design and should satisfy some constraints (the so-called calibration equations). Clearly, to each distance measure corresponds a different set of weights w_k and a different estimator.

The set of final weights w_k $(k = 1, \ldots, n)$ is found by solving the following optimization problem, well known to all survey designers:

$$\begin{cases} \min \sum_s G_k(d_k, w_k), \\ \sum_s w_k \boldsymbol{x}_k = \boldsymbol{t}_k, \end{cases} \tag{7.7}$$

where $G_k(d_k, w_k)$ is a distance function between the direct weight d_k and the final weight w_k. For the constrained optimization problem (7.7) to admit a finite solution and for this solution to be unique, the function has to satisfy precise conditions (see Deville and Särndal 1992, p. 327). Finding the solution $\boldsymbol{w}$ of the system (7.7) requires, as usual, the definition of the Lagrangian, from which we get a homogeneous system of $n + J$ equations in $n + J$ unknown variables $(\boldsymbol{\lambda}, \boldsymbol{w})$, where $\boldsymbol{\lambda}$ is the $(J \times 1)$ vector containing the Lagrange multipliers.

If the system has a solution, it can always be written in the following form, obtained from (7.7):

$$w_k = d_k F_k(\boldsymbol{x}'_k \boldsymbol{\lambda}), \tag{7.8}$$

where $d_k F_k(\cdot)$ is the inverse function of $g_k(\cdot, d_k)$. The quantity $F_k(\boldsymbol{x}'_k \boldsymbol{\lambda}) = (1/d_k) g_k^{-1}(\boldsymbol{x}'_k \boldsymbol{\lambda})$ is the factor γ_k that corrects the direct weight. This factor is a function of the linear combination of the vector $\boldsymbol{x}'_k$ containing the auxiliary variables and the J unknown values of $\boldsymbol{\lambda}$. In most practical applications $g(\cdot)$ is such that $g_k(w_k, d_k) = g_k(w_k/d_k)$ and satisfies the assumptions in Deville and Särndal (1992, p. 327).

Some examples of this kind of functions are given by Deville and Särndal (1992). However, (7.8) is not yet an implementable result, because $\boldsymbol{\lambda}$ is unknown. Deville and Särndal (1992) show that the problem can be rewritten in the form of the following system of J equations in J unknowns $\lambda_1, \ldots, \lambda_j, \ldots, \lambda_J$:

$$\boldsymbol{\phi}_s(\boldsymbol{\lambda}) = \sum_s d_k (F_k(\boldsymbol{x}'_k \boldsymbol{\lambda}) - 1) \boldsymbol{x}_k = \boldsymbol{t}_x - \hat{\boldsymbol{t}}_{x\pi}. \tag{7.9}$$

We can therefore summarize the procedure proposed by Deville and Särndal (1992) as follows.

1. Define a distance function $G_k(d_k, w_k)$.
2. Given a sample s and the function $F_k(\cdot)$ chosen at the preceding step, solve with respect to $\boldsymbol{\lambda}$ the system $\boldsymbol{\phi}_s(\boldsymbol{\lambda}) = \boldsymbol{t}_x - \hat{\boldsymbol{t}}_{x\pi}$, where the quantity on the right-hand side is known.
3. Compute the calibration estimator of $t_y = \sum_U y_k$, that is, $\hat{t}_{yw} = \sum_s w_k y_k = \sum_s d_k F_k(\boldsymbol{x}'_k \boldsymbol{\lambda}) y_k$.

Roughly speaking, it is clear that the quality of the estimates obtained by means of the estimator $\hat{t}_{yw}$ increases as the relationship between y and $\boldsymbol{x}$ gets stronger.

This procedure has been used in most surveys performed by the main national statistical institutes in the world: for example, surveys concerning business (Structural Business Statistics; EU Council Regulation No. 58/97) and families performed by Istat (Falorsi and Falorsi, 1996; Falorsi and Filiberti, 2005), as well as the American Community Survey (ACS) carried out since 2005. Among other purposes, the ACS will replace the decennial census long-form data, enabling a short-form census in 2010. This survey is based on an approach combining administrative record data with model-assisted estimation (specifically a generalized regression estimation; Fay, 2006). We recall here that the generalized regression GREG) estimator $\hat{t}_{y,GREG}$ can be rewritten in calibration form $\sum_s w_k y_k$ (Deville and Särndal, 1992; Fuller, 2002; Kott, 2006).

The Office for National Statistics (UK) devoted a considerable effort in the final years of the 20th century into the study of the most appropriate methodologies for the production of business statistics (Smith *et al.*, 2003). A 'short period' survey, the Monthly Production Inquiry of manufacturing, uses register employment as the auxiliary variable in ratio estimation. This ratio estimator is in fact a special case of a more general set of model-assisted (GREG or calibration) estimators (Deville and Särndal, 1992). It corresponds to a regression model for predicting the non-sampled observations where the intercept is set to 0 and the variance of the residuals is assumed to be proportional to the auxiliary variable (Särndal *et al.*, 1992, p. 255).

The monthly Retail Sales Inquiry uses a different form of ratio estimation known as 'matched pairs' (Smith *et al.*, 2003). More sophisticated regression estimation procedures were considered (e.g. multivariate calibration estimators) to merge and to rationalize some annual business surveys.

The US National Agricultural Statistics Service used variants of the Fuller *et al.* (1994) approach for handling undercoverage in the 2002 Census of Agriculture (see Fetter and Kott, 2003) and for adjusting an agricultural economics survey with large non-response to match totals from more reliable surveys (see Crouse and Kott, 2004).

For the 1991 and 1996 Canadian Census, calibration estimation (including regression) was used (Bankier *et al.*, 1997).

The Finnish Time Use Survey (TUS) was implemented by Statistics Finland and conducted in 1999–2000 in accordance with the Eurostat guidelines for harmonized European time use surveys. The data were collected at the household and individual levels by interviews and diaries. In the TUS, the estimation needs some special steps due to the diaries and the household sample, so that the estimators of the time use variables may be rather complicated. For instance, the allocation of diary days affects the weighting. Calibration techniques were used to balance the estimates to correspond with the auxiliary data (Väisänen, 2002).

Finally, the methodology is used at Statistics Belgium for the Labour Force Survey, the Household Budget Survey, the Time Budget Survey, the Tourism Survey, and business surveys (e.g. short-term turnover indicator).

To implement the methodology introduced by Deville and Särndal (1992), several national statistical agencies have developed software designed to compute calibrated weights based on auxiliary information available in population registers and other sources. As for this issue see the references in Estevao and Särndal (2006) and work presented at the workshop 'Calibration Tools for Survey Statisticians' that took place in Switzerland in September 2005.[1]

Going back to the solution of (7.9), it is convenient to distinguish the case when the model for the correcting factors $F_k(\boldsymbol{x}'_k\boldsymbol{\lambda})$ (and therefore $\boldsymbol{\phi}_s(\lambda)$) is a linear function of $\boldsymbol{x}$ and the case when it is not. In the first instance it is possible to solve (7.9) in closed

[1] See `http://www.unine.ch/statistics/conference`.

form, because it can be rewritten in the form $\boldsymbol{\phi}_s(\boldsymbol{\lambda}) = \boldsymbol{T}_s\boldsymbol{\lambda}$, where $\boldsymbol{T}_s$ is a symmetric positive definite $(J \times J)$ matrix. The solution is therefore given by $\boldsymbol{\lambda} = \boldsymbol{T}_s^{-1}(\boldsymbol{t}_x - \hat{\boldsymbol{t}}_{x\pi})$. In the second, instance, when the function $\boldsymbol{\phi}_s(\lambda)$ is non-linear, the solution can be found using iterative techniques, usually based on the Newton–Raphson algorithm.

To check convergence, the following criterion is commonly adopted:

$$\max_{j=1,\dots,J} \frac{|\lambda_j^{(t-1)} - \lambda_j^{(t)}|}{\lambda_j^{(t)}} < \epsilon.$$

However, in our opinion, it is preferable to use two convergence criteria. One iterates the algorithm until (i) the number of iterations does not exceed the maximum number of iterations allowed and (ii) the absolute maximum error $|t_x - \sum_s d_k F_k(\boldsymbol{x}_k'\boldsymbol{\lambda}^{(t)})\boldsymbol{x}_k|$ is larger than a tolerance threshold δ. For this parameter, Singh and Mohl (1996) suggest using a value of 0.005 or 0.01.

When choosing a distance function for use in practical applications, it is first of all necessary to consider the existence of a solution of the system $\boldsymbol{\phi}_s(\boldsymbol{\lambda}) = \boldsymbol{t}_x - \hat{\boldsymbol{t}}_{x\pi}$. Moreover, the range of the final weights $w_k = d_k F_k(\boldsymbol{x}_k')$ should be taken into consideration. Deville and Särndal (1992) report an example where the final weights w_k can be positive or negative, and further examples where certain distance measures ensure that $w_k > 0$. However, in each of the aforementioned cases the weights can be unacceptably large with respect to the initial weights $d_k = 1/\pi_k$.

A way of overcoming this drawback is given by the logit class correction discussed in the following. This function is one of the most commonly used in practice because it satisfies the range restrictions for the calibrated weights. Omitting for simplicity a multiplicative constant, the distance function is

$$G_k(d_k, w_k) = \left(\frac{w_k}{d_k} - L\right)\log\left(\frac{\frac{w_k}{d_k} - L}{1 - L}\right) + \left(U - \frac{w_k}{d_k}\right)\log\left(\frac{U - \frac{w_k}{d_k}}{U - 1}\right), \tag{7.10}$$

where L and U are two constants to be chosen such that $L < 1 < U$. The corresponding model for the correcting factors of the direct weights is the following logit correction:

$$F_k(\boldsymbol{x}_k'\boldsymbol{\lambda}) = \frac{L(U-1) + U(1-L)\exp(A\boldsymbol{x}_k'\boldsymbol{\lambda})}{U - 1 + (1-L)\exp(A\boldsymbol{x}_k'\boldsymbol{\lambda})},$$

where $A = (U - L)/((1 - L)(U - 1))$. The values of λ are computed iteratively. The distance function (7.10) gives final weights w_k always belonging to the interval (Ld_k, Ud_k).

It is worth stressing that the choice of the numerical values of L and U is not arbitrary because the condition $L < 1 < U$ must hold: some approximate rules have been developed (Verma, 1995). A rule of thumb sometimes used is to choose a starting value close to 0 for L and 'large' for U. The procedure iterates increasing L and decreasing U towards 1, and stops at the the pair (L, U) closest to 1 (according to some distance measure) and such that the optimization problem has a solution.

To conclude this brief review, we note that the methods usually studied in the literature (seven of these are in Singh and Mohl, 1996) are asymptotically equivalent to linear regression, so that the asymptotic variance of $\hat{t}_{yw}$ can be estimated as follows:

$$\hat{V}(\hat{t}_{yw}) = \sum_i \sum_j \frac{\pi_{ij} - \pi_i \pi_j}{\pi_{ij}} w_i e_i w_j e_j,$$

where the e_i are the sampling residuals $e_i = y_i - \hat{\boldsymbol{B}}' \boldsymbol{x}_i$, with $\hat{\boldsymbol{B}}' = \boldsymbol{x}'\boldsymbol{\Gamma}\boldsymbol{x}(\boldsymbol{x}'\boldsymbol{\Gamma}\boldsymbol{x})^{-1}$ and where $\boldsymbol{\Gamma} = \text{diag}(\boldsymbol{d})$ is the $n \times n$ diagonal matrix of the direct weights.

In the last decade, calibration estimation has become an important field of research in survey sampling. An excellent review of the results obtained is given by Estevao and Särndal (2006), who present some recent progress and offer new perspectives in several non-standard set-ups, including estimation for domains in one-phase sampling, and estimation for two-phase sampling.

Further developments such as calibration on imprecise data, indirect calibration, reverse calibration (Ren and Chambers, 2003) and 'hypercalibration' are treated in Deville (2005).

A large number of areas remain open for future research. We mention here further development of $GREG$ weighting in order to make this approach less affected by extreme weights and the issue of extending the calibration method to nominal and ordinal variables. The concept of model calibration (Wu and Sitter, 2001) plays a major role in the study of this topic.

7.6 Combining *ex ante* and *ex post* auxiliary information: a simulated approach

In this section we measure the efficiency of the estimates produced when the sampling design uses only the *ex ante*, only the *ex post* or a mixture of the two types of auxiliary information. Thus, we performed some simulation experiments whose ultimate aim is to test the efficiency of some of the selection criteria discussed above. After giving a brief description of the archive at hand, we detail the Istat survey based on this archive. Then we review the technical aspects of the sample selection criteria used in the simulation. Finally, we show the results.

Our frame contains $N = 2211$ slaughter-houses for which we know four variables enumerated completely in 1999, 2000 and 2001: they are respectively the total number of slaughtered (i) cattle, (ii) pigs, (iii) sheep and goats and (iv) equines.

The scatterplots of all the pairs of the four variables in 2001 are shown in Figure 7.2; the corresponding graphs for 1999 and 2000 are almost identical and therefore omitted. The main evidence is that the variables are essentially uncorrelated, as confirmed by the linear correlation coefficient, which ranges in the interval $[-0.0096, 0.0566]$. Moreover, slaughter-houses are strongly specialized and most firms are small: in particular, 38.9% of the firms slaughter only one type of animal, 24.06% two types, 21.85% three types and only 15.2% all four types.

An archive of this type has two main purposes:

- preparing the database needed to produce estimates using *ex post* the auxiliary information (the auxiliary variables are the census values of 1999 and 2000);

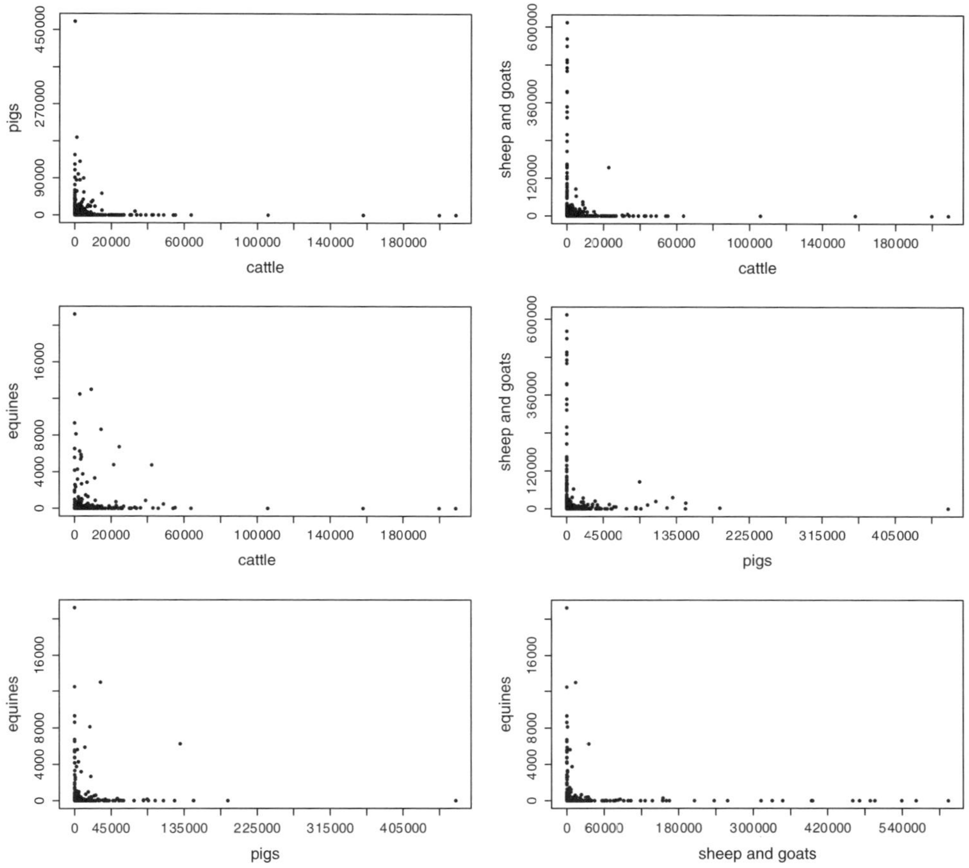

Figure 7.2 Scatterplots of the data; the unit of measurement is number of animals.

- verifying *ex post* (by computing the root mean squared errors) the goodness of the estimates produced for the year 2001 (the target year of the simulation) by means of a comparison with the known 2001 census value.

This way of proceeding can be further justified in view of the actual frequency of the slaughtering data. In parallel with the census repeated every year, Istat performs a monthly survey: the red meat slaughtering monthly survey. The aim is to obtain information on the number and weight of animals slaughtered monthly in Italy. This survey is based on a stratified sample, with a stratification by kind of slaughter-houses and geographical division, for a total of five strata. On average, the sample is of about 460 units for a population of 2211 units with the desired level of precision c set to 5% (Istat, 2007).

Our experiments, constructed as prescribed by Dorfman and Valliant (2000), simulate the monthly survey by implementing it with different selection criteria. Thus, the simulated analysis allows us to carry out comparisons with the census data by means of the empirical sample distributions of the estimates.

Table 7.1 Main results of the simulation.

Selection Criterion	Estimator	Base Year	RMSE (% of the Estimate) CAT	PIG	SH-GO	EQU
SRS	Direct	–	40.00	51.25	51.65	57.70
	Cal. Wei.	2000	14.12	25.31	27.43	26.45
	Cal. Wei.	1999	20.31	31.66	26.73	35.12
Stratified	Direct	2000	16.34	18.24	41.87	15.06
2	Direct	1999	15.93	25.90	41.12	16.83
	Cal. Wei.	2000	10.83	12.09	39.83	5.30
	Cal. Wei.	1999	12.40	21.70	38.38	11.50
Stratified	Direct	2000	6.07	8.07	10.46	7.62
5	Direct	1999	6.47	10.58	18.10	9.85
	Cal. Wei.	2000	3.50	4.77	8.99	4.81
	Cal. Wei.	1999	4.34	8.61	17.18	6.97
πps	Direct	2000	6.17	6.38	15.61	3.74
	Direct	1999	7.28	10.18	17.20	6.79
	Cal. Wei.	2000	4.52	5.04	14.87	2.57
	Cal. Wei.	1999	6.04	9.28	16.67	6.48
Balanced	Direct	2000	6.24	13.48	17.26	15.05
	Direct	1999	23.58	20.05	17.46	19.43
Bal./πps	Direct	2000	5.55	5.08	14.60	2.50
	Direct	1999	6.37	9.45	16.72	6.96

Specifically, for each scenario detailed below, we drew 2000 samples of size $n = 200$ using as *ex post* auxiliary information, whenever possible, the complete 1999 archive for the production of the 2001 estimates and the complete 2000 archive for the production of the same 2001 estimates. The aim of such a procedure is to check whether the temporal lag of the auxiliary information causes efficiency losses. The estimated quantities for 2001 are, in each replication of the experiment, the totals of the four variables considered (see Table 7.1).

For each of the selection criteria considered here, the estimates were produced by means of both the HT estimator and the Deville and Särndal (1992) calibration weighting estimator. As pointed out in Section 7.5, this is an estimator that performs a calibration of the weights of the original design and satisfies the external consistency conditions $\sum_s w_k x_k = t_k$. In other words, the sampling distributions weighted by means of the auxiliary variables are required to be equal to the known (completely enumerated) distributions.

The only exception is the balanced sampling approach, which satisfies the constraints by definition. Thus, it makes no sense to apply a calibration weighting with the estimates obtained by means of balanced sampling. If there are no total non-responses and other errors in the archive, this is a phase that does not change the weights.

The set-up of the present simulation excludes non-responses and coverage errors. It would be easy to extend the experiment in this direction. However, this would complicate the analysis of the results and would be beyond the scope of this chapter. We will therefore pursue this generalization in future works.

Before commenting on the results of the simulation, some operational details about the selection criteria are in order. The list follows the same order the first column of Table 7.1.

Simple random sampling. No relevant comment except that the direct estimate, by definition, does not use any auxiliary information; thus, it is not necessary to distinguish the two reference years (see the first line of Table 7.1).

Stratified sampling. In this case the auxiliary information plays a role in the selection of the strata *ex ante* the sample selection. Hence in Table 7.1 there are different direct estimates according to the reference year (1999 and 2000) in order to distinguish the auxiliary variables used to find the strata. Table 7.1 shows the results for two different designs: two and five strata. We have also studied the ten-strata case; the results are not reported here because the efficiency gains with respect to the five-strata case are very small. The two stratifications used have been obtained by means of the stratification trees algorithm (Benedetti *et al.*, 2008; see also Section 7.2 above), which has the advantage of combining in a single procedure the choice of the number of stratifying variables, the number of class intervals for each variable and the optimal allocation to strata.

Probability proportional to size sampling (πps). The key parameter for the selection of a πps sample is the vector of the first-order inclusion probabilities. The device used for tackling a multivariate approach (four auxiliary variables) by means of a selection criterion that is based on a single-dimensional variable is the following. We start with four vectors of first-order inclusion probabilities, one for each auxiliary variable at time t: $\pi^t_{k,i} = n^t_{k,i} / \sum_U x^t_{k,i}$, $k = 1, \ldots, 2211$; $i = 1, \ldots, 4$; $t = 1999, 2000$. In order to include all four variables in the sample selection process, we decided to switch from these four vectors to the vector $\bar{\pi}^t$ of the averages of these probabilities. In addition, notice that all the units k such that $\bar{\pi}^t_k > 1$ (which typically happens due to particularly large values), have been included in the sample. It follows that the solution proposed here to handle a multivariate survey is not a truly multivariate solution. However, the present approach is similar to Patil *et al.* (1994) who propose to base the sample selection on a single auxiliary variable deemed to be most important, whereas the remaining variables play a secondary role. Finally, the πps procedure has been carried out using SPS (Ohlsson, 1998; see also Section 7.3 above).

Balanced sampling. The balanced samples have been selected by means of the cube algorithm (Deville and Tillé, 2004; see also Section 7.4 above). The balancing constraints $\sum_s x_{k,i}/\pi_k = \sum_U x_{k,i}$ have been put on the four variables. This means that, in order to avoid the introduction of too many constraints with a relatively small archive, we do not impose the constraints per stratum. The same remarks hold for calibration weighting.

Table 7.1 displays selected results. The main conclusions concerning the use of auxiliary information can be summarized as follows. First of all, *ceteris paribus*, it is preferable to impose the balancing constraints in the balancing phase rather than *ex post*. This result is not unexpected in view of the advantages of stratification with respect to post-stratification and emerges clearly from the comparison of the root mean squared errors of simple random sampling with calibration weighting and those of the direct estimates from a balanced design.

Second, the best performances are obtained from the balanced and calibration-weighted πps selections. These two set-ups are very similar to each other and it would probably be difficult to get any significant improvement, as they are close to a minimum. From this we can draw the most important conclusion for practical applications: use estimators more sophisticated than the standard ones, trying to match them to efficient sampling designs that can also contribute to diminishing the variability of the estimates.

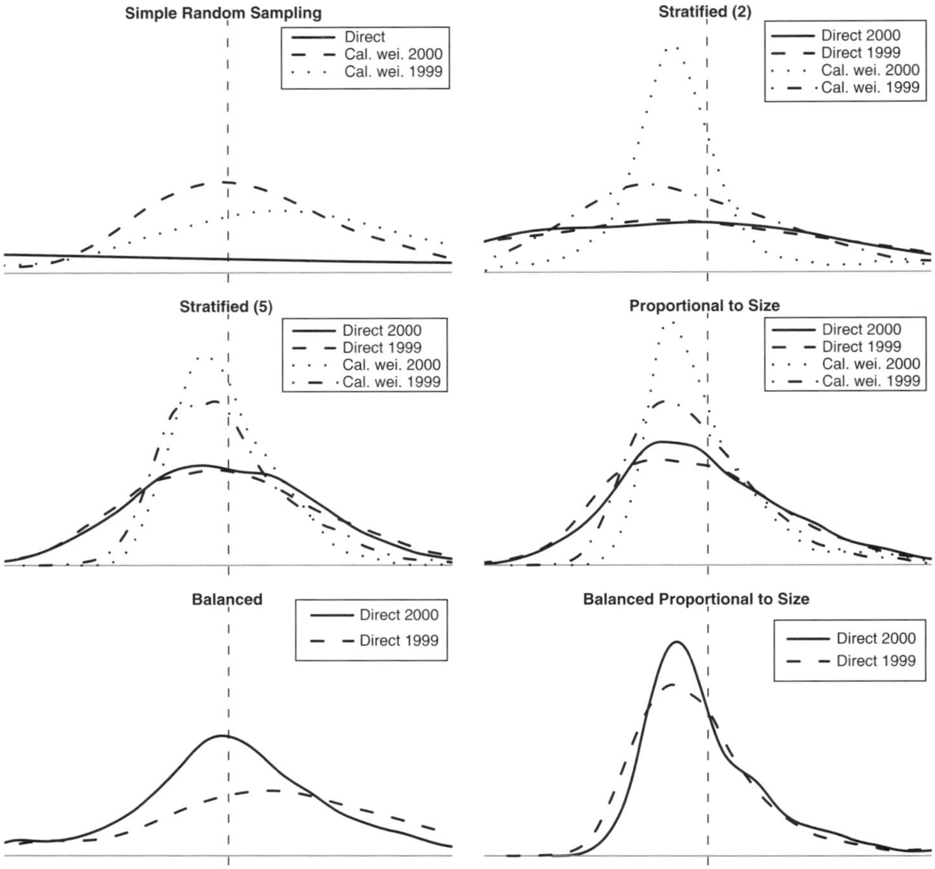

Figure 7.3 Kernel densities of the simulated distributions of the estimators.

These results are also shown in Figure 7.3, which gives, with reference to the variable 'number of slaughtered cattle', the empirical densities of the estimators used in the experiments. These densities have been obtained by means of a kernel smoother (Wand and Jones, 1995). The graphs of the other three variables give essentially the same information, and are therefore omitted.

7.7 Conclusions

In this chapter we have studied the role of auxiliary information in agricultural sample survey design. First we introduced the distinction between *ex ante* (before sample selection) and *ex post* use of the auxiliary information. In accordance with this, we reviewed three families of operational strategies, summarizing the main results in the literature and considering various topics of particular interest in the current scientific debate. The two *ex ante* strategies considered in this chapter are the construction of efficient and/or optimal stratifications and of efficient sample designs. As for the latter issue, we focused on a selection based on non-constant inclusion probabilities (πps) and on the introduction of a priori balancing constraints. Then we considered the most common (at least in the field of official statistics) strategy for the *ex post* use of auxiliary information, namely the so-called calibration weighting.

Finally, we performed some simulation experiments in order to compare, in terms of efficiency, estimates produced by designs using only *ex ante* information, designs using only *ex post* information and designs using a combination of the two.

The results appear to be of some use for practical applications. The main recommendation arising from the experiments is to use variable inclusion probabilities (πps) in combination with balancing (*ex ante* is slightly better than *ex post*).

References

Bankier, M., Houle, A.M. and Luc, M. (1997) Calibration estimation in the 1991 and 1996 Canadian censuses. *Proceedings of the Survey Research Methods Section*, American Statistical Association, pp. 66–75.

Benedetti, R., Espa, G. and Lafratta, G. (2008) A tree-based approach to forming strata in multipurpose business surveys. *Survey Methodology*, 34, 195–203.

Benedetti R., Bee, M. and Espa, G. (2010) A framework for cut-off sampling in business survey design. *Journal of Official Statistics*, in press.

Bethel, J. (1985) An optimum allocation algorithm for multivariate surveys. *Proceedings of the Surveys Research Methods Section*, American Statistical Association, pp. 209–212.

Bethel J. (1989) Sample allocation in multivariate surveys. *Survey Methodology*, 15, 47–57.

Bloch, D.A. and Segal, M.R. (1989) Empirical comparison of approaches to forming strata: Using classification trees to adjust for covariates. *Journal of the American Statistical Association*, 84, 897–905.

Breiman, L., Friedman, J.H., Olshen, R.A. and Stone, C.J. (1984) *Classification and Regression Trees*. Belmont, CA: Wadsworth International Group.

Brewer, K.R.W. (1963) A model of systematic sampling with unequal probabilities. *Australian Journal of Statistics*, 5, 5–13.

Brewer, K.R.W. (1979) A class of robust sampling designs for large scale surveys. *Journal of the American Statistical Association*, 74, 911–915.

Brewer, K.R.W. and Hanif M. (1983) *Sampling with Unequal Probabilities*. New York: Springer.

Chauvet, G. and Tillé, Y. (2005a) *Fast SAS macros for balanced sampling: user's guide*. Software manual, University of Neuchatel.

Chauvet, G. and Tillé, Y. (2005b) New SAS macros for balanced sampling. *Journées de Méthodologie Statistique*, INSEE.

Chauvet, G. and Tillé, Y. (2006) A fast algorithm for balanced sampling. *Computational Statistics*, 21, 53–61.

Chromy, J. (1987) Design optimization with multiple objectives. *Proceedings of the Survey Research Methods Section*, American Statistical Association, pp. 194–199.

Cochran, W.G. (1961) Comparison of methods for determining stratum boundaries. *Bulletin of the International Statistical Institute*, 38, 345–358.

Cochran, W.G. (1977) *Sampling Techniques*, 3rd edition. New York: John Wiley & Sons. Inc.

Crouse, C. and Kott, P.S. (2004) Evaluation alternative calibration schemes for an economic survey with large nonresponse. *Proceedings of the Survey Research Methods Section*, American Statistical Association.

Dalenius, T. (1950) The problem of optimum stratification. *Skandinavisk Aktuarietidskrift*, 34, 203–213.

Dalenius, T. (1952) The problem of optimum stratification in a special type of design. *Skandinavisk Aktuarietidskrift*, 35, 61–70.

Dalenius, T. (1957) *Sampling in Sweden. Contributions to the Methods and Theories of Sample Survey Practice*. Stockholm: Almqvist och Wiksell.

Dalenius, T. and Hodges, J.L. Jr. (1959) Minimum variance stratification. *Journal of the American Statistical Association*, 54, 88–101.

Dayal, S. (1985) Allocation of sample using values of auxiliary characteristic. *Journal of Statistical Planning and Inference*, 11, 321–328.

Deville, J.C. (2005) Calibration, past, present and future? Paper presented at the Calibration Tools for Survey Statisticians Workshop, Neuchâtel, 8–9 September.

Deville, J.C. and Särndal, C.E. (1992) Calibration estimators in survey sampling. *Journal of the American Statistical Association*, 87, 376–382.

Deville, J.C. and Tillé, Y. (2004) Efficient balanced sampling: the cube method. *Biometrika*, 91, 893–912.

Dorfman, A.H. and Valliant, R. (2000) Stratification by size revised. *Journal of Official Statistics*, 16, 139–154.

Ekman, G. (1959) An approximation useful in univariate stratification. *Annals of Mathematical Statistics*, 30, 219–229.

Estevao, V.M. and Särndal, C.E. (2006) Survey estimates by calibration on complex auxiliary information. *International Statistical Review*, 74, 127–147.

Falorsi, P.D. and Falorsi, S. (1996) Un metodo di stima generalizzato per le indagini sulle imprese e sulle famiglie. *Documenti Istat*, no. 2.

Falorsi, P.D. and Filiberti, S. (2005) GENEralised software for Sampling Estimates and Errors in Surveys (GENESEES V. 3.0). Paper presented at the Calibration Tools for Survey Statisticians Workshop, Neuchâtel, 8–9 September.

Fay, R.E. (2006) Using administrative records with model-assisted estimation for the American Community Survey. *Proceedings of the 2006 Joint Statistical Meetings on CD-ROM*, pp. 2995–3001. Alexandria, VA: American Statistical Association.

Fetter, M.J. and Kott, P.S. (2003) Developing a coverage adjustment strategy for the 2002 Census of Agriculture. Paper presented to 2003 Federal Committee on Statistical Methodology Research Conference. `http://www.fcsm.gov/03papers/fetter_kott.pdf`.

Folsom, R.E. and Singh, A.C. (2000) The generalized exponential model for sampling weight calibration for extreme values, nonresponse, and poststratification. *Proceedings of the Section on Survey Research Methods*, American Statistical Association, 598–603.

Fuller, W.A. (2002) Regression estimation for survey samples. *Survey Methodology*, 28, 5–23.

Fuller, W.A., Loughin, M.M. and Baker, H.D. (1994) Regression weighting for the 1987–88 National Food Consumption Survey. *Survey Methodology*, 20, 75–85.

Ghosh, S.P. (1963) Optimum stratification with two characters. *Annals of Mathematical Statistics*, 34, 866–872.

Glasser, G.J. (1962) On the complete coverage of large units in a statistical study. *Review of the International Statistical Institute*, 30, 28–32.

Godfrey, J., Roshwalb, A. and Wright, R. (1984) Model-based stratification in inventory cost estimation. *Journal of Business and Economic Statistics*, 2, 1–9.

Golder, P.A. and Yeomans, K.A. (1973) The use of cluster analysis for stratification. *Applied Statistics*, 18, 54–64.

Green, P.E., Frank, R.E. and Robinson, P.J. (1967) Cluster analysis in test market selection. *Management Science*, 13, 387–400.

Gunning, P. and Horgan, J.M. (2004) A new algorithm for the construction of stratum boundaries in skewed populations. *Survey Methodology*, 30, 159–166.

Hagood, M.J. and Bernert, E.H. (1945) Component indexes as a basis for stratification in sampling. *Journal of the American Statistical Association*, 40, 330–341.

Hanif, M. and Brewer, K.R.W. (1980) Sampling with unequal probabilities without replacement: a review. *International Statistical Review*, 48, 317–355.

Heeler, R.M. and Day, G.S. (1975) A supplementary note on the use of cluster analysis for stratification. *Applied Statistics*, 24, 342–344.

Hidiroglou, M.A. (1986) The construction of a self representing stratum of large units in survey design. *American Statistician*, 40, 27–31.

Hidiroglou, M.A. and Srinath, K.P. (1981) Some estimators of a population total from simple random samples containing large units. *Journal of the American Statistical Association*, 76, 690–695.

Horgan, J.M. (2006) Stratification of skewed populations: A review. *International Statistical Review*, 74, 67–76.

Istat (2007) Dati mensili sulla macellazione delle carni rosse. `http://www.istat.it/agricoltura/datiagri/carnirosse`.

Jarque, C.M. (1981) A solution to the problem of optimum stratification in multivariate sampling. *Applied Statistics*, 30, 163–169.

Julien, C. and Maranda, F. (1990) Sample design of the 1988 National Farm Survey. *Survey Methodology*, 16, 117–129.

Kish, L. (1976) Optima and proxima in linear sample designs. *Journal of the Royal Statistical Society, Series A*, 139, 80–95.

Kish, L., Anderson, D.W. (1978) Multivariate and multipurpose stratification. *Journal of the American Statistical Association*, 73, 24–34.

Kott, P.S. (2006) Using calibration weighting to adjust for nonresponse and coverage errors. *Survey Methodology*, 32, 133–142.

Lavallée, P. and Hidiroglou, M. (1988) On the stratification of skewed populations. *Survey Methodology*, 14, 33–43.

Lundström, S. and Särndal, C.E. (1999) Calibration as a standard method for treatment of nonresponse,. *Journal of Official Statistics*, 15, 305–327.

Mulvey, J.M. (1983) Multivariate stratified sampling by optimization. *Management Science*, 29, 715–723.

Neyman, J. (1934) On the two different aspects of the representative method: the method of stratified sampling and the method of purposive selection. *Journal of the Royal Statistical Society*, 97, 558–625.

Ohlsson, E. (1998) Sequential Poisson sampling. *Journal of Official Statistics*, 14(2), 149–162.

Patil, G.P., Sinha, A.K. and Taillie, C. (1994) Ranked set sampling for multiple characteristics. *International Journal of Ecology and Environmental Sciences*, 20, 94–109.

Ren, R. and Chambers, R. (2003) *Outlier Robust Imputation of Survey Data via Reverse Calibration*. S^3RI Methodology Working Papers M03/19, Southampton Statistical Sciences Research Institute, University of Southampton, UK.

Rivest, L.P. (2002) A generalization of the Lavallée and Hidiroglou algorithm for stratification in business surveys. *Survey Methodology*, 28, 191–198.

Rosén, B. (1997) On sampling with probability proportional to size, *Journal of Statistical Planning and Inference*, 62, 159–191.

Royall, R. and Herson, J. (1973) Robust estimation in finite populations I. *Journal of the American Statistical Association*, 68, 880–889.

Sadasivan, G. and Aggarwal, R. (1978) Optimum points of stratifications in bi-variate populations. *Sankhyā*, 40, 84–97.

Särndal, C.E. and Lundström, S. (2005) *Estimation in Surveys with Nonresponse*. Chichester: John Wiley & Sons, Ltd.

Särndal, C.E, Swensson, B. and Wretman J. (1992) *Model Assisted Survey Sampling*. New York: Springer.

Scott, A., Brewer, K. and Ho, E. (1978) Finite population sampling and robust estimation. *Journal of the American Statistical Association*, 73, 359–361.

Sigman R.S., Monsour N.J. (1995) Selecting samples from list frames of businesses. In B.G. Cox, D.A. Binder, B.N. Chinnappa, A. Christianson, M.J. Colledge and P.S. Kott (eds), *Business Survey Methods*, pp. 133–152. New York: John Wiley & Sons, Inc.

Singh, A.C. and Mohl, C.A. (1996) Understanding calibration estimators in survey sampling. *Survey Methodology*, 22, 107–115.

Skinner, C.J. (1999) Calibration weighting and non-sampling errors. *Research in Official Statistics*, 2, 33–43.

Smith, P., Pont, M. and Jones, T. (2003) Developments in business survey methodology in the Office for National Statistics, 1994–2000. *The Statistician*, 52, 257–295.

Sunter, A.B. (1977a) Response burden, sample rotation, and classification renewal in economic surveys. *International Statistical Review*, 45, 209–222.

Sunter, A.B. (1977b) List sequential sampling with equal or unequal probabilities without replacement. *Applied Statistics*, 26, 261–268.

Sunter, A.B. (1986) Solution to the problem of unequal probability sampling without replacement. *International Statistical Review*, 54, 33–50.

Tillé, Y. (2006) *Sampling Algorithms*. New York: Springer.

Väisänen, P. (2002) Estimation procedure of the Finnish Time Use Survey 1999–2000. Paper presented at the IATUR Annual Conference, 15–18 October, Lisbon, Portugal. http://pascal.iseg.utl.pt/cisep/IATUR/Papers/Vaisanen102f.pdf.

Verma, V. (1995) *Weighting for Wawe 1*. EUROSTAT doc. PAN 36/95.

Vogel, F.A. (1995) The evolution and development of agricultural statistics at the United States Department of Agriculture. *Journal of Official Statistics*, 11(2), 161–180.

Wand, M.P. and Jones, M.C. (1995) *Kernel Smoothing*. London: Chapman & Hall.

Wolter, K.M. (1985) *Introduction to Variance Estimation*. New York: Springer.

Wu, C. and Sitter, R.R. (2001) A model-calibration approach to using complete auxiliary information from survey data. *Journal of the American Statistical Association*, 96, 185–193.

Yates, F. (1946) A review of recent statistical developments in sampling and sampling surveys. *Journal of the Royal Statistical Society, Series A*, 109, 12–43.

Yeomans, K.A. and Golder, P.A. (1975) Further observations on the stratification of Birmingham wards by clustering: a riposte. *Applied Statistics*, 24, 345–346.

8

Estimation with inadequate frames

Arijit Chaudhuri
Indian Statistical Institute, Kolkata, India

8.1 Introduction

In India cross-checking the volume of rural credits issued by the banking system as gathered through the *Reserve Bank of India Bulletin* against the National Sample Survey results for the amount of loans reported to have been accepted by the villagers appeared to be a crucial statistical problem a couple of decades ago as the latter amounted to roughly half of the former.

Natural calamities such as floods, drought and cyclones are frequent in India, creating problems in gathering data on farm lands under cultivation and crop yields. Crop failures must be insured against. But question arises of how to fix the premium rates and amounts of subsidy to be distributed.

Irrigation is a sensitive issue, leading to difficulties for governmental assessment and reporting on seasonal differences in rice yield between irrigated and non-irrigated lands.

If frames are available and made us of, the estimation of relevant parameters is not a great problem. Here we address certain situations when adequate frames are lacking, and propose alternative estimation procedures.

8.2 Estimation procedure

8.2.1 Network sampling

It is impossible to get a complete list of rural households receiving bank loans in a district for purposes of crop cultivation, fishing, etc. in order to develop district-level estimates

Agricultural Survey Methods Edited by Roberto Benedetti, Marco Bee, Giuseppe Espa and Federica Piersimoni

for rural credits advanced by the banks. It is better to get lists of district banks from which samples may be scientifically chosen. The banks and other lending institutions may be called 'selection units' (SUs). From the sampled SUs, addresses of the account holders within the villages in the district may be gathered village-wise. The households with bank account holders are supposed to be linked to the respective SUs already chosen; these households are our 'observation units' (OUs). By exploiting these, estimation may be implemented without much difficulty.

Let $j = 1, \ldots, M$, for a known positive integer M, denote the labelled SUs with an ascertainable frame. Let N be the unknown number of households, the OUs with unascertainable labels prior to selection of the SUs, namely $i = 1, \ldots, N$ with N unknown. Let y_i be the total bank loans incurred by the ith OU in the aggregate in a fixed reference period for purposes of agriculture, fishing, fruit and vegetable cultivation in the aggregate. Then our objective is to estimate $Y = \sum_{i=1}^{N} y_i$.

Recall from Chaudhuri (2000) the theorem giving the equality of $W = \sum_{j=1}^{M} w_j$ to Y, with

$$w_j = \sum_{i \in A(j)} \frac{y_i}{m_i}, \quad j = 1, \ldots, M.$$

Here $A(j)$ is the set of OUs linked to jth SU and m_i is the number of SUs to which the ith OU is linked. Taking $t_{HT} = \sum_{j \in s} \frac{w_j}{\pi_j}$ as the standard Horvitz and Thompson (1952) estimator for W, the estimand Y is also easily estimated. Here s is a sample of SUs drawn with a probability $p(s)$ and $\pi_j = \sum_{j \in s} p(s)$ is the inclusion probability of the jth SU. An unbiased estimator for the variance of t_{HT} given by Chaudhuri and Pal (2003) is

$$v_{CP}(t_{HT}) = \sum\sum_{j<j'} \left(\frac{w_j}{\pi_j} - \frac{w_{j'}}{\pi_{j'}} \right)^2 \frac{\pi_j \pi_{j'} - \pi_{jj'}}{\pi_{jj'}} + \sum_{j \in s} \frac{w_j^2}{\pi_j^2} \beta_j.$$

Here $\pi_{jj'} = \sum_{j \in s, j'} p(s)$ is the inclusion probability of the pair of SUs (j, j') assumed positive for every $j \neq j'$ and $\beta_j = 1 + \frac{1}{\pi_j} \sum_{j2i} \pi_{ij} - \Sigma \pi_j$.

If $A(j)$ and m_i turns out to be too large, then simple random samples $B(j)$ without replacement (SRSWOR) should be taken from $A(j)$ independently across j in s to ease the burden of sampling and estimation. The union $\cup A(j)$ is a network sample and $\cup_{j \in s} B(j)$ is a constrained network sample. For the latter a revised estimator for $Y = W$ is

$$e_{HT} = \sum_{j \in s} \frac{w'_j}{\pi_j} \quad \text{with } w'_j = \sum_{i \in B(j)} \frac{y_i}{m_i}.$$

It is easy to show the unbiasedness of e_{HT} and to find an unbiased variance estimator for it.

Let C_j be the number of OUs in A_j. This number must be known or determined for every j in s. If it is too big, take from A_j an SRSWOR B_j of size d_j $(2 < d_j \leq e_j)$ for $j \in s$ such that the number $D = \sum_{j \in s} d_j$ is of a manageable size.

Let $u_j = \frac{c_j}{d_j} \sum_{i \in B_j} \frac{y_i}{m_i}$, $j \in s$. Then, $e = \frac{M}{m} \sum_{j \in s} u_j$ is an unbiased estimator of W and hence of Y.

It may be verified that

$$\hat{V}(e) = \left[\frac{M}{m}\sum_{j\in s} v(u_j)\right] + \left[\frac{\frac{M^2}{m}\left(\frac{1}{m}-\frac{1}{M}\right)}{m-1}\sum\sum_{j<k\in s}(u_j-u_k)^2\right]$$

is an unbiased estimator of $V(e)$, the variance of e. Here

$$v(u_j) = \frac{C_j^2\left(\frac{1}{d_j}-\frac{1}{C_j}\right)}{d_j-1}\sum_{i\in B_j}\left(\frac{y_i}{m_i}-\frac{\sum_{i\in B_j}\frac{y_i}{m_i}}{d_j}\right)^2.$$

8.2.2 Adaptive sampling

To estimate the bank loans received by cultivators and fishermen in a district in a specified period (say, the last calendar year), a suitable alternative method to adopt may be adaptive sampling. Here we may target the households directly in the district instead of approaching them through the banks from which the householders borrow.

Suppose, however, that we do not have a frame of households containing such a category of debtors to the banks. So, if we take a conventional sample of households, many of them may yield 'zero values' of bank loans taken by their inmates. But we may believe that a sizeable volume of loans may have been advanced collectively to the village households in the district. This is a suitable situation for adaptive sampling. Let us see how it may be implemented effectively.

Let N be the number of villages in the district of interest and let a sample of n of them be selected using the sampling scheme given by Rao *et al.* (1962). For this we need normed sizes p_i $(0 < p_i < 1, \sum_1^N p_i = 1)$ which we take as $p_i = x_i/X$, $X = \sum_1^N x_i$, with x_i as the total population in the ith village as per the latest population census of India in 2001. Let M_i be the total number of households in the ith village. Select from these M_i households an SRSWOR of m'_i households. Even if $\sum m'_i$, the total number of households actually selected, is sufficiently large, there may not be enough households selected with inmates not borrowing to a great extent from the bank for their cultivation expenses. If we stop here only to estimate $Y = \sum_1^N \sum_1^{M_i} y_{ij}$, denoting by y_{ij} the bank loans taken for the year by the inmates of the jth household of the ith village (i.e. the ijth unit), then we may employ the estimator

$$\hat{Y} = \sum_n \frac{Q_i}{p_i}\left(\frac{M_i}{m_i}\sum_{j=1}^{m_i} y_{ij}\right)$$

for Y. Here $\sum_n$ is the sum of the n random groups into which the N villages are split up; Q_i is the sum of the p_i-values over the N_i villages assigned to the ith group $(i = 1, \ldots, n)$ using the Rao *et al.* (1962) scheme. Now denote by $\sum_n\sum_n$ the n_{c_2} distinct pairs of groups over which a summing is implemented. Write $\hat{y}_i = \frac{M_i}{m_i}\sum_1^{m_i} y_{ij}$. Then the variance of $\hat{Y}$ is estimated without bias by

$$\hat{V}(\hat{Y}) = A\sum_n\sum_n Q_i Q_i\left(\frac{\hat{y}_i}{p_i}-\frac{\hat{y}_{i'}}{p_{i'}}\right)^2 + \sum_n \frac{Q_i}{p_i} M_i^2 \frac{\frac{1}{m}-\frac{1}{M_i}}{m_i-1}\sum_{j=1}^{m_i}\left(y_{ij}-\frac{\sum_1^m y_{ij}}{m_i}\right)^2.$$

Here $A = (\sum_n N_i^2 - N)/(N^2 - \sum_n N_i^2)$. The corresponding coefficient of variation, which is a relative measure of error in $\hat{Y}$ as an estimator for Y, is

$$CV = 100\frac{\sqrt{\hat{V}(\hat{Y})}}{\hat{Y}}.$$

If its value is 30% or less we may rest content. Otherwise we will need to try to improve upon it.

We proceed as follows. For every household in every village in the district let us define a neighbourhood consisting of itself and four others in the district deemed to be close to each other in a reciprocal sense in respect of the feature of being debtors to the banks of the district for cultivation purposes. In the course of conducting the field survey for the sample chosen in the manner described above, every sampled household reporting on its own bank loans is requested to identify all its neighbours. Each neighbour thus identified is then to be sampled, whether already covered or not. The process adding to the selected sample a neighbour of a selected unit is repeated until four of them are added. All such households in the district thus reached through an initial household constitute a cluster for the initial household. Any household which has not taken a bank loan will be called an 'edge' unit. Omitting all the edge units in a cluster, the residual households will constitute a 'network' for the initial household. Calling each edge unit a 'singleton' network and recognizing it as a 'genuine network', it follows (see Chaudhuri, 2000) that the 'networks' are all distinct and their union coincides with the entire population of all the households in the district. This is very useful in our estimation procedure in the absence of a 'frame'.

Let A_{ij} be the network of the ijth household and m_{ij} the number of households in A_{ij}. Then

$$t_{ij} = \frac{1}{m_{ij}}\left(\sum_{u,v \in A_{ij}} y_{uv}\right).$$

From Chaudhuri (2000) it is known that

$$T = \sum_1^N \sum_1^{M_i} t_{ij} \quad \text{exactly equals} \quad Y = \sum_1^N \sum_1^{M_i} y_{ij}.$$

Consequently to estimate Y it is enough to estimate T. Starting from the sample s for which the values y_{ij} are to be found for the units ij in s, in order to ascertain the values of t_{ij} for ij in s we have to observe the values of y_{uv} for all u, v, in A_{ij} over all ijs in s. So, this union $A(s)$ over all the u_{iv}s in A_{ij} for ijs in s is the actual sample called the adaptive sample we are to survey to ascertain the values of y_{uv} for u, v in A_{ij} for ij in s. But once we survey $A(s)$ it is a simple matter to estimate T and hence Y. Thus T is to be estimated without bias by

$$\hat{T} = \Sigma_n \frac{Q_i}{p_i}\left(\frac{M_i}{m_i}\sum_{j=1}^{m_i} t_{ij}\right),$$

which is nothing but $\hat{Y}$ with every y_{ij} in the latter replaced by t_{ij}. Obviously the variance of $(\hat{T})$ is estimated without bias by

$$\hat{V}(\hat{T}) = \hat{V}(\hat{Y})|_{y_{ij}=t_{ij}},$$

which means that $\hat{V}(\hat{Y})$ is to be evaluated by replacing every y_{ij} in the latter by the number t_{ij}.

As in network sampling, constraining the adaptive sampling often becomes a necessity if the networks A_{ij} for ij in s become so large that the adaptive sample $A(s)$ may be prohibitively expensive.

As a precaution against such a possibility we propose to proceed as follows. Let B_{ij} be an SRSWOR taken from A_{ij} with l_{ij} as the cardinality of B_{ij} so that $\sum_{ij\in s} l_{ij}$ may not exceed a predetermined number L. Then let

$$e_{ij} = \frac{1}{l_{ij}} \sum_{uv\in B_{ij}} y_{uv}.$$

Writing m_{ij} as the cardinality of A_{ij}, $l_{ij} < m_{ij}$. Writing E_r and V_r as expectation and variance operators with respect to this SRSWOR, and E_p and V_p as the same with respect to the basic two-stage sampling design already employed, let

$$E = E_p E_r \quad \text{and} \quad V = E_p V_r + V_p E_r.$$

Then $E_r(e_{ij}) = t_{ij}$ and an unbiased estimator for $V_r(e_{ij})$ is

$$v_R(e_{ij}) = \left(\frac{1}{l_{ij}} - \frac{1}{m_{ij}}\right) \frac{1}{l_{ij}-1} \sum_{uv\in B_{ij}} (y_{uv} - e_i)^2.$$

Then, instead of $\hat{T}$, the revised estimator is

$$\hat{E} = \sum_n \frac{Q_i}{p_i} \left(\frac{M_i}{m_i}\right) \sum_{j=1}^{m_i} r_{ij}.$$

Let

$$v_R(\hat{E}) = \sum_n \left(\frac{Q_i}{p_i}\right)^2 \left(\frac{M_i}{m_i}\right)^2 \sum_{j=1}^{m_i} v_R(e_{ij})$$

and $f_{ij} = e_{ij}^2 - v_R(e_{ij})$. Then $E_R(f_{ij}) = t_{ij}^2$. So,

$$\hat{V}(\hat{E}) = v_R(\hat{E}) + \hat{V}(\hat{T})|_{t_{ij}=f_{ij}} - (\hat{E})^2 - \sum_n \left(\frac{Q_i}{p_i}\right)^2 \left(\frac{M_i}{m_i}\right)^2 v_R(\hat{E})$$

may be taken as an estimator for $V(\hat{E})$.

Chaudhuri *et al.* (2005), however, considered adaptive sampling with an initial single-stage sampling. Hence their formula for the variance estimator in the case of constrained adaptive sampling turned out to be rather simple; with two-stage sampling we could not avoid complexity in our formula.

Earlier Chaudhuri (1996, 1997) and Chaudhuri and Adhikary (2000) gave alternative solutions for the main problem. Space considerations prevent us from reproducing. We also note that they employed three-stage sampling. Furthermore, they needed certain drastic assumptions to derive simple variance estimators.

References

Chaudhuri, A. (1996) Estimation of total and variance estimation in a version of unequal probability sampling in three stages. *Pakistan Journal of Statistics*, 12, 1–7.

Chaudhuri, A. (1997) On a pragmatic modification of survey sampling in three stages. *Communications in Statistics – Theory and Methods*, 26, 1805–1810.

Chaudhuri, A. (2000) Network and adaptive sampling with unequal probabilities. *Calcutta Statistical Association Bulletin*, 50, 237–253.

Chaudhuri, A. and Adhikary, A.K. (2000) Variance estimation in a specific three-stage survey sampling strategy. In A.K. Basu, J.K. Ghosh, P.K. Sen and B.K. Sinha (eds), *Perspectives in Statistical Sciences*, pp. 102–109. New Delhi: Oxford University Press.

Chaudhuri, A. and Pal, S. (2003) Systematic sampling: 'fixed' versus 'random' sampling interval. *Pakistan Journal of Statistics*, 19, 259–271.

Horvitz, D.G. and Thompson, D.J. (1952) A generalization of sampling without replacement from a finite universe. *Journal of the American Statistical Association*, 47, 663–685.

Rao, J.N.K., Hartley, H.O. and Cochran, W.G. (1962) On a simple procedure of unequal probability sampling without replacement. *Journal of the Royal Statistical Society, Series B*, 24, 482–491.

9

Small-area estimation with applications to agriculture

J.N.K. Rao
School of Mathematics and Statistics, Carleton University, Ottawa, Canada

9.1 Introduction

In the context of agriculture, the term 'small area' generally refers to a small geographical area, such as a 'tehsil' or block in India. Reliable small-area information on crop statistics is needed for formulating agricultural policies.

Crop yield statistics are generally obtained through sample surveys. For example, crop cutting experiments in sampled fields are used in India to obtain direct (or area-specific) estimates of crop yields. Data collected from sample surveys can be used to derive reliable direct estimates for large areas, such as a district, making effective use of auxiliary data. For example, in India remote sensing satellite data are currently used as auxiliary information to produce reliable direct estimates of crop areas and crop yields at the district level. Singh and Goel (2000) proposed post-stratification of the crop area on the basis of vegetation indices derived from the satellite data. The post-stratified estimators were considerably more efficient than the traditional estimators using only geographical stratification.

Direct area-specific estimates may not provide acceptable precision at the small-area level because sample sizes in small areas are seldom large enough. In some cases, sample sizes at a higher level (such as a state) also may be small (see Example 2, Section 9.3). This makes it necessary to 'borrow strength' from related areas to find indirect estimates that increase the effective sample size and thus increase the precision. Such indirect estimates are based on either implicit or explicit models that provide a link to related

Agricultural Survey Methods Edited by Roberto Benedetti, Marco Bee, Giuseppe Espa and Federica Piersimoni

small areas through supplementary data such as data from a recent census of agriculture, remote sensing satellite data and administrative records.

Indirect estimates based on implicit models include synthetic and composite estimates. Indirect estimates based on explicit models have received a lot of attention because of the following advantages over traditional synthetic and composite estimates:

1. Model-based methods make specific allowance for local variation through complex error structures in the model that links the small areas.

2. Models can be validated from the sample data.

3. Methods can handle complex cases such as cross-sectional and time series data.

4. Stable area-specific measures of variability associated with the estimates may be obtained, unlike overall measures commonly used for traditional indirect estimates.

In this chapter we provide a brief account of small-area estimation in the context of agricultural surveys.

9.2 Design issues

Efficient methods of designing surveys for use with direct estimates of large-area totals or means have received a lot of attention over the past 50 years or so. But survey design issues that have an impact on small-area statistics should also be considered. Singh *et al.* (1994) proposed several methods for use at the design stage to minimize the use of indirect small-area estimates. Those methods include (i) replacing clusters by using list frames, (ii) use of many strata to provide better sample size control at the small-area level and (iii) compromise sample allocation. They presented an excellent illustration of compromise sample size allocation in the Canadian Labour Force Survey to satisfy reliability requirements at the provincial as well as sub-provincial level. Preventive measures, such as (i)–(iii), should be undertaken at the design stage, whenever possible, to achieve adequate precision using direct estimates. Other methods for use at the design stage include the use of two or more (possibly) incomplete frames, combining data from rolling samples and integration of surveys.

Preventive measures at the design stage may significantly reduce the need for indirect estimates, but for some small areas (e.g., tehsils in India) sample sizes may not be large enough to provide adequate precision using direct estimates even after implementing such measures.

9.3 Synthetic and composite estimates

Suppose the population is divided into g large post-strata for which reliable direct estimates of the post-strata totals, $Y_{\cdot g}$, can be calculated from the survey data, where $Y_{\cdot g} = \sum_i Y_{ig}$ and Y_{ig} is the total of the characteristic of interest, y, for the units in small area i that belong to post-stratum g. Our interest is in estimat9ng the small-area totals $Y_i = \sum_g Y_{ig}$, $i = 1, \ldots, m$, using known auxiliary totals X_{ig}.

9.3.1 Synthetic estimates

A synthetic estimate of Y_i is given by

$$\hat{Y}_i^S = \sum_g X_{ig}(\hat{Y}_{\cdot g}/\hat{X}_{\cdot g}), \tag{9.1}$$

where $\hat{Y}_{\cdot g}$ and $\hat{X}_{\cdot g}$ are reliable direct estimates of post-strata totals $Y_{\cdot g}$ and $X_{\cdot g}$. The design bias of $\hat{Y}_i^S$ under repeated sampling will be small relative to the total Y_i if the ratios $R_{ig} = Y_{ig}/X_{ig}$ are homogeneous across small areas, that is, $R_{ig} \equiv R_{\cdot g} = Y_{\cdot g}/X_{\cdot g}$ for each g. Moreover, the design standard error of $\hat{Y}_i^S$ will be small relative to Y_i since it depends only on the variances and covariances of post-strata estimates $\hat{R}_{\cdot g} = \hat{Y}_{\cdot g}/\hat{X}_{\cdot g}$. Thus, the synthetic estimate $\hat{Y}_i^S$ will be reliable under the assumption $R_{ig} \equiv R_{\cdot g}$. But such an assumption may be quite strong in practice, and in fact $\hat{Y}_i^S$ can be heavily biased for small areas exhibiting strong individual effects.

The variance of $\hat{Y}_i^S$ is readily estimated, but it is more difficult to estimate the mean squared error (MSE) of $\hat{Y}_i^S$. An approximately unbiased estimate of MSE($\hat{Y}_i^S$) is given by

$$\text{mse}(\hat{Y}_i^S) = (\hat{Y}_i^S - \hat{Y}_i)^2 - v(\hat{Y}_i), \tag{9.2}$$

where $\hat{Y}_i$ is the direct estimate of Y_i and $v(\hat{Y}_i)$ is a design-unbiased estimate of variance of $\hat{Y}_i$. However, the MSE estimate (9.2) may be very unstable. Consequently, it is common practice to average them over i to get a stable estimate of MSE. But such a global measure of variability that does not vary over areas can be misleading in practice.

Example 1. Singh and Goel (2000) used a synthetic estimate of the form (9.1) to estimate crop yields in India at the tehsil level. Post-strata were formed on the basis of vegetation indices derived from the remote sensing satellite data. The survey estimates $\hat{Y}_{\cdot g}$ were obtained from crop yield surveys based on crop cutting experiments. Actually, Singh and Goel (2000) used the totals $X_{\cdot g}$ instead of $\hat{X}_{\cdot g}$ in (9.1) where $X_{\cdot g}$ is the total crop area in the gth post-stratum. On the basis of the standard error of $\hat{Y}_i^S$, their evaluation study indicated that $\hat{Y}_i^S$ is often significantly more efficient than $\hat{Y}_i$ at the tehsil level and especially at the block level. However, ignoring the bias and using the standard error may be overly optimistic.

9.3.2 Composite estimates

A simple way to balance the potential bias of synthetic estimate, $\hat{Y}_i^S$, against the instability of the direct estimate, $\hat{Y}_i$, is to take a weighted average of the two estimates. This leads to a composite estimate of the form

$$\hat{Y}_i^C = a_i\hat{Y}_i + (1 - a_i)\hat{Y}_i^S, \tag{9.3}$$

for some suitably chosen weight a_i in the range [0, 1]. Optimal weights that minimize MSE($\hat{Y}_i^S$) can be obtained, but their estimates, $\hat{a}_i$, can be very unstable as they involve mse($\hat{Y}_i^S$) given by (9.2). Purcell and Kish (1979) used a common weight, a, and then minimized the average MSE over small areas. This leads to a weight $\hat{a}$ of the form $1 - m\bar{v}/\sum_i(\hat{Y}_i^S - \hat{Y}_i)^2$, where $\bar{v} = m^{-1}\sum_i v(\hat{Y}_i)$ and m is the number of small areas. But the use of a common weight may not be reasonable if the individual variances, $V(\hat{Y}_i)$, vary considerably.

Formula (9.2) with $\hat{Y}_i^S$ replaced by $\hat{Y}_i^C$ is often used to estimate MSE($\hat{Y}_i^C$). But this MSE estimate is also very unstable.

Example 2. Eklund (1998) used a composite estimate of the form (9.3) to estimate net coverage error for the 1997 US Census of Agriculture at the state (small area) level. Survey data were used to estimate net coverage error. But the sample sizes at the state level were not large enough for the direct estimates at state level to be reliable, unlike the direct estimates at the region level. Eklund used a synthetic state estimate of the form (9.1) without post-stratification. It is given by

$$\hat{Y}_i^S = (X_i/X_R)\hat{Y}_R,$$

where $\hat{Y}_R$ is a direct estimate of the regional total Y_R, and X_i and X_R are the census totals for state i and region R of the corresponding characteristic of interest. The synthetic estimate $\hat{Y}_i^S$ was combined with the direct estimate $\hat{Y}_i$ using a composite estimate $\hat{Y}_i^C$.

Because of the instability in the MSE estimates of the form (9.2), mse($\hat{Y}_i^C$), Eklund proposed to smooth the estimates by modelling the relative mean square error, mse($\hat{Y}_i^C$)/$(\hat{Y}_i^C)^2$, using linear regression on the census state total X_i. Eklund noted some difficulties with this method.

9.4 Area-level models

We now turn to model-based methods based on small-area linking models involving random small-area effects. Such models may be broadly classified into two types: (a) area-level models, which we study in this section; (b) unit-level models, which we study in the next.

A basic area-level model that uses area-level covariates has two components. First, the direct survey estimate $\bar{y}_i$ of the ith area mean $\bar{Y}_i$, possibly transformed as $\hat{\theta}_i = g(\overline{y}_i)$, is equal to the sum of the population value $\theta_i = g(\overline{Y}_i)$ and the sampling error e_i:

$$\hat{\theta}_i = \theta_i + e_i, \quad i = 1, \ldots, m, \tag{9.4}$$

where the e_i are assumed to be independent across areas with means 0 and known variances ψ_i. The second component is a linking model that relates the θ_i to area-level covariates $\mathbf{z}_i = (z_{1i}, \ldots, z_{pi})^T$ through a linear regression model:

$$\theta_i = \mathbf{z}_i^T \boldsymbol{\beta} + v_i, \quad i = 1, \ldots, m, \tag{9.5}$$

where the model errors v_i are assumed to be independent and identically distributed with mean 0 and variance σ_v^2. The parameter σ_v^2 is a measure of homogeneity of the areas after accounting for the covariates $\mathbf{z}_i$. Combining (9.4) and (9.5), we get a mixed linear model

$$\hat{\theta}_i = \mathbf{z}_i^T \boldsymbol{\beta} + v_i + e_i, \quad i = 1, \ldots, m. \tag{9.6}$$

Using the data $\{(\hat{\theta}_i, \mathbf{z}_i),\ i = 1, \ldots, m\}$ we can obtain estimates, θ_i^*, of the realized values of θ_i from the model (9.6). A model-based estimate of $\overline{Y}_i$ is then given by $g^{-1}(\theta_i^*)$. Note that the model involves both design-based random variables, e_i, and model-based random variables, v_i.

Empirical best linear unbiased prediction (EBLUP), empirical Bayes (EB) and hierarchical Bayes (HB) methods have played a prominent role in the estimation of $\overline{Y}_i$ under model (9.6). The EBLUP method is applicable for mixed linear models and EBLUP estimates do not require a normality assumption on the random errors v_i and e_i. On the other hand, EB and HB are more generally applicable under specified distributional assumptions. HB methods lead to exact inferences via posterior distributions: $p(\theta_i|\hat{\boldsymbol{\theta}})$, where $\hat{\boldsymbol{\theta}} = (\hat{\theta}_1, \ldots, \hat{\theta}_m)^T$. EBLUP and EB estimates of θ_i are identical under normality and nearly equal to the HB estimate $E(\theta_i|\hat{\boldsymbol{\theta}})$, but measures of variability of the estimates may differ. Under EBLUP and EB we use an estimate of $\text{MSE}(\tilde{\theta}_i) = E(\tilde{\theta}_i - \theta_i)^2$ as a measure of variability of $\tilde{\theta}_i$, where the expectation is with respect to the model (9.6). On the other hand, HB uses posterior variance $V(\theta_i|\tilde{\boldsymbol{\theta}})$ as a measure of variability.

The EBLUP estimate of θ_i is a composite estimate of the form

$$\theta_i^* = \hat{\gamma}_i\hat{\theta}_i + (1 - \hat{\gamma}_i)\mathbf{z}_i^T\hat{\boldsymbol{\beta}}, \tag{9.7}$$

where $\hat{\gamma}_i = \hat{\sigma}_v^2/(\hat{\sigma}_v^2 + \psi_i)$ and $\hat{\boldsymbol{\beta}}$ is the weighted least squares estimate of $\boldsymbol{\beta}$ with weights $(\hat{\sigma}_v^2 + \psi_i)^{-1}$ obtained by regressing θ_i on $\mathbf{z}_i$: $\hat{\boldsymbol{\beta}} = (\sum_i \hat{\gamma}_i\mathbf{z}_i\mathbf{z}_i^T)^{-1}(\sum_i \hat{\gamma}_i\mathbf{z}_i\theta_i)$ and $\hat{\sigma}_v^2$ is an estimate of the variance component σ_v^2. That is, the EBLUP estimate, θ_i^*, is a weighted combination of the direct estimate, $\hat{\theta}_i$, and a regression synthetic estimate $\mathbf{z}_i^T\hat{\boldsymbol{\beta}}$ with weights $\hat{\gamma}_i$ and $1 - \hat{\gamma}_i$ respectively. The EBLUP estimate gives more weight to the direct estimate when the sampling variance, ψ_i, is small (or $\hat{\sigma}_v^2$ is large) and moves towards the regression synthetic estimate as ψ_i increases (or $\hat{\sigma}_v^2$ decreases). For the non-sampled areas, the EBLUP estimate is given by the regression synthetic estimate, $\mathbf{z}_i^T\hat{\boldsymbol{\beta}}$, using the known covariates associated with the non-sampled areas.

A lot of attention has been given to the estimation of MSE or the posterior variance, $V(\theta_i|\tilde{\boldsymbol{\theta}})$, under the HB set-up. Complex models under HB can be handled using recently developed Markov chain Monte Carlo (MCMC) methods. We refer the reader to Ghosh and Rao (1994) and Rao (1999) for some recent developments in the estimation of MSE and the computation of posterior variance.

Under model (9.7), the leading term of $\text{MSE}(\tilde{\theta}_i)$ is given by $\gamma_i\psi_i$, which shows that the EBLUP estimate can lead to large gains in efficiency over the direct estimate with variance ψ_i, when γ_i is small (or the model variance, σ_v^2, is small relative to the sampling variance, ψ_i). The success of small-area estimation, therefore, largely depends on getting good auxiliary data $\{\mathbf{z}_i\}$ that can lead to a small model variance relative to sampling variance. One should also make a thorough internal validation of the assumed model.

Sampling variances, ψ_i, may not be known in practice, in which case one often resorts to smoothing of the estimated design variances, $\hat{\psi}_i$, to get stable estimates ψ_i^*, say. The smoothed estimate ψ_i^* is then treated as a proxy for ψ_i.

Example 3. Fuller (1981) applied the area-level model (9.6) to estimate mean soybean hectares per segment in 1978 at the county level. He used the mean number of pixels of soybeans per area segment, z_{2i}, ascertained by satellite imagery, and the mean soybean hectares from the 1974 US Agricultural Census, z_{3i}, as county (area) level covariates. Survey estimates, $\overline{y}_i$, for a sample of $m = 10$ counties were obtained by sampling area segments within sampled counties.

Fuller obtained model-based estimates of the population means, $\overline{Y}_i$, for the sampled counties as well as the non-sampled counties. His model is given by

$$\overline{y}_i - z_{3i} = \beta_0 + \beta_1 z_{2i} + \beta_2 z_{3i} + v_i + e_i, \tag{9.8}$$

with known error variances $\sigma_v^2 = 25$ and $\sigma_e^2 = 18$. Note that z_{2i} and z_{3i} are known for all the 16 counties. The model (9.8) is a special case of (9.6) with $\hat{\theta}_i = \overline{y}_i - z_{3i}$ and $\psi_i = \psi = \sigma_e^2$. Fuller's estimate of $\overline{Y}_i$ for sampled counties is obtained from (9.7) as

$$\overline{y}_{iF} = g^{-1}(\theta_i^*) = \theta_i^* + z_{3i} = z_{3i} + \mathbf{z}_i^T\hat{\boldsymbol{\beta}} + \gamma(\overline{y}_i - z_{3i} - \mathbf{z}_i^T\hat{\boldsymbol{\beta}}),\ \ i = 1, \ldots, 10,$$

where $\gamma = \sigma_v^2/(\sigma_v^2 + \sigma_e^2) = 25/43$. For the non-sampled counties, $\overline{y}_{iF}^* = z_{3i} + \mathbf{z}_i^T\hat{\boldsymbol{\beta}}$, $i = 11, \ldots, 16$.

Under model (9.8), Fuller calculated the total MSE of the estimates $\overline{y}_{iF}$ for the sampled counties as 127 and for the non-sampled counties as 210. The total MSE for all 16 counties equals $127 + 210 = 337$. On the other hand, if one were to use the prior census mean z_{3i} as the estimate of $\overline{Y}_i$, then the total MSE of z_{3i} under model (9.8) is estimated as 1287, considerably larger than 337. The model-based estimates, $\overline{y}_{iF}$, outperformed the prior census predictors, z_{3i}, in terms of total MSE. They are also better than the direct estimates $\overline{y}_i$ in terms of total MSE for the sampled counties: the total MSE of the $\overline{y}_i$ is $10\sigma_e^2 = 180$ compared to 127, the total MSE of the $\overline{y}_{iF}$.

9.5 Unit-level models

A basic unit-level model assumes that the unit y-values, y_{ij}, associated with the jth population unit ($j = 1, \ldots, N_i$) in the ith area are related to unit-level covariates, $\mathbf{x}_{ij}$, for which the population mean vector $\overline{\mathbf{X}}_i$ is known. If y is a continuous response (e.g. crop yield), we assume a one-fold nested error linear regression model

$$y_{ij} = \mathbf{x}_{ij}^T\boldsymbol{\beta} + v_i + e_{ij}, \quad j = 1, \ldots, N_i;\ i = 1, \ldots, m, \tag{9.9}$$

where the random sample area effects v_i have mean 0 and common variance σ_v^2 and are independently distributed. Further, the v_i are independent of the residual errors e_{ij} which are assumed to be independently distributed with mean 0 and common variance σ_e^2 (Battese *et al.*, 1988).

If N_i is large, the population mean $\overline{Y}_i$ is approximately equal to $\overline{\mathbf{X}}_i^T\boldsymbol{\beta} + v_i$. The sample data $\{y_{ij}, \mathbf{x}_{ij},\ j = 1, \ldots, n_i;\ i = 1, \ldots, m\}$ are assumed to obey the population model (9.9). This implies that sample selection bias is absent, which is satisfied by simple random sampling within areas. For more general sampling designs, the sample data will satisfy the assumption if the selection probabilities, p_{ij}, depend only on the auxiliary variables in $\mathbf{x}_{ij}$; for example, for probability proportional to size (PPS) sample, where size is used as an auxiliary variable in model (9.9). Non-probability samples obeying model (9.9) can also be used to estimate the mean $\overline{Y}_i$ (Example 4).

We assume that $\overline{Y}_i = \overline{\mathbf{X}}_i^T \boldsymbol{\beta} + v_i$. Then the EBLUP estimate of $\overline{Y}_i$ is a composite estimate of the form

$$\overline{y}_i^* = \hat{\gamma}_i \left[\overline{y}_i + (\overline{\mathbf{X}}_i - \overline{\mathbf{x}}_i)^T \hat{\boldsymbol{\beta}}\right] + (1 - \hat{\gamma}_i)\overline{\mathbf{X}}_i^T \hat{\boldsymbol{\beta}}, \quad i = 1, \ldots, m, \tag{9.10}$$

where $\hat{\gamma}_i = \hat{\sigma}_v^2/(\hat{\sigma}_v^2 + \hat{\sigma}_e^2 n_i^{-1})$ with estimated variance components $\hat{\sigma}_v^2$ and $\hat{\sigma}_e^2$, and $\hat{\boldsymbol{\beta}}$ is the weighted least squares estimate of $\boldsymbol{\beta}$ which depends on $\hat{\sigma}_v^2$ and $\hat{\sigma}_e^2$. As the small-area sample size n_i increases, the EBLUP estimate approaches the 'survey regression' estimate $\overline{y}_i + (\overline{\mathbf{X}}_i - \overline{\mathbf{x}}_i)^T \hat{\boldsymbol{\beta}}$. On the other hand, for small n_i and small $\hat{\sigma}_v^2/\hat{\sigma}_e^2$ it tends towards the regression synthetic estimate $\overline{\mathbf{X}}_i^T \hat{\boldsymbol{\beta}}$. For the non-sampled areas, $\overline{y}_i^* = \overline{\mathbf{X}}_i^T \hat{\boldsymbol{\beta}}$.

Note that the EBLUP estimates $\overline{y}_i^*$ do not depend on the survey weights, w_{ij}, so that design consistency as n_i increases is forsaken except when the design is self-weighting, as in the case of simple random sampling. On the other hand, the EBLUP estimate under the area-level model (9.6) is design-consistent because the direct estimate $\hat{\theta}_i$, used in (9.7), is design-consistent.

The variance components σ_v^2 and σ_e^2 are estimated from the sample data, using the well-known method of fitting constants (Battese *et al.*, 1988) or the restricted maximum likelihood method (assuming normality of the errors v_i and e_{ij}).

Under model (9.9) for the sample data, the leading term of $\text{MSE}(\overline{y}_i^*)$ is given by $\gamma_i(\sigma_e^2/n_i)$, which shows that the EBLUP estimate can lead to large gains in efficiency over the survey regression estimate when γ_i is small. Note that the leading term of the MSE of the survey regression estimate equals σ_e^2/n_i.

It is possible to incorporate survey weights under unit-level models, using model-assisted estimates (Prasad and Rao, 1999). Further research on this topic would be of real practical use.

Example 4. Battese *et al.* (1988) applied the nested error regression model to estimate area under corn and soybeans for each of $m = 12$ counties in North Central Iowa, using farm interview data in conjunction with LANDSAT satellite data. Each county was divided into area segments, and the areas under corn and soybeans were ascertained for a sample of segments by interviewing farm operators; the number of sample segments, n_i, in a county ranged from 1 to 5. Unit-level auxiliary data in the form of number of pixels classified as corn and soybeans were also obtained for all the area segments, including the sample segments, in each county using the LANDSAT satellite readings. In this application, $\mathbf{x}_{ij} = (1, x_{1ij}, x_{2ij})^T$ where x_{1ij} and x_{2ij} respectively denote the number of pixels classified as corn and the number of pixels classified as soybeans in the jth area segment of the ith county, and y_{ij} denotes the number of hectares of corn (soybeans) in the jth sample area segment of the ith county.

The sample data for the second sample segment in Hardin county ($n_i = 2$) was deleted because the corn area for that segment looked erroneous in a preliminary analysis. Battese *et al.* (1988) calculated the ratio of the model-based standard error of the EBLUP estimate to that of the survey regression estimate. This ratio decreased from about 0.97 to 0.77 as the number of sample segments, n_i, decreased from 5 to 1. The reduction in standard error is considerable when $n_i \leq 3$.

The EBLUP estimates were adjusted to agree with the reliable survey regression estimate for the entire area covering the 12 counties. This adjustment produced a very small increase in the standard errors.

Battese *et al.* (1988) also reported some methods for validating the assumed model. First, they introduced quadratic terms x_{1ij}^2 and x_{2ij}^2 into the model and tested the null hypothesis that the regression coefficients associated with the quadratic terms are zero. The null hypothesis was not rejected at the 5% level. Secondly, they tested the null hypothesis that the error terms v_i and e_{ij} are normally distributed by using the transformed residuals $(y_{ij} - \hat{\alpha}_i \bar{y}_i) - (\mathbf{x}_{ij} - \hat{\alpha}_i \bar{x}_i)^T \hat{\boldsymbol{\beta}}$ with $\hat{\alpha}_i = 1 - (1 - \hat{\gamma}_i)^{1/2}$. Under the null hypothesis, the transformed residuals are independent normal with mean 0 and variance σ_e^2. The well-known Shapiro–Wilk W statistic, applied to the transformed residuals, gave p-values equal to 0.921 and 0.299 for corn and soybeans respectively, suggesting the tenability of the normality assumption.

I have recently used the HB approach and calculated 'posterior predictive' probabilities to test the model fit. Under this criterion, extreme probabilities (closer to 0 or 1) suggest poor fit and values closer to 0.5 indicate good fit. I obtained values equal to 0.507 for corn and 0.503 for corn, suggesting very good fit of the Battese–Harter–Fuller model with $\mathbf{x}_{ij} = (1, x_{1ij}, x_{2ij})^T$.

Example 5. Stasny *et al.* (1991) used a regression synthetic estimate to produce county estimates of wheat production in Kansas. They used a non-probability sample of farms, assuming a linear regression model (without the small-area effect v_i) relating wheat production of the jth farm in the ith county to predictor variables with known county totals. A measure of farm size was included as a predictor variable to account for the fact that the sample was not a probability sample. A ratio adjustment to the synthetic estimates was made to ensure that the adjusted estimates add up to the direct state estimate of wheat production obtained from a large probability sample.

9.6 Conclusions

Preventive measures at the design stage, such as those outlined in Section 9.2, may significantly reduce the need for indirect estimates. But, for many applications, sample sizes in many small areas may not be large enough to achieve adequate precision through direct estimates even after taking such measures.

We have presented model-based small-area estimation under a basic area-level model and a basic unit-level model. Various extensions of the basic models have been studied, including multivariate and time series models and logistic mixed linear models for binary responses (see Rao, 1999).

Good auxiliary information related to variables of interest, such as remote sensing satellite data related to crop yields, plays a vital role in model-based estimation. Model validation also plays an important role.

Indirect estimates of small-area totals or means, such as the EBLUP estimates, may not be suitable if the objective is to identify areas with extreme population values or to rank areas or to identify areas that fall below or above certain specified level. Ghosh and Rao (1994) and Louis (2001) reviewed some methods for handling such cases.

Finally, we should emphasize the need to formulate an overall programme that covers issues related to sample design and data development, organization and dissemination, in addition to those pertaining to methods of estimation for small areas.

References

Battese, G.E., Harter, R.M. and Fuller, W.A. (1988) An error-components model for prediction of crop areas using survey and satellite data. *Journal of the American Statistical Association*, 83, 28–36.

Eklund, B. (1998) Small area estimation of coverage error for the 1997 Census of Agriculture. *Proceedings of the Section on Survey Research Methods*, American Statistical Association, pp. 335–338.

Fuller, W.A. (1981) Regression estimation for small areas. In D.M. Gilford, G.L. Nelson and L. Ingram (eds), *Rural America in Passage: Statistics for Policy*, pp. 572–586. Washington, DC: National Academy Press.

Ghosh, M. and Rao, J.N.K. (1994) Small area estimation: an appraisal. *Statistical Science*, 9, 55–93.

Louis, T.A. (2001) Bayes/EB ranking, histogram and parameter estimation. In S.E. Ahmed and N. Reid (eds), *Empirical Bayes and Likelihood Inference*, pp. 1–16. New York: Springer.

Prasad, N.G.N. and Rao, J.N.K. (1999) On robust estimation using a simple random effect model. *Survey Methodology*, 25, 67–72.

Purcell, N.J. and Kish, L. (1979) Estimation for small domain. *Biometrics*, 35, 365–384.

Rao, J.N.K. (1999) Some recent advances in model-based small area estimation. *Survey Methodology*, 25, 175–186.

Singh, M.P., Gambino, J. and Mantel, H.J. (1994) Issues and strategies for small area data. *Survey Methodology*, 20, 3–22.

Singh, R. and Goel, R.C. (2000) Use of remote sensing satellite data in crop surveys. Technical Report, Indian Agricultural Statistics Research Institute, New Delhi.

Stasny, E.A., Goel, P.K. and Ramsey, D.J. (1991) County estimates of wheat production. *Survey Methodology*, 17, 211–225.

Part III

GIS AND REMOTE SENSING

10

The European land use and cover area-frame statistical survey

Javier Gallego[1] and Jacques Delincé[2]
[1] *IPSC-MARS, JRC, Ispra, Italy*
[2] *Institute for Prospective Technological Studies, Joint Research Centre, European Commission, Seville, Spain*

10.1 Introduction

Approaches to agricultural statistics can be classified into sampling and non-sampling methods. Non-sampling methods include farm censuses, expert eye estimates by geographic unit (enumeration areas, villages, small agricultural regions, etc.) and administrative sources (see Chapters 1 and 2 of this book). Sampling methods can be based on a list frame or an area frame. Each method has advantages and drawbacks: in general, list frame surveys are cheaper because sampled farms provide in one interview a large amount of information on crop area and yield, livestock, inputs or socio-economic variables; area frame samples are better protected against non-sampling errors due to missing or overlapping units in the frame; however non-sampling errors can also appear in area frame surveys. Martino (2003) reports a shorter delay of results and lower rates of missing data for the Italian AGRIT area frame survey than for list frame surveys in Italy, but this can vary from country to country.

An area frame survey is defined by a cartographic representation of the territory and a rule that defines how it is divided into units. In the European Land Use and Cover Area-Frame Statistical Survey (LUCAS) the population is the European Union (EU),

Agricultural Survey Methods Edited by Roberto Benedetti, Marco Bee, Giuseppe Espa and Federica Piersimoni

and the sampling frame is a representation of the EU in a Lambert azimuthal equal area projection. The frame generally matches the population quite precisely, although the concepts are slightly different. The choice of a projection, necessary for the precise definition of the frame, is not a trivial question if the territory is large. The units of an area frame can be points, transects (lines of a certain length) or pieces of territory, often named segments. When the units of the frame are points, it is often called a point frame survey; points are in principle dimensionless, but they may be defined as having a certain size for coherence with the observation rules or the location accuracy that can be achieved; LUCAS defines a point with a size of 3 m.

Several examples of agricultural area frame surveys around the world are reported in a two-volume report published by the Food and Agriculture Organization (FAO, 1996, 1998), including the June Agricultural Survey (JAS) of the US Department of Agriculture (USDA; see Chapter 11 of this book), operational since the 1930s; TER-UTI, run by the French Ministry of Agriculture, the first operational area frame survey in Europe, running since 1970 and fully operational since 1980; Morocco, the most stable area frame in Africa; and several examples in Latin America.

Surveys on a list frame are generally cheaper to conduct, especially if the list of farms contains telephone numbers or e-mail addresses, and farmers are sufficiently educated and used to filling in survey forms without the presence of a surveyor. The amount of information collected in a single interview can be important, although there is a risk of saturation by the burden of information requested by different administrative bodies. Stratification on farm size and specialization is generally very efficient. List frame surveys are an excellent solution under a few conditions:

- The list frame (often an agricultural census) matches the population well, with few missing or duplicated individuals. This generally presumes that the agricultural census is recent and had a good quality control.
- The rate of non-response is low and the answers given by the farmers are reliable.
- The precise geographic location of observations is not important because the spatial links with the environment are not a priority.

If some of the conditions above are not met, area frames can be a good solution, although they have a number of limitations:

- Part of the information needed may require an interview with the farmer, because it is not directly observable on the ground. Farms can be sampled through an area frame (Hendricks *et al.*, 1965; FAO, 1996, 1998; Gallego *et al.*, 1994), but the identification of farmers linked with the sampled points or segments may be costly, and some of the limitations mentioned for the list frame apply.
- Choosing a date on which all crops can be identified on the ground is difficult. Several visits may be necessary.
- The access to sampled points or segments may be difficult or require authorization.

Area frames may be recommended in the following cases:

- The agricultural census is old or the structure of farms is changing very quickly.

- There is a high rate of non-response or the answers provided by farmers are not considered reliable.
- The aim of the survey includes information on land not owned or managed by farms, such as common pastures.
- Geo-location of the observations is important, because they are to be analysed in a GIS environment together with digital terrain models (DTMs), soil maps, satellite images, land cover maps or other georeferenced information layers.

Multiple frame surveys can combine the advantages of list and area frames, in particular when updated lists of the largest farms are available but the quality of the lists is doubtful for smaller farms.

When area frames are based on segments (pieces of territory), a rule of thumb about their size is that the ground work per segment should last less than one working day; the average working time per segment is often 2–4 hours. If the landscape is dominated by small agricultural plots, as in most European countries, an appropriate segment size is between 10 ha and 50 ha; if plots are generally large, as in the US agricultural plains, Australia and some areas in Central Asia, a better choice may be between 200 ha and 400 ha, or even larger. Segments can be delimited by physical elements, such as roads, rivers or field boundaries (Cotter and Tomczac, 1994). This has been the choice of the USDA (see Chapter 11 of this book); the Italian AGRIT project (see Chapter 22 of this book) followed this approach between 1992 and 2001; a number of developing countries have built their area frames with this method (FAO, 1998), in many cases with the support of the USDA.

Defining an area frame of segments with physical boundaries requires an important initial investment and it was considered too expensive in the western European context when the MARS project adapted the USDA system to the EU. A cheaper solution was adopted which defined square segments on a regular grid (Gallego *et al.*, 1994); this is also the choice of the Spanish Encuesta sobre Superficies y Rendimientos de Cultivos (ESYRCE) survey (MARM, 2008). Some comparisons have been made between square segments and segments with physical boundaries: González *et al.* (1991) conclude that the standard errors are very similar, and therefore the cheaper square segments are preferable.

Point frames have been widely used for forest inventories (De Vries, 1986). For agricultural and general purpose surveys, there are several important examples in Europe, such as the French TER-UTI survey (FAO, 1998), BANCIK in Bulgaria (Eiden *et al.*, 2002), and the LUCAS survey described in this chapter. TER-UTI, BANCIK and the first version of LUCAS use a non-stratified two-stage sampling scheme, with 10–36 points per PSU or cluster. This type of frame can be seen as a frame of square segments with an incomplete observation.

The Italian AGRIT (see Chapter 22 of this book) project decided in 2002 to move from segments with physical boundaries to a sample set-up of unclustered points. The operational costs of the new AGRIT suggested the need to review the cost functions that had been used to optimize the cluster size in previous studies (Gallego *et al.*, 1999). The comparison of results and cost between the old and the new AGRIT (Martino, 2003; see chapters 22 and 13 of this book) indicates that unclustered points can provide better cost-efficiency for area frame surveys in the EU; we present further comparisons below in this chapter that support the same conclusion.

Data collection and processing are easier for points than for area segments. Ground work with segments involves delineation of fields in the segment and digitizing before computing area estimates. This may require a certain time (a few weeks to several months) for large samples and consequently delays the production of estimates. Area segments provide in exchange better information for geometric co-registration with satellite images; they also give better information on the plot structure and size; this can be useful for agri-environmental indicators, such as landscape indexes. Segments are also better adapted to combine with satellite images with a regression estimator (Carfagna, 2007; see also Chapter 13 of this book).

10.2 Integrating agricultural and environmental information with LUCAS

Land cover and land use are of increasing importance in policy design and evaluation; they constitute a key element in particular for climate change studies (Feddema *et al.*, 2005). Environmental, agricultural and regional transport policies are more and more demanding two types of land cover data: maps and statistics. A large number of land cover maps have been produced with satellite images; examples at global level are Global Land Cover 2000 (known as GLC2000), based on SPOT VEGETATION images with 1 km resolution, (Bartholomé and Belward, 2005), IGBP-DISC (Loveland *et al.*, 2000), MODIS Global Land Cover (Friedl *et al.*, 2002) or the GLOBCOVER initiative of the European Space Agency (Arino *et al.*, 2008). At regional level greater geographical detail becomes possible, using images of finer resolution; some examples are CORINE Land cover for the EU (EEA-ETC/TE, 2002), the National Land Cover Data set (NLCD) for the USA (Vogelmann *et al.*, 2001), and Africover for Africa (Kalensky, 1998)

A naive approach to land cover area estimation is simply to measure the area that has been mapped as having a given land cover type. This approach has no sampling error, but the non-sampling errors can be large. On the other hand, the accuracy requirements are different for mapping and for statistics: while 85% accuracy is often considered good for a land cover map, 15% error for land cover area estimation is usually not acceptable. Area frame sampling may be the best alternative for environmental surveys or for general land cover surveys targeting agricultural and non-agricultural issues, in particular for agri-environmental purposes. The strongest tradition of area frame sampling is probably for forest inventories (De Vries, 1986; Schreuder and Czaplewski, 1993).

LUCAS was designed as a general purpose survey integrating agricultural and environmental data. Mapping land cover and land use is not a direct aim of LUCAS, but the data provided by LUCAS have been very useful for the production and validation of land cover maps, in particular the European CORINE Land Cover map (EEA, 2006)

LUCAS has a double nomenclature: each point has a land cover code (57 classes in 2001; 68 classes in 2009) and a land use code (14 classes). In many cases land cover and land use are narrowly linked, for example if the land cover is ‘wheat’, the land use will almost certainly be ‘agriculture’. In some cases the link is less obvious and the land cover and land use codes have to be used together for meaningful statistics. For example, ‘grass’ land cover can correspond to a pasture, sport or leisure installations, grass cover in an airport or on a highway, etc. The LUCAS nomenclature is quite detailed for crops (35 classes in 2001) and relatively coarse for other land cover classes (Bettio *et al.*, 2002).

The definition of some classes takes into account the neighbourhood within a distance of 25 m; this is necessary for categories such as 'permanent grassland with sparse tree/shrub cover'.

10.3 LUCAS 2001–2003: Target region, sample design and results

In 2001 LUCAS covered 13 of the 15 countries that were members of the EU at that time (EU15). The survey in the United Kingdom and Ireland was postponed to 2002 because of the travel limitations in the countryside imposed by the outbreak of foot and mouth disease. LUCAS was also conducted in 2002 in Estonia, Hungary and Slovenia, candidate member countries at that time. The survey was carried out again in 2003 in EU15 and Hungary.

The sampling plan had a two-stage systematic design (Delincé, 2001; Bettio *et al.*, 2002). Primary sampling units (PSUs) were selected with a systematic grid of 18 km without stratification. Each PSU was a cluster of 10 points or secondary sampling units (SSUs) following a 5 × 2 rectangular pattern with a 300 m step; the SSU was defined as a cell of 3 m. National cartographic projections were used; in some countries (Spain, Italy, France) slight adaptations were made to the distances between PSUs and between SSUs within PSUs in order to improve the synergy with national area frame surveys. The sample size was slightly under 10 000 PSUs, i.e. nearly 100 000 points. The LUCAS 2001 sample is generally presented as a two-stage scheme, but it can be seen as a single-stage systematic sample in which the sampling units are clusters of 10 points with a predetermined shape. The complete sample was determined by the selection of one starting point and the sample was a single-stage systematic sample of clusters of 10 points, even if the selection of the starting point involved two steps. Therefore no second-stage term was necessary for the computation of the variance (Gallego and Bamps, 2008).

For a given land cover type c (crop in particular), the area estimate is

$$\hat{Z}_c = D\bar{y}_c = \frac{D}{\sum_j n_j} \sum_j n_j y_j = \frac{D}{n} \sum_i y_i,$$

where D is the area of the region, n_j is the number of points in PSU j (usually 10, but smaller on the country boundaries), and $y_i = 1$ if the point i has land cover c and 0 otherwise.

Estimating the variance is more problematic. Systematic sampling is more efficient than random sampling under the reasonable condition that the spatial correlation is a decreasing function of the distance, but there is no unbiased estimator of the variance under systematic sampling (Cochran, 1977; Bellhouse, 1988; Dunn and Harrison, 1993). The classical variance estimation formula for random sampling is often used for systematic sampling, but it generally overestimates the variance. Several solutions have been proposed to compute the variance by splitting the sample or combining several systematic samples (several replicates) with different starting points (Koop, 1971), but this approach can give extremely unstable estimates of the variance, and reduces the efficiency of the estimation of the mean (Gautschi, 1957). Alternative ways to estimate the variance

Table 10.1 Some area estimates of LUCAS 2003 for EU15.

Land cover	Estimated area (1000 ha)	CV (%)
Artificial area	16 070	2.16
Cereals	41 395	1.32
Common wheat	13 106	2.25
Barley	10 708	2.50
Maize	8 529	2.71
Rice	284	21.16
Root crops	2 785	4.81
Forest and wood	112 247	0.80
Shrubland	28 159	1.99
Bare land	8 898	3.72

are based on comparing each sample element with other sample elements geographically close to it. Wolter (1984) compares several estimators of this type for the one-dimensional case, some of which had been proposed by Yates (1949), Osborne (1942) and Cochran (1946). Matérn (1986) proposes similar estimators for the two-dimensional case.

The usual variance estimator of the mean for simple random sampling without replacement can be written as

$$\mathrm{var}(\bar{y}) = (1-f)\frac{s^2}{n},$$

where s^2 is an estimator of the variance of y. In our case the sampling fraction f is negligible, since the 'points' are very small.

The variance estimators for systematic sampling that we have used for LUCAS substitute the usual expression for the variance of y by a local estimate,

$$s^2 = \frac{\sum_{j \neq j'} w_{jj'}\delta_{jj'}(y_j - y_{j'})^2}{2\sum_{j \neq j'} w_{jj'}\delta_{jj'}},$$

where the weight $w_{jj'}$ is an average of the weights w_j and $w_{j'}$; $\delta_{jj'}$ is a decreasing function of the distance between j and j'. For LUCAS, the following option has been used:

$$\delta_{jj'} = \begin{cases} 1/d(j, j') & \text{if } j' \text{ is among the 8 closest clusters to } j, \\ 0 & \text{otherwise.} \end{cases} \tag{10.1}$$

If we choose $\delta_{jj'} = 1$ when $j = j'$ we have the usual variance estimator of the mean for simple random sampling. The distances $d(j, j')$ are computed between the centres of the clusters.

Table 10.1 reports some area estimates for EU15 in 2003, giving an idea of the coefficient of variation (CV) for major land cover classes and for crops that occupy a relatively small area in the EU, such as rice.

10.4 The transect survey in LUCAS 2001–2003

In each PSU of the 2001–2003 sample, a transect was defined as the 1200 m line joining the five points located at the north of the PSU. The transect was surveyed recording the intersections with linear elements (hedges, stone walls, etc.) and changes of major land cover types. Estimating the total length of linear elements is an application of the classical Buffon's needle problem (Wood and Robertson, 1998). An unbiased estimate of the total length is

$$\hat{L} = \frac{\pi D}{2\mu} \sum_j \theta_j, \tag{10.2}$$

where μ is the number of transects of length u (1200m) and θ_j is the number of intersections of transect j with the linear elements under estimation. The estimator's variance (De Vries, 1986, Chapter 13) has been adapted for systematic sampling in a similar way to the estimator of land cover:

$$\text{var}(\hat{L}) = \frac{\pi^2 D^2}{4u^2} \text{var}(\theta), \qquad \text{var}(\hat{\theta}) = \frac{\sum_{j \neq j'} w_{jj'} \delta_{jj'} (\theta_j - \theta_{j'})^2}{2(n-1) \sum_{j \neq j'} w_{jj'} \delta_{jj'}}.$$

If we eliminate the term D in (10.2) and we express u in km, we get the density of linear elements in km of length per km^2 of geographical area. Table 10.2 reports estimates by country of the density of three types of linear elements: thin green, such as hedges; wide green, such as lines of trees; and cultural elements, mainly stone walls, terrace boundaries and dikes. The main problem in estimating the total length is the application by the surveyors of the definition of the elements – for example, how long/large should a row of trees or bushes be in order to be considered a linear element?

Land cover or land use change is a major issue when the interaction between agriculture and the environment is studied. There is growing interest in the accurate quantification of deforestation for agricultural use, afforestation or abandonment of agricultural land, new built areas, etc. The area estimation of land cover change from a constant sample of points is in theory very similar to land cover area estimation.

The main difference in practice is the impact of co-location errors between the two reference dates; in 2003 the surveyor was supposed to visit the same point that had been

Table 10.2 Estimated density of linear landscape elements with LUCAS 2001 transects.

	Green $< 3m$		Green $> 3m$		Cultural	
	Density	CV (%)	Density	CV (%)	Density	CV (%)
Austria	0.84	13	0.59	15	0.11	27
Belgium	0.18	39	1.85	13	0.04	74
Germany	1.19	5	1.06	6	0.43	9
Denmark	0.74	17	1.79	10	0.02	100
Spain	0.21	14	0.06	22	0.59	11
Finland	0.02	31	0.01	63	0.01	47
France	1.45	4	1.37	4	0.43	9
Greece	0.23	17	0.35	14	1.41	19
Italy	0.62	9	0.45	9	0.63	11
Netherlands	0.64	23	0.70	22	0.13	34
Portugal	1.32	14	0.43	17	1.55	17
Sweden	0.02	34	0.01	44	0.24	14

visited in 2001, but he may have actually visited a different point. The consequence of a location mistake in one of the two visits is an apparent land cover change when there has been actually none. False land cover changes may also appear because of interpretation differences for classes with a fuzzy definition (shrubland, poor pastures, etc.). The consequent overestimation of changes was particularly serious in 2001–2003 because the 2001 observations were not made available to surveyors in 2003.

Between 2003 and 2006 the estimation of change was impossible because of the change of sampling plan. In 2009 surveyors will visit the same points as in 2006, at least for a substantial subsample, with information on the 2006 observation, including digital landscape photographs.

10.5 LUCAS 2006: a two-phase sampling plan of unclustered points

After some encouraging tests of the Joint Research Centre (JRC) in collaboration with the Greek Ministry of Agriculture in 2004 and the previous experience of the Italian AGRIT program (Martino, 2003), Eurostat decided to change sampling scheme. The new scheme used a common map projection: the Lambert azimuthal equal area recommended by the Infrastructure for Spatial Information in Europe (INSPIRE) initiative (Annoni *et al.*, 2001). This decision improved the homogeneity of the sample layout, but reduced the collaboration with member states. The area to be covered was the 25 member states at that time (EU25), although Cyprus, Malta and most islands were finally excluded.

In the first phase a systematic sample (master sample) on a square grid of about 990 000 points with a 2 km step was selected. Each point was photo-interpreted for a stratification into seven classes (Table 10.3). Around 80% of the points were photo-interpreted on aerial ortho-photos with usually 1 m resolution, and the rest on coarser resolution satellite images. Both ortho-photos and satellite images were several years old (in many cases around the year 2000); this degraded the efficiency of stratification, although it was nonetheless satisfactory for most crops. In the second phase a subsample was selected for the ground survey with a rate that depended on the stratum (Table 10.3). The subsampling method is described in the next section.

The ground survey was eventually reduced to 11 countries (Belgium, Czech Republic, Germany, Spain, France, Hungary, Italy, Luxembourg, the Netherlands,

Table 10.3 Stratification in LUCAS 2006.

Stratum	Size master sample	Subsampling rate
Arable land	253 490	50%
Permanent crops	28 655	50%
Grassland	165 142	40%
Woodland, shrubland	452 890	10%
Bare land	21 135	10%
Artificial areas	38 658	10%
Water	30 822	10%

Poland and Slovakia), covering a total area of 2 300 000 km^2 and more than 65% of the agricultural area of EU25.

10.6 Stratified systematic sampling with a common pattern of replicates

Stratified systematic sampling may be carried out as an independent systematic sample in each stratum, but doing so reduces the advantages of systematic sampling when strata are spatially mixed with each other. A better distribution of the sample was obtained in the following way.

The master sample was divided into square blocks of 9×9 points (18×18 km). The set of points with the same relative position in each block is called a replicate. Thus we have $B = 81$ replicates ranked from 1 to 81. For each stratum h we select a number b_h of replicates that corresponds to the desired sampling rate. Each point i of the master sample belongs to a stratum h and to a replicate ranked r_i. The point is selected in the second phase if $r_i < b_h$. For example, a point in replicate 25 is selected if it belongs to stratum 1 for which 50% = 40.5 replicates are chosen, but is not selected if it belongs to stratum 5. Non-integer numbers of replicates require an additional probabilistic selection rule.

If the ranking of replicates is selected at random, it may happen that selected replicates are close to each other and give redundant information because of spatial autocorrelation. To avoid sample elements too close to each other, the following procedure is applied: the first replicate is selected at random; for the second replicate a restriction is imposed to avoid replicates too close to each other. Some care is needed in defining the distance between replicates, taking into account the location of points in neighboring blocks. For example, replicate (9,9), containing all points in the last row and last column of each block, is very close to replicate (1,1).

10.7 Ground work and check survey

A total of 423 surveyors took part in the ground visits. The average number of points surveyed by each surveyor per working day ranged between 7 (Slovakia) and 23 (the Netherlands), with an average around 14–15. Surveyors were equipped with printouts of ortho-photos or satellite images, topographic maps, GPS with pre-loaded coordinates of the points to visit and a still camera. Surveyors were asked to reach the point if possible, take a picture of the location with a mark on the point, a detailed picture of the status of vegetation and four landscape pictures oriented towards north, south, east, and west. For 66.5% of the sample the observation could be made from less than 3 m; 15.2% of the points were observed from more than 100 m. In general, these points were photo-interpreted in the field, combining the information of the ortho-photo with the in-situ information.

A parallel supervision survey was carried out on 5% of the sample (around 8200 points) by an independent company. No information was provided to supervisors on the results of the main survey. Some differences were due to the different date of observation (e.g. ploughed land versus maize) or because of land cover types for which labelling was debatable (woodland versus shrubland). After editing observations with the help of in-situ pictures to eliminate pseudo-disagreements, real disagreements (wrong crop identification, wrong location, etc.) still amounted to 3%.

10.8 Variance estimation and some results in LUCAS 2006

The stratified estimator for the proportion of a given land cover c is easy:

$$\bar{y}_{st} = \sum_h w_h \bar{y}_h, \tag{10.3}$$

where w_h is the weight of stratum h estimated from the master sample and $\bar{y}_h$ is the proportion of c from the ground survey in the stratum.

Variance estimators for two-phase sampling with stratification in the first phase can be found in classical textbooks. For example, Cochran (1977, Chapter 12) gives

$$v(\bar{y}_{st}) = \sum_h w_h s_h^2 \left(\frac{1}{n' \nu_h} - \frac{1}{N} \right) + \frac{N - n}{(N-1)n'} \sum_h w_h (\bar{y}_h - \bar{y}_{dst})^2, \tag{10.4}$$

where s_h^2 is an estimate of the variance of y. For LUCAS 2006 we have adapted the formula using a local estimate of the variance:

$$s_h^2 = (1 - f_h) \frac{\sum_{i \neq j} \delta_{ij} (y_i - y_j)^2}{2 \sum_{i \neq j} \delta_{ij}},$$

where δ_{ij} is a decreasing function of the distance between i and j similar to (10.1).

Table 10.4 gives an idea of the sampling error of the estimates for the total area surveyed.

Table 10.4 Some area estimates for the 11 countries surveyed in LUCAS 2006.

Land cover	Area (1000 ha)	CV (%)
Artificial area	12 286	1.36
Cereals	42 282	0.36
Common wheat	14 149	0.71
Barley	11 556	0.79
Maize	7 850	0.96
Rice	301	3.9
Root crops	2 779	1.78
Forest and wood	57 547	0.44
Shrubland	11 293	1.43
Bare land	7 880	1.29

10.9 Relative efficiency of the LUCAS 2006 sampling plan

We have compared the variance obtained with several single-stage sampling plans.

1. Simple random sampling (srs). We make an approximation of simple random sampling using the variance of random subsamples of the available systematic sample.
2. Pure systematic. A subsample was extracted by selecting in the LUCAS 2006 sampling plan the first eight replicates in all strata. We have used the non-stratified version of (10.3) and (10.4).
3. Post-stratified sample. The systematic sample of option 2 using the information provided by the photo-interpretation of the 2 km grid. Same sampling rate for all strata.
4. The actual LUCAS 2006 sample. Two-phase systematic sampling; higher subsampling rate in agricultural strata.

The variance comparisons have been made by always using LUCAS ground observations on different subsamples (different sizes) and with the better adapted variance estimation formula in each case. The relative efficiency of the sampling approach A is computed as

$$\text{Eff}(A) = \frac{\text{var}(srs) \cdot n_{srs}}{\text{var}(A) \cdot n_A}.$$

This comparison assumes that the cost of the survey per point is the same in any of the sampling schemes and that the bias of the variance estimator for systematic sampling is similar in all cases. Table 10.5 reports the relative efficiency for major agricultural classes.

Table 10.5 Relative efficiency between different point sampling approaches compared with simple random sampling.

	Systematic	Systematic post-stratified	LUCAS 2006
Cereals	1.11	1.55	1.95
Common wheat	1.11	1.29	1.83
Durum wheat	1.43	1.84	2.60
Barley	1.15	1.35	1.88
Maize	1.21	1.44	2.06
Potatoes	1.09	1.16	1.57
Sugar beet	1.05	1.06	1.69
Sunflower	1.09	1.17	2.19
Rapeseed	1.07	1.18	1.77
Olive groves	1.63	2.97	2.63
Vineyards	1.43	2.22	3.19
Forest	1.00	1.74	0.66
Permanent grass	1.12	1.55	1.00

Table 10.6 Number of non-clustered points that give the same information as a 10-point cluster with the LUCAS 2001–2003 design.

	Equivalent points
Artificial land	4.56
Total cereals	3.27
Common wheat	4.86
Durum wheat	2.80
Barley	4.70
Maize	3.77
Sunflower	2.69
Rapeseed	6.45
Olive groves	2.52
Vineyards	3.09
Forest	1.79
Permanent grass	4.09

A comparison of Tables 10.1 and 10.4 shows that the 2006 sampling plan was much more efficient than the 2001–2003 plan, although the cost per country was slightly higher in 2006. However, there are two major differences between both sampling schemes: clustered versus non-clustered and non-stratified versus stratified. Part of the improvement comes from the stratification, as shown in Table 10.5. We have assessed the effect of clustering by comparing two non-stratified sampling approaches: the LUCAS 2001–2003 sampling plan with a cluster of 10 points every 18×18 km; and the non-stratified systematic subsample of the LUCAS 2006 sample described in option 2 above, based on the repetition of a pattern of eight points every 18×18 km. Table 10.6 gives the equivalent number of non-clustered points of a 10-point cluster (LUCAS 2001) in this sense: if we have a sample of m clusters, the equivalent number is the factor k such that a sample of $k \times m$ non-clustered points gives the same variance of the area estimator. For example, to estimate the total area of cereals the information provided by a cluster of 10 points is equivalent to the information provided by 3.27 unclustered points.

In the seven countries for which the three surveys were conducted (Belgium, Germany, Spain, France, Italy, Luxembourg and the Netherlands), the cost of the ground work per point was similar in 2001 and 2006 (around €23 per point), but it was lower in 2003 (around €16 per point), mainly because of a very low price in France. We still have to take into account that the rules for the work of enumerators were not the same in different years, but the accumulated experience suggests that the cost of one cluster of 10 points in LUCAS 2001 is similar to the cost of six to eight non-clustered points in LUCAS 2006 (Eurostat, 2007). Since the number of points equivalent to the cluster is always lower than six, except for rapeseed, we can conclude that the non-clustered sampling scheme is more efficient. This conclusion is in contrast to previous studies on the optimization of the size of sampling clusters in an area frame (Gallego *et al.*, 1999). The reason for this revision of the conclusion is that the cost function considered earlier did not take into account that the so-called fixed cost per sampling unit depends on the variable distance between units. The conclusion may change for situations in which travelling from one sampling unit (point or cluster) to the next one takes longer than in western Europe.

LUCAS 2006 surveyors spent 62% of their working time travelling, including the morning trip from home to the first point and back home in the evening. This means that there is room to optimize cost by reducing the travelling time. However, the comparison with LUCAS 2001 indicates that the time reduction due to clustering was not very efficient.

10.10 Expected accuracy of area estimates with the LUCAS 2006 scheme

Before launching any survey some idea is needed of the accuracy that can be achieved. The accuracy reached for the estimated area of land cover c mainly depends on the size D of the region and the proportion p of c. The results for each country have allowed simple linear regressions without intercept to be fitted: for agricultural classes,

$$s_{LUCAS} = 0.743 s_{ran} + \epsilon, \qquad r^2 = 0.989,$$

and for non-agricultural classes,

$$s_{LUCAS} = 1.182 s_{ran} + \epsilon, \qquad r^2 = 0.967,$$

where $s_{ran}(p) = D \times \sqrt{p(1-p)/(n-1)}$ is the standard error that would have been obtained with simple random sampling. The coefficient 0.743 confirms the good relative efficiency of the LUCAS 2006 sampling plan for crops, while the coefficient 1.182 for non-agricultural classes quantifies the loss due to the lower sampling rate in non-agricultural strata.

We can easily estimate the minimum area of a crop (or group of crops) to reach a target coefficient of variation with the LUCAS 2006 sampling plan (e.g. 5%, 2%, 1%). Table 10.7 illustrates the results for the countries covered in LUCAS 2006. We can see a certain stability of the area covered by a crop to predict a given accuracy. For example, a 5% coefficient of variation can be predicted for crops covering around 300 000–350 000 ha, unless the country is very small (Luxembourg).

Table 10.7 Approximate minimum crop area to reach a given Coefficient of Variation with the LUCAS 2006 sampling plan (areas in thousands of hectares).

			Target CV		
	Sample size	Region area	5%	2%	1%
Luxembourg	229	258	130		
Belgium	2 292	3 056	280	1160	2290
Netherlands	2 870	3 720	300	1230	2600
Slovak Rep.	3 230	4 903	340	1470	3190
Czech Rep.	5 420	7 887	320	1660	4340
Hungary	8 133	9 303	280	1400	4190
Italy	20 247	30 162	340	2110	6640
Poland	23 076	31 269	310	1880	6250
Germany	26 452	35 730	320	1790	6430
Spain	33 607	50 671	350	2030	7600
France	40 380	54 910	330	2200	6590

10.11 Non-sampling errors in LUCAS 2006

Non-sampling errors are generally more difficult to assess than sampling errors (Lesser and Kalsbeek, 1999). In this section we study the possible order of magnitude of the main sources of non-sampling errors.

10.11.1 Identification errors

The most important source of non-sampling error in an area frame survey is the identification mistakes by enumerators; this can happen as a result of:

(a) location error;

(b) incorrect identification because of inadequate enumerator training – mainly for minor crops;

(c) misinterpretation of rules to label land cover types with a fuzzy definition – this happens mainly in natural vegetation areas;

(d) unsuitable observation date – for example, if the ground visit is made when a crop has not yet emerged or has already been harvested.

Sources (a) and (b) are quantified by the check survey (Section 10.7): the 3% overall disagreement is an upper bound on the potential bias; the bias can be larger for minor crops and is substantially smaller for major crops. The rate of mistakes in LUCAS is higher than in operational area frame surveys because it has so far been a pilot survey and many enumerators had insufficient experience; a certain stability of enumerators is essential for the quality of ground data.

Source (c) refers mainly to the perception of classes with debatable labelling (e.g. shrub with sparse trees) and is not seen as critical. Source (d) has probably been important in LUCAS 2006, but its impact is difficult to assess: the will to deliver crop area estimates by 15 June led to too early a schedule of observations and as a result probably to a high number of points in fields where crops had not yet emerged or were at too early a development stage for easy identification; a colder than usual winter worsened the situation.

10.11.2 Excluded areas

Certain areas were excluded from the ground survey. We explore the possible bias introduced into the agricultural estimates by disregarding these areas, i.e. the amount of agriculture in excluded areas.

Cyprus, Malta and many other islands have been excluded. From the agricultural point of view, the most important islands excluded in the 11 countries surveyed in 2006 were the Balearic and Canary islands. Data available from other sources indicate that the arable land in these islands was around 100 000 ha in 2000, i.e. 0.7% of the arable land in Spain and 0.1% of the arable land in EU25. For permanent crops the contribution of the excluded islands is more important: around 140 000 ha, which represents 1.3% of the permanent crops in EU25.

Points in the master sample above 1200 m were considered too expensive to survey and excluded from the field work. The altitude was assessed with a DTM with a resolution

Table 10.8 Percentage of points in each stratum by altitude interval.

Strata	<1000 m	1000–1100 m	1100–1200 m	>1200 m
Arable land	98.84	0.63	0.31	0.21
Permanent crops	98.41	0.90	0.43	0.26
Permanent grassland	91.97	1.63	1.22	5.18
Wooded, shrub	91.35	2.04	1.51	5.11
Rare vegetation	74.46	1.95	1.88	21.71
Artificial land	98.61	0.49	0.29	0.60
Water, wetland	98.17	0.17	0.12	1.54
Total	93.71	1.46	1.04	3.79

of 1 km; some of the excluded points with the altitude criterion may actually have been under the 1200 m threshold. Assessing the amount of agriculture in the areas above 1200 m (according to the DTM) is difficult, but an indication can be obtained from the stratification data, available for any altitude.

Table 10.8 tells us that 0.21% of the points in stratum 1 (photo-interpreted as arable land), 0.26% of stratum 2 (permanent crops) and 5.18% of stratum 3 (grassland) are above 1200 m. These figures would give an estimate of the bias due to the elimination of such points if the accuracy of the photo-interpretation were independent of the altitude. However, the proportion of points of stratum 1 that are really arable in the ground observation decreases at higher altitude: from 67.6% below 1000 m to 55.4% (1000–1100 m) and 48.2% (1100–1200 m). We do not have data for points above 1200 m, but the data shown here strongly suggest that the proportion would in any case be below 48%. Therefore, the impact (bias) on the total arable land area will be less than 0.1% of the area of stratum 1.

In a similar way we conclude that a contribution to bias close to 0 comes from stratum 2. Stratum 3 might give an additional contribution to the bias up to 0.1% of the stratum (5.18% of points multiplied by a share that can be assumed less than 2.5%). The other strata give a smaller contribution. This is not a proper statistical estimation of bias, but supports the reasonable assumption that the bias in the arable land estimates due to the exclusion of points above 1200 m is less than 0.3% of the total area of arable land. For permanent crops, the bias comes almost exclusively from stratum 2 and can exceed 0.2% of the stratum.

The situation is different for permanent grassland. Around 5.18% of the stratum is above 1200 m, and the proportion of points that were really grassland in stratum 3 seems to be stable around 60% without a clear trend for higher altitudes. This would mean a bias of up to 3% of the stratum area.

10.12 Conclusions

At the time writing, LUCAS 2009 has been launched for 23 member states of the EU: Malta and Cyprus are still excluded from the ground survey, and Romania and Bulgaria were surveyed in 2008 and will be synchronized later with the rest of the EU. LUCAS is more focused on the application to agri-environmental indicators than to traditional agricultural statistics. LUCAS should be now carried out regularly every three years;

most of the points in the sample will remain stable between successive surveys since the estimation of land cover change matrices is one of the major targets. The sample design in 2009 is similar to the 2006 scheme, but the sampling rate will be reduced in agricultural strata and increased in the other strata. The sampling rates per stratum have been tuned separately per sub-national region using a Bethel (1989) method. The total sample size is close to 230 000 points.

The shift towards environmental concerns has also had implications for the data to be collected. Transects that had been dropped in 2006 appear again in the observation protocol, but with a different design compared with LUCAS 2001–2003. A major item introduced is the collection of soil samples from around 20 000 points that will be analysed in the laboratory (Stolbovoy *et al.*, 2005). One of the main targets is to monitor the evolution of soil organic carbon (SOC) contents required for the follow-up of the Kyoto protocol targets.

Future improvements should come from stratification updating. More recent satellite images or ortho-photos should provide a better stratification efficiency. There is still a need to compare in greater depth the approach of photo-interpretation by point, as conducted for LUCAS 2006, with a cheaper approach of simple overlay on standard land cover maps, such as CORINE Land Cover 2006 (EEA, 2007).

Acknowledgements

We are particularly grateful to the members of the different Eurostat teams that have managed LUCAS since 2001, in particular Maxime Kayadjanian, Claude Vidal, Manola Bettio, Pascal Jacques, José Lange, Thierry Maucq, Marco Fritz, Marjo Kasanko, Laura Martino and Alessandra Palmieri.

References

Annoni, A., Luzet, C., Gubler, E. and Ihde, J. (2001) Map projections for Europe. Report EUR 20120 EN, JRC, Ispra (Italy).

Arino, O., Gross, D., Ranera, F. *et al.* (2008) GlobCover: ESA service for global land cover from MERIS. *International Geoscience and Remote Sensing Symposium (IGARSS)*, Boston, 6–11 July, pp. 2412–2415.

Bartholomé, E. and Belward, A. S. (2005) GLC2000: A new approach to global land cover mapping from earth observation data. *International Journal of Remote Sensing*, 26, 1959–1977.

Bellhouse, D. R. (1988) Systematic sampling. In P. R. Krisnaiah and C. R. Rao (eds), *Handbook of Statistics*, Vol. 6, pp. 125–146. Amsterdam: North-Holland.

Bethel, J. (1989) Sample allocation in multivariate surveys. *Survey Methodology*, 15, 47–57.

Bettio, M., Delincé, J., Bruyas, P., Croi, W. and Eiden, G. (2002) Area frame surveys: aim, principals and operational surveys. In *Building Agri-environmental Indicators, Focussing on the European Area Frame Survey LUCAS*. EC report EUR 20521, pp. 12–27. http://agrienv.jrc.ec.europa.eu/publications/pdfs/agri-ind/CH0_Area_Frame_Sample_Surveys.pdf.

Carfagna, E. (2007) A comparison of area frame sample designs for agricultural statistics. *Bulletin of the International Statistical Institute*, 56th Session, 2007, Proceedings, Meeting STCPM11 on Agricultural and Rural Statistics, Lisbon, 22–29 August.

Cochran, W. (1946) Relative accuracy of systematic and stratified random samples for a certain class of populations. *Annals of Mathematical Statistics*, 17, 164–177.

Cochran, W. (1977) *Sampling Techniques*, 3rd edition. New York: John Wiley & Sons, Inc.

Cotter, J. and Tomczac, C. (1994) An image analysis system to develop area sampling frames for agricultural surveys. *Photogrammetric Engineering and Remote Sensing*, 60, 299–306.

De Vries, P. G. (1986) *Sampling Theory for Forest Inventory: A Teach-Yourself Course*. Berlin: Springer.

Delincé, J. (2001) A European approach to area frame survey. *Proceedings of the Conference on Agricultural and Environmental Statistical Applications in Rome (CAESAR)*, 5–7 June, Vol. 2, pp. 463–472. `http://www.ec-gis.org`.

Dunn, R. and Harrison, A. R. (1993) Two-dimensional systematic sampling of land use. *Applied Statistics*, 42, 585–601.

EEA (2006) The thematic accuracy of CORINE Land Cover 2000. Assessment using LUCAS. EEA Technical Report 7/2006, Copenhagen.

EEA (2007) CLC2006 Technical guidelines. EEA technical report 17/2007. `http://www.eea.europa.eu/publications/technical_report_2007_17`.

EEA-ETC/TE (2002) CORINE land cover update. Technical guidelines. `http://www.eea.europa.eu/publications/technical_report_2002_89`.

Eiden, G., Vidal, C. and Georgieva, N. (2002) Land cover/land use change detection using point area frame survey data; Application of TERUTI, BANCIK and LUCAS Data. In *Building Agri-environmental Indicators, Focussing on the European Area Frame Survey LUCAS*. EC report EUR 20521, pp. 55–74.

Eurostat (2007) LUCAS 2006 Quality Report. Standing Committee for Agricultural Statistics, 22–23 November. Document ESTAT/CPSA/522a, Luxembourg.

FAO (1996) *Multiple Frame Agricultural Surveys. Volume 1: Current Surveys Based on Area and List Sampling Methods*. FAO Statistical Development Series no. 7. Rome: FAO.

FAO (1998) *Multiple Frame Agricultural Surveys. Volume 2: Agricultural Survey Programmes Based on Area Frame or Dual Frame (Area and List) Sample Designs*. FAO Statistical Development Series no. 10. Rome: FAO.

Feddema, J. J., Oleson, K. W., Bonan, G. B., Mearns, L. O., Buja, L. E., Meehl, G. A. and Washington, W. M. (2005) The importance of land-cover change in simulating future climates. *Science*, 310, 1674–1678.

Friedl, M. A., McIver, D. K., Hodges, J. C. F., Zhang, X. Y., Muchoney, D., Strahler, A. H., Woodcock, C. E., Gopal, S., Schneider, A., Cooper, A., Baccini, A., Gao, F. and Schaaf, C. (2002) Global land cover mapping from MODIS: Algorithms and early results. *Remote Sensing of Environment*, 83, 287–302.

Gallego, J. and Bamps, C. (2008) Using CORINE land cover and the point survey LUCAS for area estimation. *International Journal of Applied Earth Observation and Geoinformation*, 10, 467–475.

Gallego, F. J., Delincé, J. and Carfagna, E. (1994) Two-stage area frame sampling on square segments for farm surveys. *Survey Methodology*, 20, 107–115.

Gallego, F. J., Feunette, I. and Carfagna, E. (1999) Optimising the size of sampling units in an area frame. In J. Gómez-Hernández, A. Soares and R. Froidevaux (eds), *GeoENV II: Geostatistics for Environmental Applications*, pp. 393–404. Dordrecht: Kluwer.

Gautschi, W. (1957). Some remarks on systematic sampling. *Annals of Mathematical Statistics*, 28, 385–394.

González, F., López, S. and Cuevas, J. M. (1991) Comparing two methodologies for crop area estimation in Spain using Landsat TM images and ground gathered data. *Remote Sensing of the Environment*, 35, 29–36.

Hendricks, W. A., Searls, D. T. and Horvitz, D. G. (1965) A comparison of three rules for associating farms and farmland with sample area segments in agricultural surveys. In S. S. Zarkovich (ed.), *Estimation of Areas in Agricultural Statistics*, pp. 191–198. Rome: FAO.

Kalensky, Z. D. (1998) AFRICOVER land cover database and map of africa. *Canadian Journal of Remote Sensing*, 24, 292–297.

Koop, J. C. (1971) On splitting a systematic sample for variance estimation. *Annals of Mathematical Statistics*, 42, 1084–1087.

Lesser, V. M. and Kalsbeek, W. D. (1999) Nonsampling errors in environmental surveys. *Journal of Agricultural, Biological, and Environmental Statistics*, 4, 473–488.

Loveland, T. R., Reed, B. C., Brown, J. F., Ohlen, D. O., Zhu, Z., Yang, L. and Merchant, J. W. (2000) Development of a global land cover characteristics database and IGBP DISCover from 1km AVHRR data. *International Journal of Remote Sensing*, 21, 1303–1330.

MARM (2008) *Encuesta sobre Superficies y Rendimientos de Cultivos. Resultados 2008*. Madrid: Ministerio de Medio Ambiente y Medio Rural y Marino. http://www.mapa.es/estadistica/pags/encuestacultivos/boletin2008.pdf.

Martino, L. (2003) The Agrit system for short-term estimates in agriculture. DRAGON Seminar Krakov/Balice, 9–11 July.

Matérn, B. (1986) *Spatial Variation*. Berlin: Springer.

Osborne, J. G. (1942) Sampling errors of systematic and random surveys of cover-type areas. *Journal of the American Statistical Association*, 37, 256–264.

Schreuder, H. T. and Czaplewski, R. L. (1993) Long-term strategy for the statistical design of a forest health monitoring system. *Environmental Monitoring and Assessment*, 27, 81–94.

Stolbovoy, V., Filippi, N., Montanarella, L., Piazzi, M., Petrella, F., Gallego, J. and Selvaradjou, S. (2005) *Validation of a EU Soil Sampling Protocol to Certify the Changes of Organic Carbon Stock in Mineral Soils (Piemonte Region, Italy)*. Report EUR 22339 EN.

Vogelmann, J. E., Howard, S. M., Yang, L., Larson, C. R., Wylie, B. K. and Van Driel, N. (2001) Completion of the 1990s national land cover data set for the conterminous United States from LANDSAT thematic mapper data and ancillary data sources. *Photogrammetric Engineering and Remote Sensing*, 67, 650–662.

Wolter, K. M. (1984), An investigation of some estimators of variance for systematic sampling. *Journal of the American Statistical Association*, 79, 781–790.

Wood, G. R., Robertson J. M. (1998) Buffon got it straight. *Statistics and Probability Letters*, 37, 415–421.

Yates, F. (1949) *Sampling Methods for Censuses and Surveys*. London: Griffin.

11

Area frame design for agricultural surveys

Jim Cotter, Carrie Davies, Jack Nealon and Ray Roberts
US Department of Agriculture, National Agricultural Statistics Service, USA

11.1 Introduction

The National Agricultural Statistics Service (NASS) has been developing, using and analysing area sampling frames since 1954 as a vehicle for conducting surveys to gather information regarding crop acreage, cost of production, farm expenditures, grain yield and production, livestock inventories and other agricultural items. An area frame for a land area such as a state or country consists of a collection or listing of all parcels of land for the area of interest from which to sample. These land parcels can be defined based on factors such as ownership or based simply on easily identifiable boundaries as is done by the NASS.

The purpose of this document is to describe the procedures used by the NASS to develop and sample area frames for agricultural surveys. The process involves many steps, which have been developed to provide statistical and cost efficiencies. Some of the key steps are as follows:

- *Stratification.* The distribution of crops and livestock can vary considerably across a state in the United States. The precision of the survey estimates or statistics can be substantially improved by dividing the land in a state into homogeneous groups or strata and then optimally allocating the total sample to the strata. The basic stratification employed by the NASS involves: (1) dividing the land into land-use strata such as intensively cultivated land, urban areas and range land, and

Agricultural Survey Methods Edited by Roberto Benedetti, Marco Bee, Giuseppe Espa and Federica Piersimoni

(2) further dividing each land-use stratum into substrata by grouping areas that are agriculturally similar.

- *Multi-step sampling.* Within each stratum, the land can be divided into all the sampling units or segments and then a sample of segments selected for a survey. This would be a very time-consuming endeavour. The time spent developing and sampling a frame can be greatly reduced by: (1) dividing the land into larger sampling units called first-step or primary sampling units (PSUs); (2) selecting a sample of PSUs and then delineating the segments only for these PSUs; and (3) selecting a sample of segments from the selected PSUs.

- *Analysis.* Several decisions are made that can have an appreciable impact on the statistical and cost efficiency. These include decisions such as the land-use strata definitions, the number of substrata, the size of the sampling units, the allocation of and the method of selecting the sample necessary to guide us in these decisions.

The major area frame survey conducted by the NASS is the June Agricultural Survey (JAS). This mid-year survey provides area frame estimates primarily for crop acreages and livestock inventories. During the survey, the interviewers visit each segment in the sample, which has been accurately identified on aerial photography, and interview each person who operates land inside the boundaries of the selected segments. With the respondent's assistance, field boundaries are identified on the photography and the acreage and crop type reported for each field in the segment. Counts of livestock within each sample segment are also obtained. This area frame information is subsequently used to provide state, regional and national estimates for crop acreages, livestock inventories and other agricultural items. Naturally, the procedures used to develop and sample area frames affect the precision and accuracy of the survey statistics.

11.1.1 Brief history

Iowa State University began construction of area frames for use in agricultural surveys in 1938. The NASS began research into the use of area sampling frames in the mid-1950s to provide the foundation for conducting probability surveys based on complete coverage of the farm sector. In 1954, area frame surveys were begun on a research basis in ten states, 100 counties with 703 ultimate sampling units or segments. These surveys were then expanded over the years and made operational in 1965 in the contiguous United States.

Changes made to the area frame methodology during the 1960s and early 1970s were mainly associated with sampling methods such as land-use stratification and replicated sampling (described in detail in Section 11.5). Technological changes were incorporated during the seventies and eighties in the form of increased computerization, use of satellite imagery, use of analytical software and development of an area frame sample management system among others.

The area frame programme has grown over the past 54 years and is now conducted in 49 states with approximately 11 000 segments being visited by data collection personnel for the major agricultural survey conducted during June of each year. Today, the NASS maintains an area frame for each state except Alaska; it also maintains an area frame for Puerto Rico. The frames are constructed one state at a time and used year after year until deemed outdated. A frame is generally utilized for 15–20 years, and when it becomes outdated, a new frame is constructed to replace it. Each year, three or four states are selected to receive a new frame. The selection of states for new frames is based on the following criteria: age of the frame, significant land-use changes, target coefficients of variance being met, and significance to the national programme.

11.1.2 Advantages of using an area frame

- *Versatility.* Since reporting units can be associated with an area of land (a sampling unit), an area frame can be used to collect data for multiple variables in one survey. For example, crop acreage, livestock, grain production and stocks, and economic data are all collected during the JAS.
- *Complete coverage.* The NASS's area frame is complete, meaning when all the sampling units are aggregated, the entire population is completely covered and every sampling unit has a known chance of being selected. The sampling units do not overlap, nor are there gaps between adjacent sampling units. This is a tremendous advantage since it provides the vehicle to generate unbiased survey estimates. Complete coverage is also useful in multiple-frame (area and list) surveys where the area frame is used to measure the degree of incompleteness of the list frame.
- *Statistical soundness.* The advantage of complete coverage combined with a random selection of sampling units is that it can provide unbiased estimates with measurable precision.
- *Non-sampling errors reduced.* Face-to-face interviews are conducted for the JAS, which generally result in better-quality data being gathered than data collected by mail or telephone. The interviewer uses an aerial photograph showing the location and boundary of the sample segment to collect data for all land within the segment boundary such as crop acreages, residential areas, and forest. If the respondent refuses to participate in the survey, or is inaccessible, the interviewer is instructed to make observations which are helpful when making non-response adjustments.
- *Longevity.* Once an area frame is constructed, it can be used year after year without having to update the sampling units. The frames can become inefficient as land use changes. However, the area frames constructed for most states last 15–20 years before they need to be replaced.

11.1.3 Disadvantages of using an area frame

- *Can be less efficient than a list frame.* If a list of farm operators can be stratified by a variable related to the survey items, it will provide greater sampling efficiency than an area frame that is stratified by land use. For example, a list frame that is stratified by peak number of cattle and calves will provide greater sampling efficiency than the area frame when estimating cattle inventory. Unfortunately, the NASS list frame does not provide 100% coverage, because of the difficulty of obtaining and maintaining a complete list of producer names, addresses, and appropriate control data. Since the area frame is a complete sampling frame, it is used to measure incompleteness in the list.

- *Cost.* An area frame can be very expensive to build and sample. New frame construction, on average, uses five full-time employees for four months per state. Also, face-to-face interviews conducted by a trained staff are also very costly.
- *Lack of good boundaries.* Although this is not a problem for most areas in the United States, it can be when building a frame in a foreign country. The importance of quality boundaries will be discussed later.
- *Sensitivity to outliers.* Because the sampling rate for the JAS is low, expansion factors are relatively high. For this reason, area frame surveys are sometimes plagued by a few 'extremely large' operations that are in sample segments. These operations can greatly distort the survey estimates. A solution to this problem is to identify all very large operations prior to the survey (special list frame) and sample them with certainty.

11.1.4 How the NASS uses an area frame

- *Acreage estimates for major commodities.* The primary focus is on corn, soybeans, winter wheat, spring wheat, durum wheat, cotton, not on list (NOL) cattle, and number of farms. The acreage for each crop is recorded for each field within the segment. These acreages are then expanded to produce an estimate of crop acreage at the state and national level.
- *Measure the incompleteness of the list.* The NASS maintains a list of farmers who operate land in the country. Because farm operations go in and out of business, the list is never complete at any given time. The JAS survey is used to find farmers who are not on the NASS's list. During the JAS data collection, the interviewer records the names of all people who operate agricultural land within each segment. These names are then checked against the NASS's list of farm operators. Those who are not present on the list are referred to as NOL. The NOL operations found during the JAS are multiplied by an expansion factor to estimate the incompleteness of the list for each state.
- *Ground truth for remotely sensed crop acreage estimates.* The crop data for each field inside the segment from the JAS are used to determine what crop spectral signatures from the satellite represent. Identified signatures are then used to classify fields throughout a state. Once the satellite data have been classified, an acreage estimate can be made for various crops grown in that state.
- *Follow-on surveys.* The NASS uses the data from the JAS for follow-on surveys such as the Objective Yield Survey where specifically cropped fields are randomly selected with probability proportional to size. These yield surveys involve making counts, measurements, and weightings of selected crops. Every 5 years additional sampling units are added to the JAS for the Agricultural Coverage Evaluation Survey (ACES). Data collected from the JAS segments, in combination with the additional ACES segments, are used to measure the completeness of the Census of Agriculture mail list. The Not on Mail List (NML) estimates are used to weight census data at the record level to produce coverage-adjusted estimates.

11.2 Pre-construction analysis

Before building a new frame, analysis is conducted to determine which states are most in need of one. Generally three to four states are selected to receive a new frame each year. Data collected from approximately 11 000 segments during the JAS is used to determine the extent to which the land-use stratification has deteriorated for each state. This involves comparing the coefficients of variation for the survey estimates of major items over the life of the frame. Typically states with the oldest frames have the highest probability of being selected. Also important is the extent to which a state contributes to the national programme for major commodities. For example, Kansas contributes approximately 20% to the national estimate for winter wheat. If it were determined that the state's JAS target coefficients of variance for winter wheat were not being met, Kansas would be likely to be selected to receive a new frame.

Once a state has been selected to receive a new frame, analysis is performed to determine the most appropriate stratification scheme to be used. Previous years' survey data are used to calculate the percentage of cultivated land in the sample segments, the average number of interviews per segment in each stratum, and the variances for important crops in each stratum. These data are used to determine the following:

Land-use strata definitions. Several land-use strata are common to all frames, including cultivated land, ag-urban, urban, and non-agricultural land. The cultivated land is divided into several strata based on the distribution of cultivation in the state. Previous years' survey data are analysed to provide information such as the percentage of cultivated land in the sample segments so that the distribution of cultivated land can be ascertained. This will help determine the number of and definition of the cultivated strata. Table 11.1 presents the land-use stratification scheme generally followed along with the codes to be used during the stratification process.

Strata 11, 12, and 20 are where the majority of cropland is present. These strata target commodities such as corn, soybeans, cotton, and wheat. In many states, strata 11 and 12 are collapsed into one stratum. The 40's strata contain less than 15% cultivation. Range and pasture land, as well as woods, mountains, desert, swampland, etc., are put into stratum 40. Cattle and other livestock operations are usually also found in stratum 40. Little to no agriculture is expected to be found in strata 31, 32, and 50. These strata are present

Table 11.1 Land-use strata codes and definitions.

Stratum	Definition
11	General cropland, 75% or more cultivated
12	General cropland, 50–74% cultivated
20	General cropland, 15–49% cultivated
31	Ag-urban, less than 15% cultivated, more than 100 dwellings per square mile, residential mixed with agriculture
32	Residential/commercial, no cultivation, more than 100 dwellings per square mile
40	Less than 15% cultivated
50	Non-agricultural, variable size segments

in all states. Stratum 31 contains dense commercial and residential areas of cities and towns. The ag-urban land in stratum 32 represents a mixture of residential and areas with the potential for agricultural activity, usually located in a band around a city, where the city blends into the rural area. Stratum 50 contains non-agricultural entities such as state and national parks, game and wildlife refuges, military installations and large airports.

These are the strata present in most states, however; adjustments may be made to the design depending on the state involved. For example, stratum 40 is often broken into two or more strata in the western states, with a special stratum for forest or desert. Also, a stratum may be added for Indian reservation land. Crop-specific strata are also used in several states to allow the opportunity to channel a sample either into, or away from, a certain area. For example, citrus strata were created in Florida. However, since an annual citrus survey conducted in Florida provides reliable estimates, the JAS is not used for citrus estimates. The citrus strata are in place to allow for a heavier sample in strata where field crops are present.

PSU and segment sizes. In the process of constructing a new area frame, all land in the selected state will be broken down into PSUs. The population of PSUs is sampled from by stratum, and the selected PSUs are further broken down into an average of six to eight segments from which one segment is chosen. This way, an entire frame need not be divided into segments, saving a tremendous amount in labour costs.

Before area frame construction can start, the sizes of the PSUs and segments must be determined. The target PSU and segment sizes are determined for each stratum based on the analysis of previous years' JAS data. The target size of the segment is determined first.

The optimum segment size for a land-use stratum depends upon a multitude of often interrelated factors such as the survey objectives, data collection costs, data variability among segments, interview length, population density, concentration of cropland, and the availability of identifiable boundaries for the segments. The segment size, which is determined in the pre-construction phase, is based on the analysis of previous years' JAS data. The target segment size varies from stratum to stratum and state to state. Table 11.2 is an example of the segment sizes per strata for a typical state.

When the PSUs in stratum 11 are broken down in this example, the resulting segments should be as close to 1 square mile as possible. The target segment size in the intensively cultivated strata (10s strata) is usually 1 square mile, with the exception of a few states where the target segment size is less. In the moderately cultivated strata (20s strata), the target segment size is typically 1–2 square miles.

Table 11.2 Target segment sizes.

Stratum	Definition	Target segment size (square miles)
11	General cropland, 75% or more cultivated	1.00
12	General cropland, 50–74% cultivated	1.00
20	General cropland, 15–49% cultivated	1.00
31	Ag-urban	0.25
32	Residential/commercial	0.10
40	Open land, <15% cultivated	2.00
50	Non-agricultural, variable size segments	PPS

The target segment size for open land strata (40s strata) varies the most. In states where good boundaries are available, the target segment size can be 1–2 square miles. In some areas (e.g. desert, mountainous, or range areas), boundaries are few and far apart. The target segment size in these areas will range from 4–8 square miles. If adequate boundaries are not available, the segments in the strata will not have a target segment size. Segment size will vary depending on available boundaries. The probability of selecting a segment is proportional to the size of the segment (PPS).

The target segment sizes in the urban and ag-urban strata are always one-tenth and one-quarter square mile, respectively. In stratum 50, the non-agricultural stratum, there is no segment size (except for states that have not received a new frame since 1985). Entities such as state parks, forests, airports, military land, etc. are placed in stratum 50. The boundary of the segment is the boundary of the entity. The probability of selecting a segment in this stratum is proportional to the size of the segment (PPS).

When determining the segment size for each stratum, the following are taken into consideration:

- *Minimize sampling variability.* Ideally the segments within a (non-PPS) substratum will be equal in size and homogeneous in terms of agricultural content to keep variance down. As the size of the segments decreases, so does the ability to delineate segments (to be discussed later) that are homogeneous with respect to the amount of cultivated land. Therefore, the sampling variability among segments increases for a given sample size.
- *Availability of boundaries.* As the size of the segments decreases, the availability of suitable boundaries also decreases. Quality boundaries are pertinent when delineating PSUs and segments. If boundaries are not available to delineate segments that are equal in size, variability may increase. Also, if poor-quality boundaries are used, the result could mean more reporting errors during the data collection phase. In highly cultivated strata, as well as the urban and ag-urban strata, quality boundaries are generally plentiful, allowing for a smaller segment size. The land in the 40s strata consists of forest, desert, range, pasture, etc. Quality boundaries in these strata are more spread apart, making a larger segment size more accommodating.
- *Data collection costs.* The interviewer must contact and interview each person operating farmland within the segment boundaries. To minimize data collection costs, the interviewer should be able to complete a segment in less than 12 hours. Research has shown that interviewers are generally able to complete an average of 3–4 interviews per segment in 12 hours.

The target segment size for a stratum will be based partly on the average number interviews per segment. In moderate to intensively cultivated strata (10s and 20s strata), farms are relative close together. A segment size of 1 square mile will result in an average of 3–4 interviews. In the lower intensively cultivated strata (40s strata), the farm operations are typically farther apart in location. The segment size in these strata can be larger and still not increase data collection costs. The target segment size in ag-urban (stratum 31), urban (stratum 32), and non-agricultural strata has no influence on data collection costs since few if any interviews are done for segments in these strata.

The target sizes of the PSUs are based on the segment size. PSUs should contain six to eight segments, so the PSU size should be about six times the segment size. PSUs that

Table 11.3 Primary sampling unit size tolerance guide.

		PSU size (sq miles)		
Stratum	Cultivation	Minimum	Target	Maximum
11	75% cultivated	1.00	6.00–8.00	9.00
12	50–75% cultivated	1.00	6.00–8.00	9.00
20	15–49% cultivated	1.00	6.00–8.00	9.00
31	Ag-urban	0.25	1.00–2.00	3.00
32	Commercial	0.10	0.50–1.00	1.00
40	>15% cultivated	4.00	20.00–24.00	36.00
50	Non-ag	–	PPS	–

are smaller than the target size will be broken down into fewer segments, and PSUs that are larger will be broken down into more segments if boundaries are available. So if a PSU in stratum 11 is only 2 square miles, it will most likely be broken down into two segments. The minimum PSU size is generally one segment. Table 11.3 is an example of the PSU size tolerance range for the target segment sizes in Table 11.2.

Once the analysis is complete and strata definitions and segment sizes are specified by the Area Frame Section (AFS), stratification will begin. After this point, the strata definitions, PSU and segment sizes are used for the life of the frame and are not changed.

11.3 Land-use stratification

The process of land-use stratification is the delineation of land areas into land-use categories (strata). The purpose of stratification is to reduce the sampling variability by creating homogeneous groups of sampling units. Although certain parts of the process are highly subjective in nature, precision work is required of the personnel stratifying the land (called stratifiers) to ensure that overlaps and omissions of land area do not occur and land is correctly stratified. The stratification unit divides the land within each county into PSUs using quality physical boundaries, then assigns them to a land-use strata (defined in the pre-construction phase). Later in area frame construction, the PSUs are further divided into segments, also using quality physical boundaries. A quality physical boundary is a permanent or, at least, long-lasting geographic feature which is easily found and identifiable by an interviewer. If an interviewer cannot accurately locate a segment in a timely manner, there is the potential for non-sampling errors to be introduced into the survey data. Also, if the field interviewer, unknowingly, does not collect data associated with all of the land inside the sampled area or collects data for an area outside of that selected, then survey results will be biased. Quality boundaries include highways, roads, railroads, rivers, streams, canals, and section lines.

The stratifier breaks down all the land within each county into PSUs. The stratifiers locate the best physical boundaries and draw off PSUs as close to the target PSU size (defined in the pre-construction phase) as possible. They use the following materials to locate boundaries:

- *Topographic quadrangle maps.* Produced by the US Geological Survey (USGS), digital raster graphic maps are scanned images of USGS 1:100 000 scale topographic maps.

- *Tele Atlas data.* Produced by Dynamap, these accurate and complete digital vector data provide the NASS with an accurate map base on which to verify boundaries during the frame construction process.
- *National Agriculture Imagery Program (NAIP).* This is operated by the Farm Service Agency (FSA), which acquires one- and two-metre digital ortho-imagery during the agricultural growing seasons in the continental USA. Coverage provides approximately 20% one-metre and 80% two-metre. The NAIP imagery is used to verify boundaries during the frame construction process and has improved the accuracy of the frame. These data are also used when generating the photo enlargements used for the JAS data collection.

Simultaneously, they use the following materials to classify the PSUs into strata:

- *Satellite imagery.* Satellite imagery is derived from digital data collected by sensors aboard satellites. The AFS currently uses imagery from the LANDSAT 7 satellite. Satellite imagery is used primarily to ascertain where the cultivated areas and the pasture areas are present in a county.
- *Cropland Data Layer.* This is an agriculture specific land cover geospatial product developed by the Spatial Analysis Research Section (SARS). It is used during the frame construction process as an additional layer to assist the stratifier in isolating agriculture. It is also used to assist in isolating crop specific strata in a number of states.

Once all of the PSUs in the county have been delineated and classified into strata, the PSU identification number is attached. This is done automatically in ArcGIS9. The PSUs are numbered beginning in the upper right-hand corner and winding through the county in a serpentine fashion. Figure 11.1 shows an example of the numbering scheme. The first number is the stratum and the second is an incremental PSU number.

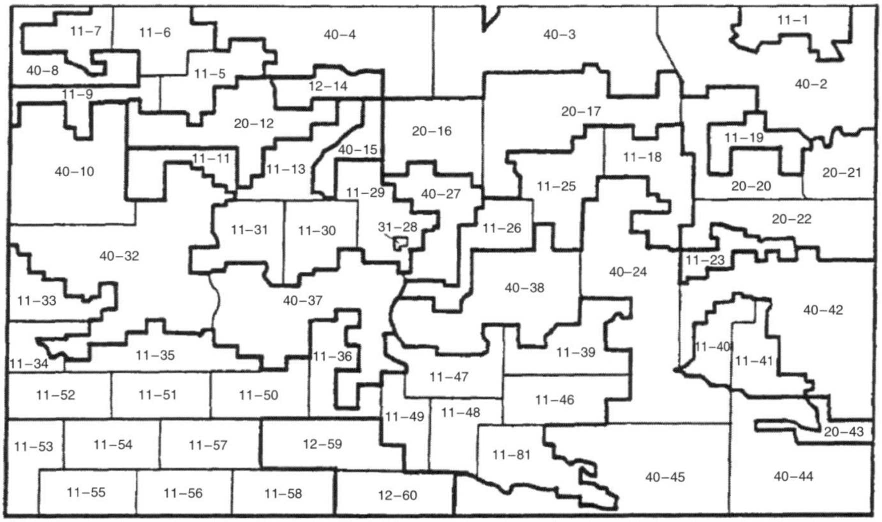

Figure 11.1 PSU ordering in a serpentine manner.

11.4 Sub-stratification

There is a further level of stratification which is applied to the frame. Sub-stratification is the process used to divide the population of sampling units within each stratum equally into categories (substrata). These substrata do not have a definition associated with them like strata do (e.g. 50% or more cultivated). Sampling units are placed into substrata based on likeness of agricultural content and, to a certain extent, location. Sub-stratification activities include ordering the PSUs, ordering the counties, calculating the number of sampling units in the strata, determining the number of substrata, and placing the sampling units into substrata.

Recall in Section 11.3 that when the stratifier completes stratification for a county, the PSUs within the county are ordered automatically in ArgGIS9. The PSUs are numbered beginning in the upper right-hand corner and winding through the county in a serpentine fashion. This ordering plays a role in creating the substrata. Once stratification is complete and all PSUs within each county are ordered, the counties are ordered by an area frame statistician. This county ordering is based on a multivariate cluster analysis of county level crop and livestock data. The purpose of cluster analysis is to group counties into clusters or groups which generally have the same overall agricultural make-up.

Figure 11.2 exhibits the county ordering used in the Pennsylvania area sampling frame. Note that in all but one instance, the ordering proceeds from one county into an adjacent county. The reason for the exception along the southern border of the state is that Somerset County is more agriculturally similar to Fulton County than the adjacent Bedford County. The county ordering need not be continuous. If the counties in one corner of the state are very similar to those in another corner, the ordering can skip across several counties. The starting point of the ordering is somewhat arbitrary, so a logical starting point would be any corner of the state. However, if the cluster analysis indicates a clear distinction between two groups of counties, it may be advantageous to start in one area and end in the other.

The county ordering 'links' the PSU ordering within each county together. In the example above, the PSU ordering for the state begins with the PSU ordering in the first

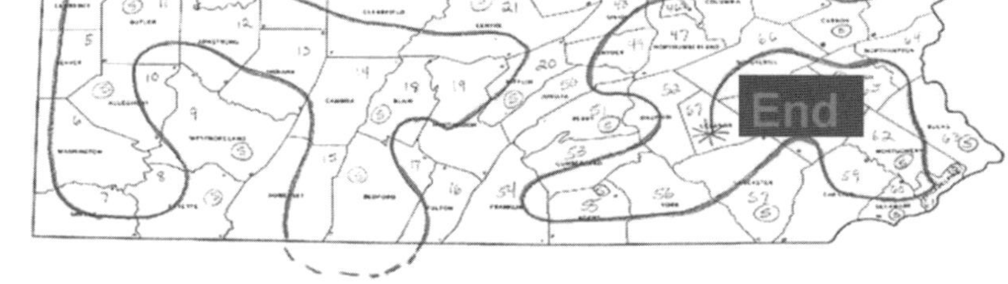

Figure 11.2 County ordering used for the Pennsylvania area frame.

county, Erie County. The PSUs ordered in the second county, Crawford county, go next in the ordering, and so on. When the ordering 'enters' a county from the west or the south, the order of the PSUs in the county is reversed. PSUs within a county are ordered by arbitrarily starting in the northeast corner of the county. Therefore, reversing the order will ensure a fairly continuous ordering of PSUs from one county to the next.

Since sampling units (not PSUs) are placed into substrata, the population of sampling units needs to be calculated for each stratum. Only PSUs that are chosen for the sample are physically broken down into sampling units or segments. However, the number of potential sampling units must be determined for all PSUs in the population in order to calculate the population of sampling units. The number of sampling units varies from PSU to PSU depending on the PSUs size. The number of sampling units for any given PSU in the strata is

$$N_{ik} = \frac{S_{ik}}{S_i},$$

where N_{ik} is the number of potential sampling units in the kth PSU from the ith land-use stratum, rounded to the nearest whole number, S_{ik} is the size of the kth PSU from the ith land-use stratum, and S_i is the target size of sampling units in the ith land-use stratum.

For example, if a PSU is 8.3 square miles in size, and the target segment size is 1.0 square miles, then the number of potential sampling units that could be delineated from the PSU would be eight. The number of potential sampling units is determined for all PSUs in the population of each land-use stratum. Then the total number of sampling units is

$$N_i = \sum_{i=1}^{k} N_{ik},$$

where N_i is the number of sampling units in the ith land-use stratum, and N_{ik} is the number of sampling units in the kth PSU from the ith land-use stratum.

After the population of segments has been determined for each stratum, the number of substrata for each land-use stratum is established. Several factors are considered in the determination, including experience with sampling frames in other states, the number of sample segments and replicates within each stratum and the degree of homogeneity among the sampling units within the various strata. Generally, the higher the intensity of cultivation and variation in crops, the higher the number of substrata relative to the sample size.

Table 11.4 is the sample design for Pennsylvania. The land-use stratum definition and target segment size are determined in the pre-construction phase. The number of sampling units for each stratum is calculated after stratification is complete. The sample size is determined by the sample allocation.

The NASS employs a concept called replicated sampling which provides several key benefits in the estimation process (described in Section 11.5). Approximately 20% of the replicates are rotated out of the sample each year with new replicates taking their place. The number of substrata is the sample size divided by the number of replicates, as is illustrated in Table 11.4.

In this example, there are 2800 sampling units in stratum 13. So 700 sampling units will go into each of the four substrata. The first 700 sampling units in the ordering within the stratum will go into substratum 1. The next 700 sampling units in the ordering within the stratum will go into substratum 2, and so on. In many cases the substrata

Table 11.4 Pennsylvania sample design showing the number of substrata and replications.

Stratum	Land-use stratum	Target segment size (sq mi)	Number of sampling units	Number of substrata	Number of replications	Sample size
13	>50% cult.	1.00	2800	4	6	24
20	15–49% cult.	1.00	17084	14	7	98
31	Ag-urban	0.25	8284	1	5	5
32	Commercial	0.10	1814	1	2	2
40	<15% cultivated	2.00	11344	8	6	48
50	Non-ag	PPS	45	1	2	2

'break' will 'split' the last PSU (some sampling units will be in one substratum, the rest in the next substratum). Each substratum will contain the same number of sampling units, except the last, which may contain slightly more or fewer than the others due to rounding. For example, the 17 084 sampling units in stratum 20 are divided into 14 substrata. The first 13 substrata will contain 1220 units and the last will contain 1224.

Sub-stratification is implemented to reduce variability in sampling units. The land-use stratification is based on the percentage of cultivation. Therefore, while the majority of the segments within a stratum may be intensely cultivated, the agricultural make-up of the segments may differ depending on the location of the segments within the state. Ordering the population of PSUs according to agricultural content will yield greater precision in the estimates for individual commodities. Sub-stratification is particularly effective in areas of intensive cultivation where cropland content varies across the state. Utilizing substrata in grazing or range strata contributes very little to reducing variance except possibly for cattle. Therefore, more substrata are used in the intensely cultivated strata as compared to the range or lightly-cultivated strata.

11.5 Replicated sampling

NASS's area frames have been sampled using a replicated design since 1974. Replicated sampling is characterized by the selection of a number of independent subsamples or replicates from the same population using the same selection procedure for each replicate. Each replicate is therefore an unbiased representation of the population.

A replicate for NASS's area frame sample design is a random sample of land areas (segments) selected within a land-use stratum. The sub-stratification within each land-use stratum has been incorporated into the sampling process to improve the sampling efficiency and the sample dispersion. Therefore, a replicate is more specifically defined as a simple random sample of one segment from each substratum in a land-use stratum.

The first segment randomly selected in each substratum in a land-use stratum is designated as replicate 1, the second segment selected from each substratum is designated as replicate 2, and so forth. The number of replicates is the same for each substratum in a given land-use stratum. Therefore, the number of sample segments in a land-use stratum is simply the product of the number of replicates and the number of substrata in the land-use stratum,

$$n_i = r_i \cdot s_i,$$

where n_i is the number of segments in the sample for the ith land-use stratum, r_i is the number of replicates for each substratum in the ith land-use stratum and s_i is the the number of substrata in the ith land-use stratum.

Suppose, for example, we want to select a replicated sample of two replicates from a land-use stratum consisting of three substrata with ten segments in each substratum. Then the total sample size for the land-use stratum would be $n_i = r_i \cdot s_i = 2 \cdot 3 = 6$ segments, as illustrated in Table 11.5. Notice that a simple random sample of one segment is selected

Table 11.5 Replicated sampling process for a land-use stratum.

		Replicate	
Substratum	Segment	1	2
1	1		
	2		
	3		
	4		
	5		
	6		×
	7		
	8	×	
	9		
	10		
2	11		
	12		
	13		
	14		
	15		
	16		
	17		
	18		×
	19	×	
	20		
3	21		
	22		
	23		
	24		
	25		
	26	×	
	27		
	28		×
	29		
	30		

in each substratum for a replicate so that the number of sample segments in a replicate is simply the number of substrata.

The number of replicates certainly does not need to be the same in each substratum. Sometimes it may be advantageous to vary the number of replicates in the substrata for a land-use stratum. For example, if a crop is localized to a few counties in a state and greater precision is desired for data pertaining to this crop, then the sampling variance could be reduced for this crop by increasing the number of replicates in the substrata corresponding to these counties.

There are six reasons why the NASS uses replicated sampling:

1. *Sample rotation.* A sample rotation scheme is used to reduce respondent burden caused by repeated interviewing, avoid the expense of selecting a completely new area sample each year, and provide reliable measures of change in the production of agricultural commodities from year to year through the use of the ratio estimator. Sample rotation is accomplished each year by replacing segments from specified replicates in each land-use stratum with newly selected segments. Approximately 20% of the replicates in each land-use stratum are replaced annually. The sample design does not rotate exactly 20% of the segments because the number of replicates is not always a multiple of 5. To illustrate how replicated sampling simplifies the sample rotation process, Table 11.6 shows the numbering scheme for a hypothetical land-use stratum with five replicates in each of eight substrata. The first digit in the five-digit segment number represents the year the segment rotated into the sample, e.g. 50001 entered in 2005. The remaining four digits are simply unique numbers. The sample rotation in 2010 will be performed by replacing the segments in the 50000 series (replicate 1), which have been in the sample for 5 years, with segments numbered 00001, 00002,. . ., 00008. In 2011, the segments will be replaced from replicate 2 since the 60000 series would have completed its five-year sample cycle. In 2012, the segments from replicate 3 will be replaced and so forth.

2. *Methodology research.* Replicated sampling provides the capability to test alternative survey procedures or evaluate current methodology since different replicates can be assigned to the research and operational methods. For example, if there are a total of ten replicates in a land-use stratum and there is a need to compare two

Table 11.6 Replicated sampling process for a land-use stratum.

	Replicate				
Substratum	1	2	3	4	5
1	50001	60009	70017	80025	90033
2	50002	60010	70018	80026	90034
3	50003	60011	70019	80027	90035
4	50004	60012	70020	80028	90036
5	50005	60013	70021	80029	90037
6	50006	60014	70022	80030	90038
7	50007	60015	70023	80031	90039
8	50008	60016	70024	80032	90040

approaches to asking a particular question, then five replicates could be assigned to each method. The test statistic could then be easily derived using the means or totals from each replicate for each approach. Some examples of survey procedures that might be tested are different questionnaire designs and alternative interviewing approaches.

3. *Quality assurance.* Replication also facilitates quality assurance analysis by allowing data comparisons among years in order to determine if significant differences in survey processes exist over time. For example, segment sizes can readily be compared among replicates to determine if the average size and the variability in size differ significantly from year to year. If so, this may indicate that the manual procedures for delineating segments (to be discussed later) need to be reviewed.

4. *Sample management.* Replication allows easy management of the sample due to the replicate numbering scheme. This simplifies the process of designating a subsample of segments for one-time or repetitive surveys, increasing or decreasing the sample size in a land-use stratum to improve sampling efficiency, and identifying segments to be rotated out of the area frame sample. For example, replicates are added every 5 years for the ACES survey to estimate the completeness of the Census mail list.

5. *Variance estimation.* Replicated sampling provides a simple, unbiased method for estimating the sampling variance using replicate means or totals. The NASS estimates the sampling variance for agricultural surveys using the sub-stratification design rather than replicate totals. However, replicate totals are sometimes used for variance and covariance estimation to simplify multivariate statistical analysis in research studies. The benefit of using replicate totals to estimate the sampling variance is most pronounced in underdeveloped countries where a computer facility or the necessary statistical software is not available.

6. *Rotation effects.* Replication readily provides the vehicle for evaluating sample rotation effects. Rotation effects are defined as the impact on survey data resulting from the number of years a segment has been in the sample. The NASS has a five-year rotation process which permits replicate totals to be compared for segments in the sample from one to five years.

11.6 Sample allocation

The area frame sample is used to collect data on a wide range of agricultural items such as crop acreages, livestock inventories and economic data. Therefore, the allocation of the sample across states and within states to the land-use strata is extremely important. The NASS evaluates optimum allocations of the sample to obtain the most precision in the major survey estimates for a given budget. The number of sample segments allocated to each land-use stratum and state depends on factors such as the average data collection cost per segment in each stratum, the variability of the data in each stratum resulting from the intensity and diversity of agriculture, the total number of segments or land area in each stratum, and the importance of the state's agriculture relative to the national agricultural statistics programme.

An optimum sample allocation to the land-use strata is generated for each of the most important agricultural survey items (univariate) and for all of the important commodities considered simultaneously (multivariate). These important commodities include corn, soybeans, cotton, winter wheat, spring wheat, durum wheat, number of farms, and NOL cattle. The allocations are evaluated not only from an area frame perspective but also from a multiple frame point of view where the area frame is used to measure the incompleteness in the list frame. Finally, optimum allocations are conducted at the national, regional, and state levels to assess the allocations at the various inference levels.

The NASS places the most importance on the multivariate optimum allocation for the area frame non-overlap estimates at the state level since it is important to provide useful statistics at the state level. Adjustments are made to this sample allocation to improve the precision of the regional and national estimates without seriously hindering the precision levels for the states. Minor adjustments to the optimum allocation are also made to provide a multiple of five replicates in each stratum to simplify the sample rotation process and to protect against the impact of outliers by not allowing the sampling rate to be too small in a stratum, e.g. 1 in 750 segments.

The optimum allocation of a sample for multi-purpose surveys can be viewed as a problem in convex programming. An iterative, nonlinear programming algorithm is used to provide the univariate and multivariate optimum sample allocations for the area frames. The algorithm is guaranteed to converge to the optimum solution. A brief description of the multivariate sample allocation model follows.

Suppose each of the j survey items, $1 \leq j \leq p$, from the p selected survey items must satisfy the constraint

$$\operatorname{var}(\hat{Y}_j) \leq v_j,$$

where $\operatorname{var}(\hat{Y}_j)$ is the estimated sampling variance for the jth survey total, and v_j is the desired or target sampling variance for the jth survey total. Assume the cost function

$$C(x) = \sum_{i=1}^{l} a_{ij} c_i n_i = \sum_{i=1}^{l} a_{ij} \frac{c_i}{x_i},$$

where c_i is the average cost per segment in the ith land-use stratum, n_i is the number of sample segments in the ith land-use stratum, l is the number of land-use strata, and $x_i = 1/n_i$ with $n_i \geq 1$. The problem then reduces to minimizing the cost function subject to the constraints

$$\sum_{i=1}^{l} a_{ij} x_i \leq 1, \qquad 1 \leq j \leq p,$$

where $a_{ij} = N_i^2 s_{ij}^2 / (v_j + \sum_{i=1}^{l} N_i s_{ij}^2)$, s_{ij}^2 is the square of the standard deviation for the jth survey item in the ith land-use stratum and N_i is the number of segments in the ith land-use stratum. The nonlinear algorithm iteratively finds the intersection between $A_k = \{x : C(x) = k\}$ for fixed values of k, and $F = \{x : a_j' x \leq 1\}$. The intersection is the optimal solution. Experience has shown that the program converges rapidly to the optimal solution.

Given this allocation model, the input for the model is generated as follows:

- The average cost per segment for each land-use stratum, c_i, is estimated by having the interviewers keep time records during field work.

Table 11.7 Number of segments in the area frame sample, 2008.

State	Number of segments	State	Number of segments
Alabama	236	Nebraska	473
Arizona	118	Nevada	26
Arkansas	342	New Hampshire	10
California	404	New Jersey	48
Colorado	267	New Mexico	124
Connecticut	8	New York	96
Delaware	23	North Carolina	319
Florida	100	North Dakota	420
Georgia	290	Ohio	220
Idaho	148	Oklahoma	335
Illinois	401	Oregon	194
Indiana	264	Pennsylvania	179
Iowa	452	Rhode Island	8
Kansas	487	South Carolina	119
Kentucky	189	South Dakota	395
Louisiana	249	Tennessee	334
Maine	32	Texas	1120
Maryland	61	Utah	69
Massachusetts	12	Vermont	21
Michigan	145	Virginia	179
Minnesota	393	Washington	267
Mississippi	298	West Virginia	66
Missouri	383	Wisconsin	219
Montana	316	Wyoming	53

- The population counts that are calculated after the stratification process.
- The desired sampling variance for the estimated total of each item, v_j, is established by the AFS after consultation with others in the NASS.
- The square of the standard deviation, s_{ij}^2, for the jth item in the ith land-use stratum is estimated using the previous two years' survey data.

The area frame sample allocations among and within states are evaluated periodically to determine if a reallocation of the sample is worthwhile. The sample allocations among the 48 states for 2008 are shown in Table 11.7.

11.7 Selection probabilities

There are two methods for selecting the ultimate sampling unit or segment – equal and unequal selection. Which method is used depends on the availability of adequate boundaries for segments. If good boundaries are plentiful so that segments can be made approximately the same size within a land-use stratum, then segments are selected with equal probability. If adequate boundaries are not available, then unequal probability of selection is used since segment sizes are allowed to vary greatly in order to ensure easily identifiable segment boundaries.

The use of unequal selection probabilities is restricted to the non-agricultural stratum in area frames developed since 1985 and to some open land strata (40s strata) in some western states. In all other land-use strata in the USA, equal probability of selection is used. About 96% of the approximately 11 000 segments in the area frame sample are selected based on the equal probability of selection method.

The probability expressions for equal and unequal probability of selection will now be derived in the context of the NASS's area frame design. These expressions provide the statistical foundation for area frame sampling.

11.7.1 Equal probability of selection

A two-step procedure is used to select sample segments from the selected PSUs when selection probabilities are equal. Recall that the number of segments delineated within the selected PSU depends on the size of a PSU. The number of segments in a PSU is simply the total area of the PSU divided by the target (desired) segment size for the land-use stratum in which the PSU has been stratified. This quotient is rounded to the nearest integer since fractional segments are not allowed. For example, if a PSU in an intensively cultivated stratum is 7.1 square miles and the target segment size is 1.0 square mile, then the number of segments for the PSU is seven.

1. A sample of PSUs ia selected within each substratum in a given land-use stratum. Selection is done randomly, with replacement, with probability proportional to the number of segments in the PSU. That is, the probability of selecting the kth PSU in the jth substratum from the ith land-use stratum is

$$P(A_{ijk}) = \frac{N_{ijk}}{N_{ij}},$$

where A_{ijk} is the kth PSU in the jth substratum from the ith land-use stratum, N_{ijk} is the number of sampling units (segments) in the kth PSU from the jth substratum in the ith land-use stratum, and N_{ij} is the number of sampling units (segments) in the jth substratum from the ith land-use stratum.

2. After the sample of PSUs is drawn, each selected PSU is divided into the required number of segments. This step involves randomly selecting a segment with equal probability from the selected PSU. That is, the probability of selecting the mth segment given that the kth PSU was selected from the jth substratum in the ith land-use stratum is

$$P(B_{ijkm}|A_{ijk}) = \frac{1}{N_{ijk}},$$

where B_{ijkm} is the mth segment in the kth PSU from the jth substratum and the ith land-use stratum. Therefore, the unconditional probability of selecting the mth segment in the kth PSU from the jth substratum in the ith land-use stratum is

$$P(B_{ijkm}) = P(A_{ijk}P(B_{ijkm}|A_{ijk}) = \frac{N_{ijk}}{N_{ij}} \cdot \frac{1}{N_{ijk}} = \frac{1}{N_{ij}}.$$

Therefore, all sampling units within a given substratum in a land-use stratum have an 'equal' probability of selection using the two-step selection procedure. This

Table 11.8 Selection probabilities for the two-step procedure.

PSU	Number of segments in PSU	$P(A_{ijk})$	$P(B_{ijkm} \mid A_{ijk})$	$P(B_{ijkm})$
1	2	2/40	1/2	1/40
2	3	3/40	1/3	1/40
3	5	5/40	1/5	1/40
4	6	6/40	1/6	1/40
5	7	7/40	1/7	1/40
6	8	8/40	1/8	1/40
7	9	9/40	1/9	1/40

fact is illustrated in Table 11.8 for a hypothetical substratum with seven PSUs. This table shows the number of required segments in each PSU, the probability of selecting each PSU, $P(A_{iik})$, the probability of selecting a segment given the PSU was selected, $P(B_{ijkm}|A_{ijk})$, and the unconditional probability of selecting a segment in the PSU, $P(B_{ijkm})$. Notice that the unconditional selection probability is the same for all segments, as previously stated.

11.7.2 Unequal probability of selection

PSUs are selected with unequal probability in less cultivated strata (40s strata) in some western states and in the non-agricultural stratum (stratum 50) for states receiving a new area frame since 1985. This type of selection is performed because adequate boundaries are not available in these areas to draw off segments of approximately the same size. The probability of PSU selection in these strata is proportional to its size (PPS). In PPS strata, the PSUs are not further broken down into segments. Therefore, the PSU and segment are synonymous. The probability of selecting the kth PSU in the jth substratum from the ith land-use stratum is

$$P(A_{ijk}) = \frac{S_{ijk}}{S_{ij}},$$

where A_{ijk} is the kth PSU in the jth substratum from the ith land-use stratum, S_{ijk} is the size (in acres) of the kth PSU in the jth substratum from the ith land-use stratum, and S_{ij} is the size (in acres) of the jth substratum in the ith land-use stratum.

The selection probabilities for all situations encountered during the sampling process have now been formulated. The expansion factor or weight assigned to each segment to expand the survey data to population totals is derived from these selection probabilities. The expansion factor for a segment in a substratum is simply the inverse of the product of the probability of selection for the segment and the number of segments in the sample for the substratum,

$$e_{ijm} = \frac{1}{p_{ijm} n_{ij}},$$

where e_{ijm} is the expansion factor for the mth segment in the jth substratum and the ith land-use stratum, p_{ijm} is the probability of selecting the mth segment in the jth substratum from the ith land-use stratum, n_{ij} is the number of segments or replicates in the sample for the jth substratum in the ith land-use stratum.

11.8 Sample selection

The procedures used to select the area frame samples will be described in this section for the equal and unequal probability of selection methods.

11.8.1 Equal probability of selection

Recall that a two-step selection procedure is followed when segments are selected with equal probability. The first step is PSU selection. An SAS program is run which uses the selection probabilities discussed in the previous section to select the chosen PSUs. The program creates a listing of all chosen PSUs.

Personnel in the sample selection unit break down the chosen PSUs that have equal probability of selection into segments in ArcGIS9. NAIP photography is used because it provides valuable detail in terms of land use and availability of boundaries. Three criteria are followed when delineating segments using aerial photography in order to control the total survey error (non-sampling errors and sampling variability):

- Use the most permanent boundaries available for each segment so that reporting problems during the data collection phase caused by ambiguous boundaries will be minimized.
- Create segments that are as homogeneous as possible with respect to agricultural content. Since crop types are generally not distinguishable on the aerial photography, homogeneity is usually based on the amount of cultivated land. This criterion reduces the sampling variability among segments in a given substratum.
- Choose boundaries so that the size of each segment is as close to the target segment size as practical. Deviations from the target size as large as 25% are permitted to satisfy the first two criteria. This criterion, like the second, helps control sampling variability.

After the required number of segments has been delineated for a selected PSU, the segments are automatically numbered in ArcGIS9. Then one segment is chosen at random also in ArcGIS9.

11.8.2 Unequal probability of selection

Recall that PSUs selected with unequal probability in less cultivated strata (40s strata) in some western states and in the non-agricultural stratum (stratum 50) because adequate boundaries are not available in these areas to draw off segments of approximately the same size. The probability of PSU selection in these strata is proportional to its size (PPS). PSUs in PPS strata vary in size and are not broken down further. Because the PSU and segment are one in the same, the sample selection unit reviews the boundaries identified by the stratification unit.

Once the chosen PSUs with equal probability of selection are broken down into segments and the boundaries of chosen PSUs with unequal probability of selection are reviewed, the sample is prepared for data collection. A $24' \times 24'$ image of each segment ($8'$ = one mile scale) is printed onto Kodak photographic paper. This printout is used to collect data for all land inside the segment boundary.

11.9 Sample rotation

As mentioned earlier, the NASS uses a five-year rotation scheme for the sample segments. Rotation is accomplished by replacing segments from specified replicates within a land-use stratum with newly selected segments. Preferably, the number of replicates is a multiple of 5 to provide a constant workload for sample selection and preparation activities in the AFS and for data collection work in the state offices. Naturally, instances occur when the number of replicates is not a multiple of 5, especially in urban, commercial, and non-agricultural strata where the sample size is small (usually two replicates).

Table 11.9 illustrates how the replicates are rotated over a five-year cycle (2008–2012) for different numbers of replicates. If a land-use stratum has two replicates, the segments in replicate 1 will be replaced with all new segments in 2010 and will stay in the sample for 5 years. In 2015, the segments in replicate 1 will be replaced again. Likewise, new segments will rotate into replicate 2 in 2011 and 2016. No segments rotate into or out of the sample in the years in between (2012, 2013, 2014). If a stratum has five replicates, then the segments in one replicate are replaced with new segments each year which is 20% of the sample.

All segments are not in the sample exactly 5 years as has been implied. Segments from the first and last four years of an area frame's life are in the sample less than 5 years, as shown in Table 11.10. This table presents the rotation cycle for an area frame assuming a 20-year life and, for simplicity, five replicates in each land-use stratum.

The national area frame sample size is approximately 11 000 segments. The total number of segments rotated each year is approximately 3000. This results from an average of 800 segments being selected for new area frames (except in census years when no states receive new frames) and about 2200 segments being selected based on a 20% rotation of the remaining 15 000 or so segments. Therefore, approximately 27% of the national area frame sample is based on newly selected segments each year.

Table 11.9 Rotation of replicates depending upon the number of replicates.

	Year				
Number of replicates	2008	2009	2010	2011	2012
2			1	2	
3			1	2	3
4	4	1		2	3
5	4	5	1	2	3
6	4	5,6	1	2	3
7	4,7	5,6	1	2	3
8	4	5	1,6	2,7	3,8
9	4,9	5	1,6	2,7	3,8
10	4,9	5,10	1,6	2,7	3,8
11	4,9	5,10	1,6,11	2,7	3,8
12	4,9	5,10	1,6,11	2,7,12	3,8
13	4,9	5,10	1,6,11	2,7,12	3,8,13
14	4,9,14	5,10	1,6,11	2,7,12	3,8,13
15	4,9,14	5,10,15	1,6,11	2,7,12	3,8,13

Table 11.10 Rotation cycle for a 20-year period assuming five replicates in the stratum.

Year	Replicates																				
2009	1	2	3	4	5																
2010		2	3	4	5	1															
2011			3	4	5	1	2														
2012					4	5	1	2	3												
2013					5	1	2	3	4												
2014						1	2	3	4	5											
2015							2	3	4	5	1										
2016								3	4	5	1	2									
2017									4	5	1	2	3								
2018										5	1	2	3	4							
2019											1	2	3	4	5						
2020												2	3	4	5	1					
2021													3	4	5	1	2				
2022														4	5	1	2	3			
2023															5	1	2	3	4		
2024																1	2	3	4	5	
2025																	2	3	4	5	1
2026																		3	4	5	1	2
2027																			4	5	1	2
2028																				5	1	2

11.10 Sample estimation

This final section will briefly discuss the approaches used to estimate agricultural production with an area frame sample of segments. The NASS uses two area frame estimators, namely the closed and weighted segment estimators. Both require that the interviewer collect data for all farms that operate land inside each segment. (A farm is defined to be all land under one operating arrangement with gross farm sales of at least $1000 a year.) The portion of the farm that is inside the segment is called a tract. The interviewer draws the boundaries of each tract on the photo enlargement, accounting for all land in the segment.

When an interviewer contacts a farmer, the closed segment approach requires that the interviewer obtain data only for that part of the farm within the tract. For example, the interviewer might ask about the total number of hogs on the land in the tract. The most common uses of the closed segment estimator are to estimate crop acreages and livestock inventories. An interviewer accounts for all land in each tract by type of crop or use and for all livestock in the tract. The main disadvantage of the closed segment estimator arises when the farmer can only report values for the farm rather than for a tract which is a subset of the farm. For example, 'How many tractors do you own?' can only be answered on a farm basis. Thus, the closed segment estimator is not applicable for many agricultural items. Economic items and crop production are two major examples which farmers find difficult or impossible to report on a tract basis.

The weighted segment estimator, by contrast, does not have this limitation. It can be used to estimate all agricultural characteristics, which is a major advantage for this estimator. The weighted segment approach requires that the interviewer obtain data on the entire farm. For example, the interviewer would ask about the total number of hogs on all land in the farm. Using the weighted segment approach, the interviewer obtains farm data for each tract, but these farm data are weighted. The weight used by the NASS is the ratio of tract acres to farm acres.

Suppose the following situation occurs for a specific farm: tract acres = 10, farm acres = 100, hogs on the tract = 20, and hogs on the farm = 40. The closed segment value of number of hogs would be 20, and the weighted segment value would be $40 \cdot 10/100 = 4$.

When estimating survey totals and variances for these estimators, segments can be treated as a stratified sample with random selection within each substratum. The formulas for each of the three estimators can be described by the following notation. For some characteristic, Y, of the farm population, the sample estimate of the total for the closed segment estimator is

$$\hat{Y}_c = \sum_{i=1}^{l} \sum_{j=1}^{s_i} \sum_{k=1}^{n_{ij}} e_{ijk} y_{ijk},$$

where l is the number of land-use strata, s_i is the number of substrata in the ith land-use stratum, n_{ij} is the number of segments sampled in the jth substratum in the ith land-use stratum, e_{ijk} is the expansion factor or inverse of the probability of the selection for the kth segment in the jth substratum in the ith land-use stratum,

$$y_{ijk} = \begin{cases} \sum_{i=1}^{f_{ijk}} t_{ijkm} & \text{if } f_{ijk} > 0, \\ 0 & \text{if } f_{ijk} = 0, \end{cases}$$

where f_{ijk} is the number of tracts in the kth segment, jth substratum, and ith land-use stratum, and t_{ijkm} is the tract value of the characteristic Y for the mth tract in the kth segment, jth substratum, and ith land-use stratum.

The weighted segment estimator would also be of the same form,

$$\hat{Y}_w = \sum_{i=1}^{l} \sum_{j=1}^{s_i} \sum_{k=1}^{n_{ij}} e_{ijk} y_{ijk},$$

except that

$$y_{ijk} = \begin{cases} \sum_{i=1}^{f_{ijk}} a_{ijkm} y_{ijkm} & \text{if } f_{ijk} > 0, \\ 0 & \text{if } f_{ijk} = 0, \end{cases}$$

where a_{ijkm} is the weight for the mth tract in the kth segment, jth substratum, and ith land-use stratum.

The following weight is currently in use:

$$a_{ijkm} = \frac{\text{tract acres for the } m\text{th tract}}{\text{farm acres for the } m\text{th tract}}.$$

The precision of an estimate can be measured by the standard error of the estimate. An estimate becomes less precise as the standard error increases. Given the same number of segments to make an estimate, weighted segment estimates are usually more precise than closed segment estimates. For both estimators, the formula for the sampling variance can be written as

$$\text{var}(\hat{Y}) = \sum_{i=1}^{l} \sum_{j=1}^{s_i} \frac{1 - 1/e_{ij}}{1 - 1/n_{ij}} \sum_{k=1}^{n_{ij}} (y'_{ijk} - y'_{ij.})^2,$$

where $y'_{ijk} = e_{ij} y_{ijk}$ and $y'_{ij.} = (1/n_{ij}) \sum_{k=1}^{n_{ij}} y'_{ijk}$. The standard error is then

$$\text{se}(\hat{Y}) = \sqrt{\text{var}(\hat{Y})}.$$

In closing, research into non-sampling errors associated with these estimators has shown that the closed estimator, when applicable, is generally the least susceptible to non-sampling errors. The closed segment estimator is much relied on for NASS's area frame surveys, and the weighted segment estimator is the most used for multiple-frame surveys where the area frame is only used to measure the incompleteness in the list frame.

11.11 Conclusions

The JAS is an annual area frame survey conducted by the NASS to gather agricultural data such as crop acreages, cost of production, farm expenditures, grain yield and production, and livestock inventories. The JAS provides estimates for major commodities, including corn, soybeans, winter wheat, spring wheat, durum wheat, cotton, NOL cattle, and number of farms. The JAS also provides measurement of the incompleteness of the NASS list frame, provides ground truth data to verify pixels from satellite imagery, and serves as a base for the NASS's follow-on surveys. The development and implementation of an area frame requires several steps, including dividing all land into PSUs, classifying the PSUs into strata and substrata, sampling the PSUs, dividing the chosen PSUs into segments, and randomly selecting a segment from each chosen PSU. While area frames can be costly and time-consuming to build and sample, the importance of the results from the JAS justify the effort.

12

Accuracy, objectivity and efficiency of remote sensing for agricultural statistics

Javier Gallego[1], Elisabetta Carfagna[2] and Bettina Baruth[1]

[1]*IPSC-MARS, JRC, Ispra, Italy*

[2]*Department of Statistical Sciences, University of Bologna, Italy*

12.1 Introduction

Agricultural statistics emerged during the 1970s as one of the most promising applications of satellite images. A significant number of papers were published in the 1980s and 1990s on different ways to combine accurate data from a sample of sites (usually by in-situ observation) with less accurate data, covering the region of interest (classified satellite images). The number of satellites and Earth observation (EO) sensors in orbit is growing (Table 12.1), but the efforts to marry image processing with agricultural statistics seem fraught with difficulties, with a few honourable exceptions.

In 2008 the world saw an agricultural commodities crisis with high price volatility and, in particular, soaring food prices. The crisis had multiple causes besides the bad climatic conditions (decreasing worldwide cereal stocks, development of bio-fuels, changes in the diet in emerging countries, expected changes due to the impact of climate change on agriculture, etc.). The crisis increased the awareness that constant agricultural monitoring is important. Remote sensing has an acknowledged potential to support the monitoring system, but has so far played a subordinate role for agricultural statistics

Agricultural Survey Methods Edited by Roberto Benedetti, Marco Bee, Giuseppe Espa and Federica Piersimoni

Table 12.1 Main characteristics of some optical multispectral satellite sensors.

Satellite/sensor	Resolution	Channels	Swath (km)	Year
Ikonos (multispectral)	4 m	4	11	2000
Quickbird (multispectral)	2.44 m	4	16.5	2001
IRS-LISS-IV	6	5	24	2003
SPOT 5 HRG	2.5–10	5	60	2002
SPOT 4 HRVIR	20	4	60	1998
Aster	15–90	14	60	1999
REIS (RapidEye)	5	5	77	2008
CBERS2-HRC	20	5	113	2003
IRS LISS-III	23–70	5	141	2003
Landsat TM-ETM	30 m	6	180	1982+
Landsat MSS	60 × 80 m	4	180	1972
IRS-P6 AwiFS	56–70	4	700	2003
MERIS	300	15	1150	2002
SPOT-Vegetation	1 km	4	2250	2002
MODIS	250–1000 m	36	2330	1999+
NOAA(16-17)-AVHRR3	1 km	6	3000	2000+
Planned launches				
Landsat 8 (LDCM)-OLI	15–60	8	180	2012
SENTINEL	10–60	12	200	2012

within operational services, both for area and yield estimates. The aim of this chapter is to give a description of the current situation, with some reflections on the history of a few major programmes in different parts of the world and on the best practices document of the Global Earth Observation System of Systems (GEOSS, 2009). The focus is on the distinction between approaches that can be considered operational and topics at a research level.

Most of the chapter focuses on the use of remote sensing for crop area estimates. Some indications are given on the accuracy assessment of land use/cover maps that can be used for agricultural information. The last part concentrates on quantitative approaches of crop yield forecasting, giving some examples of operational systems where remote sensing plays a role. Yield forecasting makes a forward estimate of the harvest for the current cropping season.

12.2 Satellites and sensors

The main characteristics of sensors for use in agricultural statistics are as follows:

- *Spectral resolution.* Most agricultural applications use sensors that give information for a moderate number of bandwidths (four to eight bands). Near infrared (NIR), short wave infra-red (SWIR) and red are particularly important for measuring the activity of vegetation; red-edge bands (between red and NIR) also seem to be promising for crop identification (Mutanga and Skidmore, 2004). Panchromatic (black and white) images are very useful for the identification of the shape of fields, but their ability to discriminate crops is limited to cases in which the spatial pattern of individual plants can be caught, generally permanent crops (vineyards, orchards, olive trees). A question that remains open for research is the additional value of hyperspectral sensors (Im and Jensen, 2008; Rama Rao, 2008).

- *Swath and price.* The financial resources devoted to agricultural statistics are often insubstantial. Only systems able to cover large areas at a low price have the chance to become sustainable. The swath is also strongly linked with the frequency of possible image acquisition which may be important in following the phenological cycle of different crops.
- *Spatial resolution.* The appropriate spatial resolution mostly depends on the size of parcels. A useful rule of thumb for images to be suitable for crop area estimation in an agricultural landscape is that most pixels should be pure (i.e. fully inside a plot); only a minority of pixels is shared by several plots. Sub-pixel analysis techniques are available for area estimation, but they have not yet proved to be implementable. Coarse resolution images, in which most pixels are mixed, are mainly used for vegetation monitoring and yield forecasting.

12.3 Accuracy, objectivity and cost-efficiency

In this section we discuss the main properties that we would like to find in an estimation method, in particular for agricultural statistics.

The term *accuracy* corresponds in principle to the idea of small bias, but the term is often used in practice in the sense of small sampling error (Marriott, 1990). Here we use it to refer to the total error, including bias and sampling error. Non-sampling errors are generally more difficult to measure or model (Lessler and Kalsbeek, 1999; Gentle *et al.*, 2006), and remote sensing applications are no exception. In this chapter we give some indications as to how to guess the order of magnitude of non-sampling errors.

Objectivity is also difficult to measure. All methods entail some risk of subjectivity: in an area frame survey with in-situ observations, enumerators may be influenced by the observation rules, in particular for land cover types with a fuzzy definition (grassland, rangeland, natural vegetation, grass or crops with sparse trees, etc.); surveyors may also make systematic mistakes in the recognition of minor crops, and this will result in a biased estimate. Some idea of the non-sampling errors can be obtained by visiting a subsample with independent enumerators. Unlike sampling errors, non-sampling errors often grow with the sample size because controlling the surveyors becomes more difficult.

How biased and subjective are estimates obtained from remote sensing? Satellite images are very objective, but the method of extracting information from them has a certain degree of subjectivity, depending on the method applied for the analysis. We do not aim to give a general answer to this difficult question, but we give indications for a few cases.

Methods combining ground surveys with remote sensing provide a tool to assess the *cost-efficiency* of remote sensing through the comparison of variances (Allen, 1990, Taylor *et al.*, 1997; Carfagna, 2001), although the straightforward comparison can be completed with considerations on the added value of the crop maps produced by image classifications.

A key financial benchmark is the cost of current surveys. For example, FAO (2008) reports an average expenditure per country in Africa in 2007 of US$657 000, including estimation of crop area and yield, means of production and socio-economic information. Remote sensing applications to agricultural statistics can be sustainable in the long term if their total cost can be budgeted without endangering the feasibility of surveys that cannot be substituted by satellite technology. With this narrow margin for the cost-efficiency of EO approaches, the free data policy that is being adopted for some medium-high resolution images (LDMC-OLI or CBERS-HRC) seems essential. In Europe the financial margin is wider; for example, the Spanish Ministry of Agriculture spends €1.3 million per year on an area frame survey with in-situ observation of around 11 000 segments of 49 ha.

12.4 Main approaches to using EO for crop area estimation

Carfagna and Gallego (2005) give a description of different ways to use remote sensing for agricultural statistics; we give a very brief reminder.

Stratification. Strata are often defined by the local abundance of agriculture. If a stratum can be linked to one specific crop, the efficiency is strongly boosted (Taylor *et al.*, 1997).

Pixel counting. Images are classified and the area of crop c is simply estimated by the area of pixels classified as c. An equivalent approach is photo-interpretation and measurement of the area classified into each crop or land cover category. The equivalent for fuzzy or sub-pixel classification on coarse resolution images is to sum the area estimated for crop c in each pixel.

Pixel counting or similar estimators have no sampling error if the image coverage is complete, but they have a bias that is approximately the difference between the error of commission ϕ and the error of omission ψ of the image classification (Sielken and Gbur, 1984). The relative bias b_c for crop c due to misclassification can be written as

$$b_c = \frac{\lambda_{+c} - \lambda_{c+}}{\lambda_{c+}} = \phi_c \frac{\lambda_{+c}}{\lambda_{c+}} - \psi_c, \tag{12.1}$$

where λ_{+c} is the area classified as crop c and λ_{c+} is the area of c in ground observations.

Since there is no particular reason for compensation between errors of commission and omission, the potential risk of bias is of the same order of magnitude as the classification error. An illustration of the potential bias of pixel counting is given by Hannerz and Lotsch (2008), who compare the total agricultural area in Africa identified by six land cover maps obtained by image classification; the total agricultural area derived from each of the maps ranges between 644 000 km^2 for the MODIS land cover map created by Boston University (Friedl *et al.*, 2002) and the 3 257 000 km^2 of GLCC IFPRI (Global Land Cover Characterization, International Food Policy Research Institute: Wood *et al.*, 2000). The most consistent figures are found for Egypt, for which the agricultural area mapped ranges between 22 000 and 37 000 km^2. The main source of these extreme differences is probably the coarse resolution images used for global land cover maps, inadequate for the size of African agricultural fields.

12.5 Bias and subjectivity in pixel counting

In most classification algorithms the analyst can tune parameters to obtain a more balanced classification. For example, in the traditional maximum likelihood supervised classification, a priori probabilities can be adjusted to obtain a higher or lower number of pixels in each class. Tuning parameters can reduce the bias in pixel counting, but introduces a certain margin for subjectivity. Other algorithms, for example the decision tree classifier implemented in See5 (Quinlan, 1996), can be seen as 'black boxes'; the user provides input data but cannot explicitly tune parameters. However, changing the proportion of classes in the input data will change the number of pixels in each output class. In practice, the bias of pixel counting can become a risk of subjectivity. If users disregard the issue of objectivity-subjectivity, they can be happy because the estimated areas correspond to expected values, but the algorithm becomes somewhat less interesting.

12.6 Simple correction of bias with a confusion matrix

A reasonable reaction to the risk of bias may be to compute a confusion matrix and correct the pixel counting estimates with the difference between errors of commission and omission. However, some care is needed to ensure that the estimated confusion matrix is unbiased. To illustrate the possible effects of incorrect estimation of the confusion matrix, we give an example using data from a photo-interpretation carried out for the stratification of the LUCAS survey (Chapter 10 of this book). Table 12.2 shows the raw confusion matrix between photo-interpretation and ground observations in LUCAS. This table simply reports the number of points which have been photo-interpreted as c and belong to class c' in the ground survey. It does not take into account that the non-agricultural strata have been subsampled at a rate 5 times lower than the agricultural strata. Let us consider the 'forest and woodland' class. The raw (apparent) error of commission is 16.4% and the raw error of omission is 28.5%. We could try to compensate by using equation (12.1), that is, by multiplying the pixel counting estimate by $\lambda_{c+}/\lambda_{+c} = 22\,735/19\,980 = 1.138$. Let us look now at Table 12.2b, in which the cells have been weighted with the inverse of the sampling probability to obtain an unbiased estimate of the confusion matrix. The error of commission is now 21% while the error of omission is 8%. Therefore it would be more correct to multiply the pixel counting estimate by $\lambda_{c+}/\lambda_{+c} = 88\,471/99\,900 = 0.886$. Thus using the unweighted confusion matrix would lead us, in this example, to a multiplicative bias of $1.138/0.886 = 1.285$, that is, an overestimation by approximately 28.5%. This example gives a simplified illustration of the perverse effects of an incorrect estimation of the confusion matrix if the sampling plan is not taken into account. If there is no sampling plan, the correction becomes much more difficult.

12.7 Calibration and regression estimators

Calibration and regression estimators combine more accurate and objective observations on a sample (e.g. ground observations) with the exhaustive knowledge of a less accurate

Table 12.2 Unweighted and weighted (unbiased) confusion matrix in LUCAS photo-interpretation for stratification.

(a) Unweighted

	Ground observations					
Photo-interpretation	Arable	Permanent Crops	Permanent Grass	Forest & Wood	Other	Total
Arable land	67313	1751	17597	2035	2760	91456
Permanent crops	651	9516	546	573	287	11573
Permanent grass	4940	658	26969	3693	4244	40504
Forest & wood	308	185	1962	16248	1277	19980
Other	195	47	299	186	2925	3652
Total	73407	12157	47373	22735	11493	167165
Error of commission (%)	**32.9**	**16.9**	**28.6**	**16.4**	**6.3**	
Error of omission (%)	**8.3**	**21.7**	**43.1**	**28.5**	**74.7**	

(b) Weighted

	Ground observations					
Photo-interpretation	Arable	Permanent Crops	Permanent Grass	Forest & Wood	Other	Total
Arable land	67313	1751	17597	2035	2760	91456
Permanent crops	651	9516	546	573	287	11573
Permanent grass	4940	658	26969	3693	4244	40504
Forest & wood	1540	925	9810	81240	6385	99900
Other	975	235	1495	930	14625	18260
Total	75419	13085	56417	88471	28301	261693
Error of commission (%)	**32.0**	**15.7**	**24.0**	**21.1**	**12.8**	
Error of omission (%)	**10.7**	**27.3**	**52.2**	**8.2**	**48.3**	

or less objective source of information, or co-variable (classified images). A characteristic property of these estimators is that they do not inherit the bias of the co-variable; therefore they can be used to correct the bias due to pixel counting.

The suggestion given above of correcting estimates using the difference between the errors of commission and omission can be seen as a primitive approach to a calibration estimator. There are two main types of calibration estimators, often somewhat confusingly referred to as 'direct' and 'inverse' (for a discussion see Gallego, 2004):

$$\hat{\lambda}_{dir}(g) = P_g \Lambda_c, \qquad \hat{\lambda}_{inv}(g) = (P_c')^{-1} \Lambda_c,$$

where Λ_c is the column vector with the number of pixels classified into each class c, $P_g(g, c) = \lambda_{gc}/\lambda_{g+}$ and $P_c(g, c) = \lambda_{gc}/\lambda_{+c}$ for the sample.

The regression estimator (Hansen *et al.*, 1953; Cochran, 1977) has been used for crop area estimation since the early days of satellite EO (Hanuschak *et al.*, 1980):

$$\bar{y}_{reg} = \bar{y} + b(\bar{X} - \bar{x}),$$

where $\bar{y}$ and $\bar{x}$ are the sample means of the ground observations and the image classification, $\bar{X}$ is the population mean for the image classification and b is the angular coefficient of the regression between y and x. Regression estimators can be used for other variables besides crop area; for example, Stehman and Milliken (2007) use them to estimate the evapotranspiration of irrigated crops for water management in the lower Colorado basin. Ratio estimators (Lathrop, 2006) can be seen as a particular case of regression estimators. Small-area estimators (Battese *et al.*, 1988) are more complex but can be seen also as members of the family.

12.8 Examples of crop area estimation with remote sensing in large regions

12.8.1 US Department of Agriculture

The early Large Area Crop Inventory Experiment (LACIE: Heydorn, 1984) analysed samples of segments (5 × 6 nautical miles), defined as pieces of Landsat MSS images, and focused mainly on the sampling errors. It soon became clear (Sielken and Gbur, 1984) that pixel counting entailed a considerable risk of bias linked to the errors of commission and omission. Remote sensing was still too expensive in the 1980s (Allen, 1990), but the situation changed in the 1990s with the reduction of cost, both for image purchasing and processing (Hanuschak *et al.*, 2001).

The National Agricultural Statistical Service (NASS) crop area estimation programme has evolved a consistent core methodology based on the regression estimator, developing in several respects (Boryan *et al.*, 2008; Johnson, 2008):

- *Software.* The in-house PEDITOR application has been substituted with a combination of commercial software: ERDAS, ARC-GIS, See5 and SAS.
- *Ground data.* Administrative data from FSA-CLU (Farm Service Agency-Common Land Units) are used to train the image classification, while the sample survey data from the June Enumerative Survey (JES) provide the main variable for the regression estimator.
- *Images.* TM images, now in a difficult situation, have been substituted with AWiFS, suitable for the US landscape. MODIS images make a minor contribution.

USDA-NASS distributes the cartographic results of image classification as ‘cropland data layers’ that should not be confused with crop area estimates (Mueller and Seffrin, 2006).

The USDA Foreign Agricultural Service (FAS) makes intensive use of satellite images for worldwide agricultural monitoring. The FAS does not follow any specific method for crop area estimates; it rather makes an audit of data provided by agricultural attachés of the embassies (Taylor, 1996) with a ‘convergence of independent evidence’ decision support system that combines image analysis with other sources of information (Van Leeuwen *et al.*, 2006).

12.8.2 Monitoring agriculture with remote sensing

The Monitoring Agriculture with Remote Sensing (MARS) project of the European Union was launched in the late 1980s with two major crop area estimation activities. The 'regional crop inventories' (Taylor *et al.*, 1997) borrowed the USDA-NASS scheme, with area frame ground survey as main variable and classified images as co-variable for a regression correction. The cost-efficiency threshold was achievable with Landsat-TM images. Unfortunately the availability of TM images became problematic and the positive assessment resulted in few operational applications.

The 'rapid estimates of crop area changes' (called Activity B or Action 4) of the MARS project was an attempt to produce crop area estimates without a ground survey in the current year. Unsupervised classifications (pixel clustering and visual labelling) were performed on a set of 60 sites of 40 km × 40 km. For most sites four SPOT-XS images were acquired. Confusion matrices could not be computed because many clusters received generic labels such as 'light reddish turning to light greenish'. An unpublished assessment by the JRC of the method, carried out in 1994, indicated that the margin for subjectivity could be of the order of ±10% to ±30% for major crops. An analysis of the errors computed showed that the contribution of images was debatable (Gallego, 2006).

12.8.3 India

FASAL (Forecasting Agricultural output using Space and Land-based observations) is part of the very ambitious Indian remote sensing applications program (Navalgund *et al.*, 2007). Parihar and Oza (2006) mention a complex combination of econometric models with meteorological data and satellite images for agricultural assessment. Dadhwal *et al.* (2002) report a large number of studies conducted by the Indian Space Research Organisation (IRSO) with different types of images and algorithms. Area estimation often seems to be carried out by simple pixel counting. The classification accuracy reported is often above 90% and therefore the margin for subjectivity of pixel counting is within ±10%.

12.9 The GEOSS best practices document on EO for crop area estimation

GEOSS is an initiative to network earth observation data. GEOSS promotes technical standards for different aspects of remote sensing. One of them is dedicated to the use of remote sensing for crop area estimation (GEOSS, 2009) that gives a simplified summary of the methods and of kinds of satellite data that can be used in operational projects and those that still require research:

- *Satellite data.* Synthetic aperture radar (SAR) images cannot be considered for crop area estimation, except for paddy rice. Coarse or moderate resolution optical images can be used only for landscapes with very large fields.
- *Approaches.* Pure remote sensing approaches are acceptable in two cases: when a ground survey is not possible (because local authorities are not involved or for

security reasons) or when the accuracy requirements are looser than the errors of commission/omission. Crop area forecasting several months before harvest is not possible in general, but interesting indications can be given in some cases. For example, Ippoliti-Ramilo *et al.* (2003) identify the area prepared for planting annual crops in Brazil with TM images with an accuracy around 90%, thus the margin for subjectivity is less than 10%. This is useful if the uncertainty as to cultivated area is higher than 10%.

12.10 Sub-pixel analysis

Standard image classification attributes one class to each pixel. This is often known as the sharp or hard approach. Alternative soft or sub-pixel methods are not new but are receiving more and more attention. Soft classifications can have at least three different conceptual bases: probabilistic, fuzzy or area-share (Pontius and Cheuk, 2006). In the probabilistic conception each pixel belongs to a class with a certain probability. The fuzzy conception corresponds to a vague relationship between the class and the pixel; it is very attractive for classes with an unclear definition, but difficult to use for area estimation. In the area-share conception, classes have a sharp definition and the classification algorithm estimates the proportion x_{ik} of pixel i that belongs to class k.

The area-share conception is closer to the idea of crop area estimation; the pixel counting estimator can be easily adapted:

$$\hat{X}_k = \sum_i x_{ik}.$$

Woodcock and Gopal (2000) give an extension of the direct calibration estimator for fuzzy classifications, but its use in practical situations is difficult and it does not seem to have been used often. Regression estimators are easier to apply, but we could not find any application in the literature. A promising correlation ($r^2 = 0.54$) is reported by Verbeiren *et al.* (2008) for Maize in Belgium between the sub-pixel classification of SPOT-VEGETATION images and ground data. If this type of correlation is confirmed in countries running agricultural area frame surveys, the relative efficiency of regression estimators would be above 2, with excellent chances of becoming cost-efficient.

12.11 Accuracy assessment of classified images and land cover maps

When classified images or land cover maps are directly used to produce agricultural statistics, assessing the accuracy of the derived statistics is closely linked to the accuracy assessment of land cover maps. A few paragraphs on this issue may be worthwhile here.

Land cover/use map classes frequently include pure and mixed classes concerning agriculture, depending on the purposes of the project and the characteristics of the agriculture in the area, particularly the field size. A crucial question is how the accuracy is defined and measured. We believe that the accuracy of a digital map is the result of two different kinds of quality assessment: quality control and validation.

When polygons are delineated by photo-interpretation, quality control should be carried out by repeating the production process for a sample of polygons, with the same basic material and the same procedure. On the other hand, validation of a land cover map means the assessment of the level of agreement with a representation of reality considered more reliable. To validate a land cover map, a sample is compared with the corresponding ground truth. If suitable ground truth cannot be acquired, it may be substituted with EO data with a more detailed scale.

Insufficient resources aretoo often devoted to the validation of digital maps. Few land cover products undergo quality control and validation using statistical sampling (for a review see Carfagna and Marzialetti, 2009b); Strahler *et al.* (2006) stress the need for a validation plan and sample design to be included in every funded effort to map global land cover.

The level of accuracy needed for agriculture is often higher than for other kinds of land cover. An 85% accuracy is generally considered satisfactory, but 15% inaccuracy in the extent of agricultural land is usually insufficient for the management of agricultural markets or for food security management.

The quality of a digital map can be evaluated and improved at the same time, in particular when requirements cannot be predefined, by adopting a quality control method proposed by Carfagna and Marzialetti (2009a) based on an adaptive sequential sample design. This method is always more efficient than stratified sampling with proportional allocation and is also more efficient than the procedure proposed by Thompson and Seber (1996, pp. 189–191), who suggested stratified random sampling in two or, more generally, k phases. A similar approach, the two-step adaptive procedure with permanent random numbers, can be adopted for validation (Carfagna and Marzialetti, 2009b).

Congalton and Green (1999) classify the historical evolution of validation of image classification in three stages: (a) 'it looks nice'; (b) 'the area classified in each class seems correct'; (c) confusion matrices. The two first options are considered unacceptable; nevertheless the difficulties of extending confusion matrices with all their properties to sub-pixel classification have forced part of the remote sensing community to go back to the second stage. The third stage needs to be subdivided. An additional complication in assessing the possible bias appears because measuring the classification accuracy is not always a straightforward operation for several reasons: mixed pixels, fuzziness in the nomenclature, coexistence of classification errors with co-location inaccuracy, and spatial correlation between the training and the test set. Sources of bias in accuracy assessment are studied by Verbyla and Hammond (1995), Verbyla and Boles (2000) and Hammond and Verbyla (1996).

Benchmarking with alternative methods is essential in order to assess how operational a method can be. For example, a sub-pixel classification of MODIS images based on neural networks for winter crop area estimation by province in Spain achieved $r^2 = 0.88$ between official statistics and the MODIS-derived estimates (Eerens and Swinnen, 2009), both referring to 2006. This looks like an excellent result, but not if we consider that we get $r^2 = 0.92$ between official statistics and the area of the class 'arable land' in the CORINE 2000 Land cover map (Nunes de Lima, 2005). This means that a six-year-old land cover map still gives a better indication of the extent of winter crops by province than a full-time series of MODIS images of the current year; a great deal of progress is needed before we can claim that sub-pixel analysis is useful in this context.

12.12 General data and methods for yield estimation

The best estimates of crop yield are still based on traditional survey techniques. The main disadvantage is that results usually become known several months after harvest. Most traditional survey techniques are based on yield (or evaluation of potential yield) reported by a statistical sample of farmers. Alternatively, objective yield surveys are conducted by doing plant counts and fruit measurements within a sample of parcels. Objective methods are applied for instance by USDA/NASS and Statistics Canada (USDA/NASS, 2006; Wall *et al.*, 2008). In the EU, member states have a certain freedom to choose their methods, as long as these meet common quality requirements. Eurostat collects and harmonizes data to produce comparable statistics for the EU. For statistical purposes, in general the net harvested yield is considered, derived by deducting harvesting and other losses from the biological yield.

The major reasons for interest in bringing in alternative methods such as remote sensing in the yield forecasting are related to cost-effectiveness and timeliness. A combination of ground observations and alternative methods should also lead to gains in accuracy.

12.13 Forecasting yields

A baseline forecast can be produced as a simple average or trend of historical yield statistics. Various kinds of information are needed to improve the baseline forecast by capturing the main parameters that influence crop yield, including data on weather and weather anomalies, observations on the state of the crops and on relevant environmental conditions and reports on developments of crop markets (Piccard *et al.*, 2002). Statistics on past regional yields are always necessary to calibrate the forecasting model.

The maximum yield of a crop is primarily determined by its genetic characteristics and how well the crop is adapted to the environment. Environmental requirements of climate, soil and water for optimum crop growth vary with the crop and variety (Doorenbos, 1986). The main factors for yield are temperature and precipitation (Wall *et al.*, 2008).

The yield forecasting process usually integrates time series of historical yield statistics and crop yield indicators, which can come from remote sensing, biophysical models, field measurements, etc. These indicators are used to parameterize a forecasting model explaining the relation using a best-fit criterion. The model parameters are derived at several time steps during the growing season and the forecast model is then applied to forecast the current season's yield.

To identify the yield–indicator relationship, some statistical forecasting models are parametric, for example performing multiple regressions between crop yield statistics and crop indicators. Others can be seen as non-parametric, for example those using neural networks (Prasad *et al.*, 2006), in which parameters are hidden; or scenario analysis, which looks for similar years that are used to forecast.

Many forecasting methods make use of a mixture of data sources and approaches (Bouman, 1992; Doraiswamy *et al.* 2003, 2004; Fang *et al.*, 2008; Manjunath *et al.*, 2002; Rojas, 2007; Weissteiner *et al.*, 2004). Over time, more information becomes available, which needs to be assimilated into the model to increase the accuracy of yield estimates (Wall *et al.*, 2008). Understanding the physical and biological basis of the relationship between indicators and the final yield is important for obtaining a higher accuracy; unfortunately, such relations are often weak, complex and insufficiently known. The remedy often proposed is to add variables in an empirical model (Rasmussen, 1998).

12.14 Satellite images and vegetation indices for yield monitoring

The use of remote sensing within the crop yield forecasting process has a series of requirements for most of the applications.

- Information is needed for large areas while maintaining spatial and temporal integrity.
- Suitable spectral bands are needed that characterize the vegetation to allow for crop monitoring, whether to derive crop indicators to be used in a regression or to feed biophysical models.
- High temporal frequency is essential to follow crop growth during the season.
- Historical data (both images and statistical data on yields) are required for comparison and calibration of the methodology
- A quick processing system is critical for short-term forecasts of the ongoing season in order to produce timely and useful figures.

The need to cover a large area with frequent images narrows the usable sensors to coarse and medium resolution data (NOAA AVHRR, SPOT-VEGETATION, MODIS TERRA/ACQUA, IRS P6 AWIFS, ENVISAT, MERIS), as demonstrated in the literature (Shi *et al.*, 2007; Salazar *et al.*, 2008). The need for coarse resolution for a satisfactory description of the phenological behaviour prevents individual parcels from being seen in most agricultural landscapes. Techniques are available to weigh the signal according to the presence of crops based on area statistics (Genovese *et al.* 2001), or to perform a so-called yield correlation masking (Kastens *et al.*, 2005). Alternatively, the signal can be unmixed based on crop-specific classification of high and very high remote sensing imagery (Dong *et al.*, 2008; Zurita Milla *et al.*, 2009), but spatial resolution remains a strong limitation in an analysis that looks for crop-specific information.

Few studies concentrate on high- and very high-resolution satellites (e.g. Kancheva *et al.*, 2007) as the only source of information due to the limitations of area coverage and temporal frequency. Radar imagery plays a subordinate role (Blaes and Defourny, 2003).

Remote sensing data integrate information on different factors that influence the yield. The yield indicator most frequently used is the normalized difference vegetation index (NDVI: Myneni and Williams, 1994), which uses the red (R_r) and near infrared (R_n) reflectance:

$$NDVI = \frac{R_n - R_r}{R_n + R_r}.$$

The formula is based on the fact that chlorophyll absorbs red radiation whereas the mesophyll leaf structure scatters the near infrared (Pettorelli *et al.*, 2005). Compared to the individual reflectances, the NDVI is less sensitive to external influences related to the atmosphere, viewing geometry, and the state of the underlying soil. It has been criticized (Pinty *et al.*, 1993), but remains the most popular vegetation index.

The relationship between the NDVI and vegetation productivity is well established, and the spectral growth profiles derived from vegetation indices such as the NDVI show close physiological analogy with canopy-related determinants of yield (Kalumbarme *et al.*, 2003). However, the regression based approaches of the NDVI for yield forecasting only work for a given region and the same range of weather conditions where it was developed (Doraiswamy *et al.*, 2005). Apart from the NDVI, a series of related vegetation indices such as the Vegetation Condition Index (Kogan, 1990) are applied within similar regression based approaches showing moderate errors for the yield prediction (Salazar *et al.*, 2008). Other vegetation indices, such as the global environmental monitoring index (GEMI: Pinty and Verstraete, 1992) have been used more for natural vegetation than for yield forecasting.

Wall *et al.* (2008) review the most common approaches to the use of vegetation indices in crop yield forecasting models developed in the past two decades. They conclude that the most accurate yield estimates from remote sensing data have been reported using regression analysis and extensive multi-temporal data sets.

As for all the remote sensing derived products, a careful treatment of data is needed to avoid a distortion of the results, as remote sensing data are affected by systematic and random errors (Kalumbarme *et al.*, 2003). Pettorelli *et al.* (2005) give an overview of post-processing steps such as smoothing and interpolation and pertinent measures that can be assessed from NDVI time series in the context of ecological studies that can also be applied for crop yield forecasts.

Remote sensing can be also used to retrieve biophysical parameters, such as the fraction of absorbed photosynthetically active radiation (FAPAR) or leaf area index to supply inputs to crop growth models (Dowraiswamy *et al.*, 2005; Yuping, 2008; Fang *et al.*, 2008; Hilker *et al.*, 2008). Dorigo *et al.* (2007) give a summary of different strategies for assimilating remote sensing data in biophysical models.

12.15 Examples of crop yield estimation/forecasting with remote sensing

12.15.1 USDA

The USDA publishes monthly crop production figures for the United States and the world. At NASS, remote sensing information is used for qualitative monitoring of the state of crops but not for quantitative official estimates. NASS research experience has shown that AVHRR-based crop yield estimates in the USA are less precise than existing surveys.[1] Current research is centred on the use of MODIS data within biophysical models for yield simulation.

The Foreign Agricultural Service (FAS) regularly assesses global agricultural conditions. Images from geostationary satellites such as GOES, METEOSAT, GMS and from microwave sensors such as SSM/I are used to derive precipitation estimates, as input for models which estimate parameters such as soil moisture, crop stage, and yield. NASA

[1] `http://www.nass.usda.gov/Surveys/Remotely_Sensed_Data_Crop_Yield/index.asp`

and the University of Maryland provide FAS with data from instruments on several satellites: NASA's Aqua and Terra, TOPEX/Poseidon, Jason and Tropical Rainfall Measuring Mission.

NOAA-AVHRR and MODIS images are used to monitor vegetation conditions with bi-weekly vegetation index numbers such as the NDVI (Reynolds, 2001). Other satellite data sources include TM, ETM+ and SPOT-VEGETATION. These are often only visually interpreted. A large imagery archive allows incoming imagery to be compared with that of past weeks or years. When a new image is processed and loaded, key information is automatically calculated and stored in a database. Most of the information is available through the FAS Crop Explorer[2] providing easy-to-read crop condition information for most agricultural regions in the world. With these data, producers, traders, researchers, and the public can access weather and satellite information useful for predicting crop production worldwide.

12.15.2 Global Information and Early Warning System

The Global Information and Early Warning System (GIEWS) was established by the FAO in the early 1970s and is the leading source of global information on food production and food security. The system continually receives economic, political and agricultural information from a wide variety of sources (UN organizations, 115 governments, 4 regional organizations and 61 non-governmental organizations). Over the years, a unique database on global, regional, national and subnational food security has been maintained, refined and updated.

In many drought-prone countries, particularly in sub-Saharan Africa, there is a lack of continuous, reliable information on weather and crop conditions. For this reason, GIEWS, in collaboration with FAO's Africa Real Time Environmental Monitoring Information System (ARTEMIS), has established a crop monitoring system using near real-time satellite images. Data received directly by ARTEMIS from the European METEOSAT satellite are used to produce cold cloud duration (CCD) images for Africa every 10 days. These provide a proxy estimate for rainfall. ARTEMIS maintains an archive of CCD images dating back to 1988, which allows GIEWS's analysts to pinpoint areas suffering from low rainfall and drought by comparing images from the current season to those from previous years or the historical average. Similarly, since 1998, the Japan Meteorological Agency has been providing the FAO with 10-day estimated rainfall images for Southeast Asia computed from data received from the Japanese GMS satellite. In addition to rainfall monitoring, GIEWS makes extensive use of NDVI images that provide an indication of the vigour and extent of vegetation cover. These allow GIEWS analysts to monitor crop conditions throughout the season. Data obtained from NOAA satellites are processed by the NASA Goddard Space Flight Center to produce 10-day, 8-kilometre resolution vegetation images of Africa, Latin America and the Caribbean. The FAO, in collaboration with the Joint Research Centre (JRC) of the European Commission, has access to 10-day real-time images from the SPOT-4 VEGETATION instrument. These cover the whole globe at 1-kilometre resolution and are suitable for crop monitoring at subnational level.

[2] http://www.pecad.fas.usda.gov/cropexplorer/

12.15.3 Kansas Applied Remote Sensing

The Kansas Applied Remote Sensing (KARS) Program was established in 1972 by NASA and the state of Kansas. In 1998 it became the Great Plains Regional Earth Science Applications Center. This focuses on the assessment of grassland condition and productivity, monitoring and projecting crop production and yield, and monitoring changes in land use and land cover. Yield forecasting is based on statistical modelling of NOAA-AVHRR NDVI data.

12.15.4 MARS crop yield forecasting system

Since 1993 the MARS project of the JRC has been running a crop yield forecasting system for the quantitative assessment of the major European crops in all EU member states. The system has also been used outside the EU since 2000. It is based on:

- low-resolution, high-frequency remote sensing products – NOAA-AVHRR, SPOT VEGETATION and MODIS – received every 10 days with world coverage. An archive of remote sensing data has also been maintained with data for Europe since 1981.
- meteorological information received daily from European Centre for Medium-Range Weather Forecasts models and from more than 4000 synoptic stations in Europe and Africa. Other parameters are derived from METEOSAT satellite data. More than 30 years of data have been archived.
- additional geospatial information – soil maps, land cover/land use maps, phenology information and agricultural statistics.

Depending on the region and on data availability, different crop monitoring procedures are implemented. During the growing season, crop qualitative assessments are produced looking at anomalies between crop development profiles of the current year and historical profiles. Before the end of the season, for some countries, quantitative crop production estimates are computed using a simple NDVI regression model or a multiple regression with water satisfaction indices (Nieuwenhuis *et al.*, 2006) and NDVI. The model is calibrated with historical agricultural statistics. In addition to the remote sensing approaches, the MARS Crop Yield Forecasting System uses several simulation crop growth models: WOFOST, LINGRA and WARM. Results are published in monthly or bimonthly climatic and crop monitoring bulletins for Europe with quantitative yield forecasts by country and monthly national and regional crop monitoring bulletins for food-insecure regions including qualitative and, for some countries, quantitative estimates (`http://mars.jrc.it/mars/`).

References

Allen, J.D. (1990) A look at the Remote Sensing Applications Program of the National Agricultural Statistics Service. *Journal of Official Statistics*, 6, 393–409.

Battese, G.E., Harter, R.M. and Fuller, W.A. (1988) An error-components model for prediction of county crop areas using survey and satellite data. *Journal of the American Statistical Association*, 83, 28–36.

Blaes, X. and Defourny, P. (2003) Retrieving crop parameters based on tandem ERS 1/2 interferometric coherence images. *Remote Sensing of Environment*, 88, 374–385.

Boryan, C., Craig, M. and Lindsay, M. (2008) Deriving essential dates of AWiFS and MODIS data for the identification of corn and soybean fields in the U.S. heartland. Pecora 17 – The Future of Land Imaging. Going Operational, Denver, Colorado, 18–20 November.

Bouman, B.A.M. (1992) Accuracy of estimating the leaf area index from vegetation indices derived from crop reflectance characteristics: a simulation study. *International Journal of Remote Sensing*, 13, 3069–3084.

Carfagna, E. (2001) Cost-effectiveness of remote sensing in agricultural and environmental statistics. In *Proceedings of the Conference on Agricultural and Environmental Statistical Applications in Rome* (CAESAR), 5–7 June, Vol III, pp. 617–627. http://www.ec-gis.org/document.cfm?id = 427&db = document.

Carfagna, E. and Gallego, F.J. (2005) Using remote sensing for agricultural statistics. *International Statistical Review*, 73, 389–404.

Carfagna, E. and Marzialetti, J. (2009a) Sequential design in quality control and validation of land cover data bases. *Journal of Applied Stochastic Models in Business and Industry*, 25, 195–205.

Carfagna, E. and Marzialetti, J. (2009b) Continuous innovation of the quality control of remote sensing data for territory management. In P. Erto (ed.), *Statistics for Innovation*, pp. 172–188. New York: Springer.

Cochran, W. (1977) *Sampling Techniques*, 3rd edition. New York: John Wiley & Sons. Inc.

Congalton, R.G. and Green, K. (1999) *Assessing the Accuracy of Remotely Sensed Data: Principles and Practices*. Boca Raton, FL: Lewis.

Dadhwal, V.K., Singh, R.P., Dutta, S. and Parihar, J.S. (2002) Remote sensing based crop inventory: A review of Indian experience. *Tropical Ecology*, 43, 107–122.

Dong, Q., Eerens, H. and Chen, Z. (2008) Crop area assessment using remote sensing on the North China Plain. In *Proceedings ISPRS*, Vol. 37, Part B8, pp. 957–962.

Doorenbos, J. (ed.) (1986) *Yield response to water*. FAO irrigation and drainage paper, 33.

Doraiswamy, P.C., Moulin, S., Cook, P.W. and Stern, A. (2003) Crop yield assessment from remote sensing. *Photogrammetric Engineering and Remote Sensing*, 69, 665–674.

Doraiswamy, P.C., Hatfield, J.L., Jackson, T.J., Akhmedov, B., Prueger, J. and Stern, A. (2004), Crop condition and yield simulations using landsat and MODIS. *Remote Sensing of Environment*, 92, 548–559.

Doraiswamy, P.C., Sinclair, T.R., Hollinger, S., Akhmedov, B., Stern, A. and Prueger, J. (2005) Application of MODIS derived parameters for regional crop yield assessment. *Remote Sensing of Environment*, 97, 192–202.

Dorigo, W.A., Zurita-Milla, R., de Wit, A.J.W., Brazile, J., Singh, R. and Schaepman, M.E. (2007) A review on reflective remote sensing and data assimilation techniques for enhanced agroecosystem modeling. *International Journal of Applied Earth Observation and Geoinformation*, 9, 165–193.

Eerens, H. and Swinnen, E. (2009) Crop mapping with MODIS 250m over Europe: first results of the hard and soft classifications. MARSOP3 interim report, JRC Ispra.

Fang, H., Liang, S., Hoogenboom, G., Teasdale, J. and Cavigelli, M. (2008) Corn-yield estimation through assimilation of remotely sensed data into the CSM-CERES-maize model. *International Journal of Remote Sensing*, 29, 3011–3032.

FAO (2008) *The State of Food and Agricultural Statistical Systems in Africa - 2007*, RAF publication 2008E. Accra: FAO Regional Office for Africa.

Friedl, M.A., McIver, D.K., Hodges, J.C.F., Zhang, X.Y., Muchoney, D., Strahler, A.H., Woodcock, C.E., Gopal, S., Schneider, A., Cooper, A., Baccini, A., Gao, F. and Schaaf, C. (2002) Global land cover mapping from MODIS: Algorithms and early results. *Remote Sensing of Environment*, 83, 287–302.

Gallego, F.J. (2004) Remote sensing and land cover area estimation. *International Journal of Remote Sensing*, 25, 3019–3047.

Gallego, F.J. (2006) Review of the main remote sensing methods for crop area estimates. In B. Baruth, A. Royer and G. Genovese (eds), *Remote Sensing Support to Crop Yield Forecast and Area Estimates ISPRS archives*, Vol. 36, Part 8/W48, pp. 65–70. http://www.isprs.org/publications/PDF/ISPRS_Archives_WorkshopStresa2006.pdf.

Genovese, G., Vignolles, C., Nègre, T. and Passera, G. (2001) A methodology for a combined use of normalised difference vegetation index and CORINE land cover data for crop yield monitoring and forecasting. A case study on Spain. *Agronomie*, 21, 91–111.

Gentle, J., Perry, C. and Wigton, W. (2006) Modeling nonsampling errors in agricultural surveys. In *Proceedings of the Section on Survey Research Methods*, American Statistical Association, 3035–3041.

GEOSS (2009) Best practices for crop area estimation with Remote Sensing. Ispra, 5–6 June 2008. http://www.earthobservations.org/geoss.shtml.

Hammond, T.O. and Verbyla, D.L. (1996) Optimistic bias in classification accuracy assessment. *International Journal of Remote Sensing*, 17, 1261–1266.

Hannerz, F. and Lotsch, A. (2008) Assessment of remotely sensed and statistical inventories of African agricultural fields. *International Journal of Remote Sensing*, 29, 3787–3804.

Hansen, M.H., Hurwitz, W.N. and Madow, W.G. (1953) *Sample Survey Methods and Theory*. New York: John Wiley & Sons, Inc.

Hanuschak, G., Hale, R., Craig, M., Mueller, R. and Hart, G. (2001) The new economics of remote sensing for agricultural statistics in the United States. *Proceedings of the Conference on Agricultural and Environmental Statistical Applications in Rome* (CAESAR), 5–7 June, Vol. 2, pp. 427–437.

Hanuschak, G.A., Sigman, R., Craig, M.E., Ozga, M., Luebbe, R.C., Cook, P.W. *et al.* (1980) Crop-area estimates from LANDSAT: Transition from research and development timely results. *IEEE Transactions on Geoscience and Remote Sensing*, GE-18, 160–166.

Heydorn, R.P. (1984) Using satellite remotely sensed data to estimate crop proportions. *Communications in Statistics – Theory and Methods*, 23, 2881–2903.

Hilker, T., Coops, N.C., Wulder, M.A., Black, T.A. and Guy, R.D. (2008) The use of remote sensing in light use efficiency based models of gross primary production: A review of current status and future requirements. *Science of the Total Environment*, 404, 411–423.

Im, J. and Jensen, J.R. (2008) Hyperspectral remote sensing of vegetation. *Geography Compass*, 2, 1943–1961.

Ippoliti-Ramilo, G.A., Epiphanio, J.C.N. and Shimabukuro, Y.E. (2003) Landsat-5 thematic mapper data for pre-planting crop area evaluation in tropical countries. *International Journal of Remote Sensing*, 24, 1521–1534.

Johnson, D.M. (2008) A comparison of coincident landsat-5 TM and resourcesat-1 AWiFS imagery for classifying croplands. *Photogrammetric Engineering and Remote Sensing*, 74, 1413–1423.

Kalubarme, M.H., Potdar, M.B., Manjunath, K.R., Mahey, R.K. and Siddhu, S.S. (2003) Growth profile based crop yield models: A case study of large area wheat yield modelling and its extendibility using atmospheric corrected NOAA AVHRR data. *International Journal of Remote Sensing*, 24, 2037–2054.

Kancheva, R., Borisova, D. and Georgiev, G. (2007) Spectral predictors of crop development and yield. In *Proceedings of the 3rd International Conference on Recent Advances in Space Technologies* (RAST 2007), article number 4283987, pp. 247–251. Piscataway, NJ: Institute of Electrical and Electronics Engineers.

Kastens, J.H., Kastens, T.L., Kastens, D.L.A., Price, K.P., Martinko, E.A. and Lee, R. (2005) Image masking for crop yield forecasting using AVHRR NDVI time series imagery. *Remote Sensing of Environment*, 99, 341–356.

Kogan, F.N. (1990) Remote sensing of weather impacts on vegetation in non-homogeneous areas. *International Journal of Remote Sensing*, 11, 1405–1419.

Lathrop, R. (2006) The application of a ratio estimator to determine confidence limits on land use change mapping. *International Journal of Remote Sensing*, 27, 2033–2038.

Lessler, V.M. and Kalsbeek, W.D. (1999) Nonsampling errors in environmental surveys. *Journal of Agricultural, Biological, and Environmental Statistics*, 4, 473–488.

Manjunath, K.R., Potdar, M.B. and Purohit, N.L. (2002) Large area operational wheat yield model development and validation based on spectral and meteorological data. *International Journal of Remote Sensing*, 23, 3023–3038.

Marriot, F.H.C. (1990) *A Dictionary of Statistical Terms*, 5th edition. Harlow: Longman.

Mueller, R. and Seffrin, R. (2006) New methods and satellites: A program update on the NASS cropland data layer acreage program. In B. Baruth, A. Royer and G. Genovese (eds), *Remote Sensing Support to Crop Yield Forecast and Area Estimates ISPRS archives*, Vol. 36, Part 8/W48, pp. 97–102. http://www.isprs.org/publications/PDF/ISPRS_Archives_WorkshopStresa2006.pdf.

Mutanga, O. and Skidmore, A.K. (2004) Narrow band vegetation indices overcome the saturation problem in biomass estimation. *International Journal of Remote Sensing*, 25, 3999–4014.

Myneni, R.B. and Williams, D.L. (1994) On the relationship between FAPAR and NDVI. *Remote Sensing of Environment*, 49, 200–211.

Navalgund, R.R., Jayaraman, V. and Roy, P.S. (2007) Remote sensing applications: An overview. *Current Science*, 93, 1747–1766.

Nieuwenhuis, G.J.A., de Wit, A.J.W., van Kraalingen, D.W.G., van Diepen, C.A. and Boogaard, H.L. (2006) Monitoring crop growth conditions using the global water satisfaction index and remote sensing. Paper presented to ISPRS Conference on Remote Sensing: From Pixels to Processes, Enschede. http://www.itc.nl/external/isprsc7/symposium/proceedings/TS23_4.pdf.

Nunes de Lima, M.V. (ed.) (2005) *CORINE Land Cover Updating for the Year 2000. Image2000 and CLC2000: Products and Methods*, Report EUR 21757 EN. Ispra, Italy: JRC. http://www.ec-gis.org/sdi/publist/pdfs/nunes2005eur2000.pdf.

Parihar, J.S. and Oza, M.P. (2006) FASAL: An integrated approach for crop assessment and production forecasting. *Proceedings of SPIE, the International Society for Optical Engineering*, 6411, art. no. 641101.

Pettorelli, N., Vik, J.O., Mysterud, A., Gaillard, J.M., Tucker, C.J. and Stenseth, N.C. (2005) Using the satellite-derived NDVI to assess ecological responses to environmental change. *Trends in Ecology and Evolution*, 20, 503–510.

Piccard, I., van Diepen, C.A., Boogaard, H.L., Supit, I., Eerens, H. and Kempeneers, P. (2002) Other yield forecasting systems: description and comparison with the MARS Crop Yield Forecasting System. Internal report, JRC Ispra.

Pinty, B. and Verstraete, M.M. (1992) GEMI: A non-linear index to monitor global vegetation from satellites. *Vegetatio*, 101, 15–20.

Pinty, B., Leprieur, C. and Verstraete, M. (1993) Towards a quantitative interpretation of vegetation indices. Part 1: Biophysical canopy properties and classical indices. *Remote Sensing Reviews*, 7, 127–150.

Pontius Jr., R.G. and Cheuk, M.L. (2006) A generalized cross-tabulation matrix to compare soft-classified maps at multiple resolutions. *International Journal of Geographical Information Science*, 20, 1–30.

Prasad, A.K., Chai, L., Singh, R.P. and Kafatos, M. (2006) Crop yield estimation model for Iowa using remote sensing and surface parameters. *International Journal of Applied Earth Observation and Geoinformation*, 8, 26–33.

Quinlan, J.R. (1996) Improved use of continuous attributes in C4.5. *Journal of Artificial Intelligence Research*, 4, 77–90.

Rama Rao, N. (2008) Development of a crop-specific spectral library and discrimination of various agricultural crop varieties using hyperspectral imagery. *International Journal of Remote Sensing*, 29, 131–144.

Rasmussen, M.S. (1998) Developing simple, operational, consistent NDVI-vegetation models by applying environmental and climatic information. Part II: Crop yield assessment. *International Journal of Remote Sensing*, 19, 119–139.

Reynolds, C. (2001) Input data sources, climate normals, crop models, and data extraction routines utilized by PECAD. Paper presented to Third International Conference on Geospatial Information in Agriculture and Forestry, Denver, Colorado, 5–7 November.

Rojas, O. (2007) Operational maize yield model development and validation based on remote sensing and agro-meteorological data in Kenya. *International Journal of Remote Sensing*, 28, 3775–3793.

Salazar, L., Kogan, F. and Roytman, L. (2008) Using vegetation health indices and partial least squares method for estimation of corn yield. *International Journal of Remote Sensing*, 29, 175–189.

Shi, Z., Ruecker, G.R., Mueller, M. *et al.* (2007) Modeling of cotton yields in the Amu Darya river floodplains of Uzbekistan integrating multitemporal remote sensing and minimum field data. *Agronomy Journal*, 99, 1317–1326.

Sielken R.L., Gbur E.E. (1984) Multiyear, through the season, crop acreage estimation using estimated acreage in sample segments. *Communications in Statistics – Theory and Methods*, 23, 2961–2974.

Stehman, S.V. and Milliken, J.A. (2007) Estimating the effect of crop classification error on evapotranspiration derived from remote sensing in the lower Colorado river basin, USA. *Remote Sensing of Environment*, 106, 217–227.

Strahler, A.S., Boschetti, L., Foody, G.M., Friedl, M.A., Hansen, M.C. and Herold, M. (2006) Global land cover validation recommendations for evaluation and accuracy assessment of Global Land Cover Maps. EUR Report 22156, JRC Ispra.

Taylor, J., Sannier, C., Delincé, J, Gallego, F.J. (1997) Regional crop inventories in Europe assisted by remote sensing: 1988–1993. Synthesis Report, EUR 17319 EN, JRC Ispra.

Taylor, T.W. (1996) Agricultural analysis for a worldwide crop assessment. In *Proceedings of the SPOT Conference*, Paris, 15–18 April, pp. 485–488.

Thompson, S.K. and Seber, G.A.F. (1996) *Adaptive Sampling*. New York: John Wiley & Sons, Inc.

USDA/NASS (2006) The yield forecasting program of NASS. SMB Staff Report 06-01. `http://www.nass.usda.gov/Education_and_Outreach/Understanding_Statistics/yldfrcst2006.pdf`.

Van Leeuwen, W.J.D., Hutchinson, C.F., Doorn, B., Sheffner, E. and Kaupp, V.H. (2006) Integrated crop production observations and information system. In *Proceedings of the International Geoscience and Remote Sensing Symposium* (IGARSS), pp. 3506–3508.

Verbeiren, S., Eerens, H., Piccard, I., Bauwens, I. and Van Orshoven, J. (2008) Sub-pixel classification of SPOT-VEGETATION time series for the assessment of regional crop areas in Belgium. *International Journal of Applied Earth Observation and Geoinformation*, 10, 486–497.

Verbyla, D.L. and Boles, S.H. (2000) Bias in land cover change estimates due to misregistration. *International Journal of Remote Sensing*, 21, 3553–3560.

Verbyla, D.L. and Hammond, T.O. (1995) Conservative bias in classification accuracy assessment due to pixel-by-pixel comparison of classified images with reference grids. *International Journal of Remote Sensing*, 16, 581–587.

Wall, L., Larocque, D. and Léger, P. (2008) The early explanatory power of NDVI in crop yield modelling. *International Journal of Remote Sensing*, 29, 2211–2225.

Weissteiner, C.J., Braun, M. and Kühbauch, W. (2004) Regional yield predictions of malting barley by remote sensing and ancillary data. *Proceedings of SPIE, the International Society for Optical Engineering*, 5232, 528–539.

Wood, S., Sebastian, K. and Scherr, S.J. (2000) *Pilot Analysis of Global Ecosystems: Agroecosystems*. Washington, DC: World Resources Institute and International Food Policy Research Institute.

Woodcock, C.E. and Gopal, S. (2000) Fuzzy set theory and thematic maps: Accuracy assessment and area estimation. *International Journal of Geographical Information Science*, 14, 153–172.

Yuping, M., Shili, W., Li, Z. *et al.* (2008) Monitoring winter wheat growth in North China by combining a crop model and remote sensing data. *International Journal of Applied Earth Observation and Geoinformation*, 10, 426–437.

Zurita-Milla, R., Kaiser, G., Clevers, J.G.P.W., Schneider, W. and Schaepman, M.E. (2009) Downscaling time series of MERIS full resolution data to monitor vegetation seasonal dynamics. *Remote Sensing of Environment*, 113, 1874–1885.

13

Estimation of land cover parameters when some covariates are missing

Roberto Benedetti[1] and Danila Filipponi[2]

[1]*Department of Business, Statistical, Technological and Environmental Sciences, University 'G. d'Annunzio' of Chieti-Pescara, Italy*

[2]*Istat, National Institute of Statistics, Rome, Italy*

13.1 Introduction

Land cover information is growing more and more important for the implementation and evaluation of environmental policies, and high precision is expected of it. AGRIT is an Italian point frame sample survey the purpose of which is to produce area estimates of the main crops, with a predetermined sample error for the different crops (see also Chapter 22 of this book).

When auxiliary information is available, the design-based regression estimator (Hansen *et al.*, 1953; Cochran, 1977) is a classical technique used to improve the precision of a sample estimator. This technique has been widely applied to improve the efficiency of crop area estimates since the early days of satellite image data (Allen, 1990; Flores and Martnez, 2000; see also Chapter 12 of this book). However, there is an increasing demand for land cover estimates over small areas (usually districts or provinces) and the regression estimator is not sufficiently precise to produce small-area estimates of land cover type due to the small sample sizes in the small area considered. In this chapter, to increase the precision of the small-area estimates we use a

Agricultural Survey Methods Edited by Roberto Benedetti, Marco Bee, Giuseppe Espa and Federica Piersimoni

model-based approach defining statistical models to 'borrow strength' from related small areas in order to increase the effective sample size (for further details see Chapter 9 of this book). Small-area estimates have optimal statistical properties. The choice of good auxiliary information, linked to variables of interest, is very important for the success of any model-based method. Here, two different aspects associated with the quality of the auxiliary variable are considered, where the available information is given by remote sensing satellite data covering the national agricultural area.

The location accuracy between the ground survey and satellite images and the difficulties in improving this accuracy through geometrical correction have been considered one of the main problems in relating remote sensing satellite data to crop areas or yields, mainly in point frame sample surveys where the sampled point represents a very small portion of the territory (see Chapter 10 of this book). Moreover, outliers and missing data are often present in the satellite information; they are mainly due to cloudy weather that does not allow identification or correct recognition of crop areas from the digital images acquired.

In this chapter, the first problem is addressed by using the basic area-level model to improve the land cover estimate at a small-area level (Rao, 2003). Thus, the small-area direct estimator is related to area-specific auxiliary variables, that is, number of pixels classified in each crop type according to the satellite data in each small area. The missing data problem is addressed by using a multiple imputation

The following section gives some background on the AGRIT survey, describing the sampling design and the direct estimator used. Suection 13.2.1 gives some details of the area-specific models used. Spatial autocorrelation amongst the small-area units is also considered in order to improve the small-area estimators. In Section 13.3 the problem of missing values in the auxiliary variable is addressed; in particular, we give a brief review of the missing data problem and a description of the multiple imputation approach. Finally, the concluding sections summarize and discuss some results.

13.2 The AGRIT survey

13.2.1 Sampling strategy

The AGRIT is a survey designed to obtain areas and yields estimates of the main crops in accordance with the LUCAS nomenclature of land cover (Table 13.1) valid at the European level. The survey covers the entire area of Italy, about 301 300 km^2, and for the main categories of the nomenclature high precision is expected.

A stratified two-phase sampling design is used. In the first phase the reference frame is defined. The Italian territory is partitioned into N quadrants of 1000 m size. In each quadrant n' units are selected using an aligned spatial systematic sample: one point is randomly selected within each quadrant and the remaining $n' - 1$ are placed in the same position within the other quadrants. Using aerial photos each point is then classified according to a land use hierarchical nomenclature with a total of 25 entries. The sampling frame obtained in the first phase is divided into 103 provincial domains and stratified, using the nomenclature of land use, into the following four classes: (1) arable land; (2) permanent crops; (3) permanent grassland; (4) non-agricultural land (urban, forest, water and wetland, etc.). Clearly, the first three classes are those relevant to agriculture. In each Italian province a sample is then selected that is stratified by class of land use for

Table 13.1 LUCAS nomenclature of land cover for which area estimates are produced.

Cereals	Soft wheat
	Durum wheat
	Barley
	Maize
Root crops	Sugar beet
Non-permanent industrial crops	Sunflower
	Soya
Dry pulses, vegetables	Tomatoes

a total of about 150 000 sampled points. The observation unit is the point, i.e. the smallest part of the territory homogeneous and identifiable from the aerial photo. Theoretically, the point has no dimension, but it is self-evident that the observations of land cover are defined at a metric scale. This leads to the need to assign an operational dimension to each point. Thus, the observation unit is defined as a circle of 3 m radius around its theoretical position, thus having a total area of about 30 m^2. If the land characteristics change in the area considered, the operational dimension is extended to 700 m^2 (i.e. a circle with a radius of 15 m) and multiple registrations, proportional to each different land cover, are assigned to the points concerned.

A ground survey, based on direct observations of the crops in the portion of territory, is carried out for all the points belonging to the layers of agricultural interest (strata 1–3), while for points belonging to stratum 4 the information recorded is limited to that obtained by ortho-photo interpretation.

The aim of the sampling strategy is to produce area estimates of the main crops at a national, regional and provincial level. In order to define the problem formally, let us denote by A the area of the entire territory, A_d the domain d $(d = 1, \ldots, D)$ in a partition of the territory and c $(c = 1, \ldots, C)$ the crops for which estimates of land use are provided. The parameters of interest are $\bar{Y}_A^c$, and $Y_A^c = A\bar{Y}_A^c$ respectively denoting the percentage and total of land cover for crop c for the entire territory, and $\bar{Y}_d^c$ and $Y_d^c = A_d\bar{Y}_d^c$ respectively denoting the percentage and total of land cover for crop c in domain d.

Simple random sampling is assumed for the first phase, since it behaves like aligned spatial systematic sampling, and stratified sampling for the second phase. Thus, in the first phase sampling, n' represents the total number of sampled points, n'_h the number of sampled points in stratum h $(h = 1, \ldots, H)$ and $w_h = n'_h/n'$ is the proportion of sampled points belonging to stratum h, while in the second phase sampling, n represents the total number of points and n_h the number of sampled points in stratum h. Finally, y_{ih}^c represents the proportion of land cover for crop c observed at point i in stratum h.

In accordance with the formula for two-stage sampling when simple random sampling is carried out in the first stage and stratified sampling in the second stage, the Horwitz–Thompson (HT) estimator of $\bar{Y}_A^c$ is given by

$$\bar{y}_A^c = \sum_{h=1}^{H}\left(w_h \sum_{i=1}^{n_h} \frac{y_{ih}^c}{n_h}\right) = \sum_{h=1}^{H} w_h \bar{y}_h^c,$$

while the unbiased estimator of the corresponding variance is

$$v(\bar{y}_A^c) = \sum_{h=1}^{H} \frac{w_h^2 s_h^2}{n_h} + \frac{1}{n'} \sum_{h=1}^{H} w_h (\bar{y}_h^c - \bar{y}^c)^2,$$

where $s_h^2 = \sum_{i=1}^{n_h} \frac{(y_{ih}^c - \bar{y}_h^c)^2}{n_h - 1}$.

Once the percentage of land cover for crop c is estimated, the estimate of the total $Y_A^c = A\bar{Y}_A^c$ is straightforward and is given by $\hat{y}_A^c = A\bar{y}_A^c$, while the unbiased estimator of the corresponding variance is $v(\hat{y}_A^c) = A^2 v(\bar{y}_A^c)$ and $CV(\hat{y}_A^c) = \sqrt{v(\hat{y}_A^c)}/\hat{y}_A^c$. If the parameters of interest are $\bar{Y}_d^c$ and $Y_d^c = A_d \bar{Y}_d^c$, that is, the percentage and total of land cover for crop c in domain d, the previous theory is applicable by defining a domain membership variable $y_{d.ih}^c$, where $y_{d.ih}^c = y_{ih}^c$ if the unit $i \in A_d$ and $y_{d.ih}^c = 0$ otherwise.

13.2.2 Ground and remote sensing data for land cover estimation in a small area

In the AGRIT project remote sensing satellite data covering the national agricultural area for the spring and summer periods are available. Remote sensing satellite data provide a complete spectral resolution of an area that can be used to classify the area by crop types. The availability of such information does not eliminate the need for ground data, since satellite data do not always have the accuracy required to estimate the different crop areas. This can be used as auxiliary information to improve the precision of the direct estimates. In this framework, the design based regression estimator has often been used to improve the efficiency of land cover estimates for a large geographical area when classified satellite images can be used as auxiliary information. However, the regression estimator is not sufficiently precise to produce small-area estimates of land cover type due to the small sample sizes in the small area considered.

Here, to increase the precision of the small-area estimates a model-based approach is followed, defining a statistical model to 'borrow strength' from related small areas to increase the effective sample size. The model suggested for estimating main crops at provincial level is the basic area-level model, and we refer the reader to Chapter 9 of this book for a comprehensive review. Let us denote by $z_d = (z_{d1}, z_{d2}, \ldots, z_{dp})$ the $(1 \times p)$ vector containing the number of pixels classified in each crop type according to the satellite data in the small area $d, d = 1, \ldots, D$, where $D = 103$ is the number of Italian provinces. It seems natural to assume that the area covered by a crop Y_d^c, in the small area d is in some way linked to z_d. A model-based estimate of Y^c $(D \times 1)$, eventually transformed as $\theta^c = g(Y^c)$, is the empirical best linear unbiased predictor (EBLUP) based on the linear mixed model:

$$\hat{\theta}^c = \beta^c Z + v + e, \tag{13.1}$$

where $\hat{\theta}^c = g(\hat{Y}^c)$ is a D-component vector of the direct survey estimators of θ^c, $\beta(D \times 1)$ is the vector of regression parameters, $Z(D \times p)$ is the matrix of covariates, $e(D \times 1)$ are the sampling errors assumed to be independent across areas with mean 0 and variance matrix equal to $R = diag(\psi_1, \psi_2, \ldots, \psi_m)$ where the ψ_i are the known sampling variances corresponding to the dth small area, and v $(D \times 1)$ are the model errors assumed to be independent and identically distributed with mean 0 and variance matrix equal to

$G = \sigma_v^2 I$. Independence between v and e is also assumed. For this model the EBLUP estimator of the dth small-area total of land cover for crop c, $\theta_d^{\star c}$, is the weighted average of the direct surveys estimator $\hat{\theta}_d^c$ and the regression synthetic estimator $z_d^T \hat{\beta}^c$, where the weights depend on the estimator of the variance component $\hat{\sigma}_v^2$. The EBLUP estimator of θ_d^c and estimation of the mean square error (MSE) of the EBLUP are described by Rao (2003). The small-area characteristics usually have spatial dependence. The spatial autocorrelation amongst neighbouring areas can be introduced to improve the small-area estimation. Here, in order to take into account the correlation between neighbouring areas we use conditional spatial dependence among random effects (Cressie, 1991). An area-level model with conditional spatial dependence among random effects can be considered an extension of model (13.1) where all the parameters have the same meaning as previously explained, and the model errors v are assumed with mean 0 and covariance matrix $G = \sigma_v^2(I - \rho W)^{-1}$. W $(D \times D)$ is a known proximity matrix that indicates the interaction between any pair of small areas. The elements of $W \equiv [W_{ij}]$ with $W_{ij} = 0$ for all i are binary values, $W_{ij} = 1$ if the jth small area is physically contiguous with the ith small area and $W_{ij} = 0$ otherwise. The constant ρ is called the spatial autoregressive coefficient and is a measure of the overall level of spatial autocorrelation.

The spatial model 'borrows strength' from related small areas by using two parameters: the regression parameters and the spatial autoregressive coefficient. By setting $\rho = 0$ we obtain (13.1). Considering the spatial model as a special case of generalized mixed models, the BLUP of θ^c and the MSE of the BLUP can be easily obtained as

$$\theta^{\star c}\left(\rho, \sigma_v^2\right) = Z\hat{\beta}\left(\rho, \sigma_v^2\right) + \lambda\left(\rho, \sigma_v^2\right)\left[\hat{\theta} - Z\hat{\beta}\left(\rho, \sigma_v^2\right)\right],$$

$$MSE\left[\theta^{\star c}\left(\rho, \sigma_v^2\right)\right] = g_1\left(\rho, \sigma_v^2\right) + g_2\left(\rho, \sigma_v^2\right),$$

where

$$g_1\left(\rho, \sigma_v^2\right) = \lambda\left(\rho, \sigma_v^2\right) R,$$

$$g_2\left(\rho, \sigma_v^2\right) = RV^{-1}\left(\rho, \sigma_v^2\right) Z\left(Z^T V^{-1}(\rho, \sigma_v^2)\, Z\right)^{-1} Z^T V^{-1}\left(\rho, \sigma_v^2\right) R,$$

and

$$\hat{\beta}\left(\rho, \sigma_v^2\right) = \left[Z^T V^{-1}\left(\rho, \sigma_v^2\right) Z\right]^{-1} Z^T V^{-1}\left(\rho, \sigma_v^2\right)\hat{\theta},$$

$$V\left(\rho, \sigma_v^2\right) = \sigma_v^2 A^{-1}\left(\rho, \sigma_v^2\right) + R,$$

$$\lambda\left(\rho, \sigma_v^2\right) = \sigma_v^2 A^{-1}\left(\rho, \sigma_v^2\right) V^{-1}\left(\rho, \sigma_v^2\right),$$

$$A\left(\rho, \sigma_v^2\right) = (I - \rho W).$$

The BLUP of θ^c and the MSE of the BLUP are both functions of the parameter vector (ρ, σ_v^2) which is unknown and needs to be estimated. Assuming normality, the parameters (ρ, σ_v^2) can be estimated by a maximum likelihood (ML) or restricted maximum likelihood (REML) procedure. Therefore, the EBLUP, $\theta^{\star c}(\rho, \sigma_v^2)$, and the naive estimator of the MSE are obtained by replacing the parameter vector (ρ, σ_v^2), with its estimator $(\hat{\rho}, \hat{\sigma}_v^2)$. The naive estimator of the MSE of the EBLUP underestimates the MSE, since it does not take into account the additional variability due to the estimation of the parameters. If $(\hat{\rho}, \hat{\sigma}_v^2)$ is a REML estimator then an approximation of $MSE[\theta^{\star c}(\hat{\rho}, \hat{\sigma}_v^2)]$ is given by Prasad and Rao (1990).

13.3 Imputation of the missing auxiliary variables

13.3.1 An overview of the missing data problem

Let us denote by $\hat{Y}$ $(D \times C)$ the matrix of estimates of the areas covered by crop types and by Z $(D \times C)$ the matrix containing the number of pixels classified by crop types according to the satellite data in each small area. $\hat{Y}$ is considered fully observed, while satellite images often haver missing data. Outliers and missing data in satellite information are mainly due to cloudy weather that does not allow the identification or correct recognition of what is being gown from the acquired digital images. Availability of good auxiliary data is fundamental to the success of any model-based method and therefore attention has been given to its control and imputation.

Missing values are a commonly occurring complication in any statistical analysis and may represent an obstacle in the way of making efficient inferences for data analysts who rely on standard software. In the presence of missing data, many statistical programs discard those units with incomplete information (case deletion or listwise deletion). However, unless the incomplete case can be considered representative of the full population the statistical analysis can be seriously biased, although in practice it is difficult to judge how large the bias may be. Moreover, even if the hypothesis of no differences between the observed and non-observed data holds, discarding the units with missing information is still inefficient, particularly in multivariate analysis involving many variables. The only good feature of the case deletion method is its simplicity, and therefore the method is often used and might be acceptable when working with a large data set and a relatively small amount of missing data. Simulations and real case studies showing the implications of the case deletion method for parameter bias and inefficiency have been reported by Barnard and Meng (1999), Brown (1994), and Graham *et al.* (1996). For further references on case deletion methods, see Little and Rubin (1987, Chapter 3).

In general, when missing values occur there are three main concerns for survey methodologists: (i) bias, because the complete case cannot be representative of the full population; (ii) inefficiency, because a portion of the sample has not been observed; and (iii) complications for the data analysis. The development of statistical methods to address missing data problems has been a dynamic area of research in recent decades (Rubin, 1976; Little and Rubin, 1987; Laird, 1988; Ibrahim, 1990; Robins *et al.*, 1994; Schafer, 1997).

Rubin (1976) developed missing data typologies widely used to classify the nature of missingness and to which we refer to describe and evaluate the different imputation methods. Adopting Rubin notation, denote by Y the complete data matrix, partitioned $Y = (Y_{obs}, Y_{mis})$, where Y_{obs} is the observed part and Y_{mis} the missing part, and by R a matrix of response indicators of the same dimension of the data matrix, that identifies what is observed and what is missing (i.e. $R_j = 1$ if the jth element of Y is observed and $R_j = 0$ otherwise). According to Rubin (1976), missingness should be considered as a probabilistic phenomenon and therefore R regarded as a set of random variables having an unknown joint distribution generally referred to as the *distribution of missingness*. Rubin (1976) defined different typologies for these distributions. The missingness mechanism is referred to as completely at random (MCAR) if

$$P(R/Y) = P(R),$$

that is, the missingness is not related to the data. In general, it is more realistic to assume that missingness mechanism is missing at random (MAR), or

$$P(R/Y) = P(R/Y_{obs}),$$

that is, the probability of missingness depends on the observed data, but not on the missing data. MCAR and MAR are also called *ignorable non-response*. If the probability of missingness $P(R/Y)$ cannot be simplified, that is, the distribution depends on the unobserved values Y_{mis}, the missing data are said to be missing not at random (NMAR). NMAR is also called *non-ignorable non-response*.

Another important classification of missing data is based upon the pattern of non-response. If the data can be arranged in matrix form, where the rows are the units and the columns the observed variables, then there are several patterns of missing data. Given a set of variables $Y_1, Y_2, \ldots, Y_p$, if the items can be ordered in such a way that there is a hierarchy of missingness (i.e. if Y_j is missing then $Y_{j+1}, \ldots, Y_p$ are missing as well), the missingness is said to be *monotone*. In general, in many realistic settings, the data set may have an arbitrary pattern, in which any variable may be missing for any unit.

Until the 1980s, missing values were mainly treated by single imputation. This is the practice of filling in missing values on incomplete records using some established procedure, and it is often used as a method to address missing data. Various single-imputation procedures have been developed in the missing data literature (Madow *et al.*, 1983; Schafer and Schenker, 2000). For a complete review of single-imputation methods and further references, see Little and Rubin (1897).

The single-imputation approach is potentially more efficient than case deletion, because no units are discarded, and has the virtue of simplicity, since the imputation produces a complete data set that can be analysed with standard methods. It is emphasized that the idea of filling in the incomplete record does not imply replacing missing information with a prediction of the missing value. In fact it is more important to preserve the distribution of the variable of interest and relationships with other variables than to predict the missing observation accurately, since the goal of the statistical procedure is to make inference about a population of interest.

However, even if single imputation saves the marginal and joint distributions of the variables of interest, it generally does not reflect the uncertainty due to the fact the missing values have been filled in by imputed values; in other words, single imputation does not take into account the fact that the imputed values are only a guess. This may have very serious implications especially if there are very many missing observations. The problems related to single-imputation methods are well documented in Rubin (2004). In general, some single-imputation methods perform better then others in terms of bias reduction, even if the non-response bias is never completely eliminated. The main problem of all single-imputation methods is the undercoverage of confidence intervals. If a nominal 95% confidence interval is considered, then the actual coverage is much lower and this is valid for all types of imputation methods and types of missingness. The low performance in terms of coverage is due to the fact that single-imputation methods tend to understate the level of uncertainty. This problem is solved by multiple-imputation methods.

13.3.2 Multiple imputation

One approach to the problem of incomplete data, which addresses the question of how to obtain valid inference from imputed data, was proposed by Rubin (1976) and explained in detail by Rubin (2004). Rubin introduces the idea of multiple imputation (MI) and he described the MI approach as a three-step process. First, instead of imputing a single value for each missing datum, $m > 1$ likely values are drawn from the predictive distribution $P(Y_{mis}/Y_{obs})$ in order to reflect the uncertainty about the true value to impute. Second, m possible alternative versions of the complete data are produced by substituting the ith simulated value, $i = 1, \ldots, m$, in the corresponding missing data. The m imputed data sets are analysed using standard procedures for complete data. Finally, the results are combined in such a way as to produce statistical inferences that properly reflect the uncertainty due to missing values; for example, confidence intervals with the correct probability coverage. Reviews of MI have been published by Rubin (1996), Schafer (1997, 1999) and Sinharay *et al.* (2001).

MI is an attractive approach in the analysis of incomplete data because of its simplicity and generality. From an operational prospective MI, like single imputation, allows surveys statisticians to adopt standard statistical procedures without additional complications. Moreover, from a statistical prospective MI is a device for representing missing data uncertainty allowing valid statistical inference.

Rubin (2004) presents a procedure that combines the results of the analysis and generates valid statistical inferences. Let Q be a scalar parameter. It could be a mean, correlation or regression coefficient. Let $\hat{Q} = \hat{Q}(Y_{obs}, Y_{mis})$ be the statistics to be used to estimate Q if no data were missing and $U = U(Y_{obs}, Y_{mis})$ the squared standard error. The method assumes that the sample is large so that the inferences for Q can be based on the assumption that $(Q - \hat{Q}) \sim N(0, U)$.

If unobserved values exist, then m independent simulated versions $Y_{mis}^{(1)}, \ldots, Y_{mis}^{(m)}$ are generated and m different data sets are analysed as if they where complete. From this we calculate the m estimates for Q and U, that is, $\hat{Q}^{(i)} = \hat{Q}^{(i)}(Y_{obs}, Y_{mis})$ and $U^{(i)} = U^{(i)}(Y_{obs}, Y_{mis})$, $i = 1, \ldots, m$. Then the combined point estimate for Q from multiple imputation is the average of the m complete-data estimates

$$\bar{Q} = m^{-1} \sum_{i=1}^{m} \hat{Q}^{(i)},$$

and the estimate of the uncertainty associated with $\bar{Q}$ is given by

$$T = \bar{U} + (1 + m^{-1})B,$$

where $\bar{U} = m^{-1} \sum_{i=1}^{m} \hat{U}^{(i)}$ is the within-imputation variance, which is the average of the m complete-data estimates, and $B = (m-1)^{-1} \sum_{i=1}^{m} (\hat{Q}^{(i)} - \bar{Q})^2$ is the between-imputation variance, which is the variance among the m complete data estimates.

Tests and confidence intervals when Q is a scalar are based on $\bar{Q} \pm kT^{1/2}$ where k is the percentile of a t distribution with $\nu = (1-m)\left[1 + \frac{\bar{U}}{(1+m^{-1})B}\right]^2$ degrees of freedom. This procedure is very general, no matter which imputation method has been used to complete the data set. When there is no missing information, the m complete-data estimates $\hat{Q}^{(i)}$ would be identical and $T = \bar{U}$. A measure of the relative increase in variance due to non-response (Rubin, 2004) is given by $r = (1 + m^{-1})B/\bar{U}$ and the estimated rate of missing information for Q is approximately $\lambda = r/(1 + r)$.

The validity of the method, however, depends on how the imputations are carried out. In order to obtain valid inferences the imputation cannot be generated arbitrarily, but they should give on average reasonable predictions for the missing data, while reflecting the uncertainty about the true value to impute. Rubin recommends basing MI on Bayesian arguments: a parametric model is specified for the complete data, a prior distribution is applied for the unknown parameters and then m independent draws are simulated from the conditional distribution of Y_{mis} given Y_{obs}. In general, suppose that the complete data follow a parametric model $P(Y/\theta)$, where $Y = (Y_{obs}, Y_{mis})$ and θ has a prior distribution. Since

$$P(Y_{mis}/Y_{obs}) = \int P(Y_{mis}/Y_{obs}, \theta)P(\theta/Y_{obs})d\theta,$$

the repeated imputation for Y_{mis} can be obtained by first simulating m independent plausible values of the unknown parameter from the observed data posterior $\theta^{(i)} \sim P(\theta/Y_{obs})$ and then drawing the missing data from the conditional predictive distribution $Y_{mis}^{(i)} \sim P(Y_{mis}/Y_{obs}, \theta)$, $1 = 1, \ldots, m$.

The computation necessary for creating MIs depends on the parametric model and on the pattern of missingness. For data sets with monotone missing patterns that assume multivariate normality the observed data posterior is a standard distribution and therefore MIs can be performed through formulas. For data sets with arbitrary missing patterns, special computational techniques such as Markov chain Monte Carlo (MCMC) methods are used to impute all missing values or just enough missing values to make the imputed data sets have monotone missing patterns. MCMC methods for continuous, categorical, and mixed multivariate data are described by Schafer (1997). Most of the techniques currently available for generating MIs assume that the distribution of missingness is ignorable, that is, they assume that the missing data are MAR.

13.3.3 Multiple imputation for missing data in satellite images

In this context an MI procedure has been used to generate repeated imputation of the missing values, with the imputation model chosen based on the following observations. The imputation model needs to be general enough to lead to valid inference in later analysis. This means that the imputation model should preserve the relationships among variables measured on a subject, that is, the joint distribution in the imputed values. For this reason, it is not necessary to distinguish between dependent and independent variables, but the variables can be treated as a multivariate response. A model that preserves the multivariate distributions of the variable will also preserve the relations of any variables with the others. Distinctions between response and covariate can be made in a subsequent analysis.

Here, it is seems reasonable to suppose that the area $\hat{Y}^c$ $(D \times 1)$ covered by a crop c, $c = 1, \ldots, C$, is somehow related to Z^c $(D \times 1)$ and therefore to impute missing values in the explanatory variables Z^c, it is assumed that the random vectors $(\hat{Y}^c, Z^c)$ have a multivariate normal distribution, that is, $(\hat{Y}^c, Z^c) \sim N(\mu^c, \Sigma^c)$, where (μ^c, Σ^c) are unknown parameters. Since information about (μ^c, Σ^c) is not available an improper prior is applied. The MCMC method is then used to obtain m independent draws from the predictive distribution.

13.4 Analysis of the 2006 AGRIT data

The aim of the AGRIT survey is to produce area estimates of the main crops, as given in Table 13.1, at a national, regional and provincial level. Direct estimates of the area LCT_c covered by crop type c ($c = 1, \ldots, C$), estimated standard errors (SE) and coefficients of variation (CV) were obtained as described in Section 13.2 for the 103 Italian provinces. The auxiliary variable is the number of pixels classified as crop type c according to the satellite data in the small area d, $d = 1, \ldots, D$, with $D = 103$.

Table 13.2 Mixed models.

Model 1. Direct estimator	$\hat{y}_d = \sum_{h=1}^{H}\left(w_{d.h}\sum_{i=1}^{n_{d.h}}\frac{y_{d.ih}}{n_{d.h}}\right)$
Model 2. Area-level model	$\hat{y}_d = \beta_0 + \beta_1 z_d + v_d + e_d$ $v \sim N(0, \sigma_v^2 I)$
Model 3. Area-level model with spatial correlation	$\hat{y}_d = \beta_0 + \beta_1 z_d + v_d + e_d$ $v \sim N(0, G)$ $G = \sigma^2[\exp(-W/\theta)]$

Table 13.3 Estimates of parameters for small-area models of LCT.

Crops	R^2	% missing	β_0	β_1	σ_v^2	β_0	β_1	ρ	σ_v^2
			Model 2			Model 3			
Durum wheat	0.94	26.1	113.6	0.09	1064.6	98.2	0.09	13.5	1069.9
Soft wheat	0.95	6.5	118.2	0.10	193.4	182.8	0.09	70.4	240.5
Barley	0.93	11.4	280.5	0.09	21.7	291.8	0.09	30.4	22.7
Maize	0.97	0.5	87.7	0.08	358.5	87.7	0.08	0.0	358.5
Sunflower	0.97	7.1	50.4	0.09	8.1	50.4	0.09	0.2	8.1
Soya	0.96	16.0	−139.3	0.09	29.6	−170.9	0.09	59.4	33.5
Sugar beet	0.94	13.0	23.2	0.09	5.6	23.2	0.09	0.9	5.6
Tomatoes	1.00	20.7	−3.8	0.08	0.9	−3.8	0.08	1.0	0.9

Table 13.4 Distribution of the LRT model 2 versus model 3 for the $M = 10$ data sets.

Crops	Min	Q_1	Median	Q_3	Max
Durum wheat	0.00	0.00	0.00	8.90	13.97
Soft wheat	0.00	11.95	15.97	18.54	22.73
Barley	0.31	0.96	1.99	2.90	4.63
Maize	0.00	0.00	0.00	0.00	0.00
Sunflower	0.00	0.00	0.00	0.00	8.12
Soya	0.16	0.31	0.43	1.35	3.67
Sugar beet	0.00	0.00	0.00	0.00	0.00
Tomatoes	0.00	0.00	0.00	0.00	0.00

When the explanatory variable Z^c has missing values we generate $M = 10$ complete data sets $(\hat{y}^c, Z_1^c), \ldots, (\hat{y}^c, Z_M^c)$ by imputing one set of the plausible values according to the procedure describe in Section 13.3. The choice of $M = 10$ is justified by Rubin (2004).

Two different mixed models, as described in Table 13.2, were used to improve the estimates of LCT_c. In the area-level model with correlation (model 3) a spatially continuous covariance function $G = \sigma^2[\exp(-W/\theta)]$ was introduced, where the elements of W are the distances between the ith and jth small areas. This choice can be motivated by a greater stability of the correlation parameter θ, with respect to the spatial autoregressive coefficient ρ, for the M imputed data sets. Table 13.3 presents the estimated parameters for the simple mixed effect model and for the mixed effect model with spatial correlation. The R^2 coefficients between the LCT estimates and the auxiliary variables, and the percentage of missing data for each crop type are also given in the table.

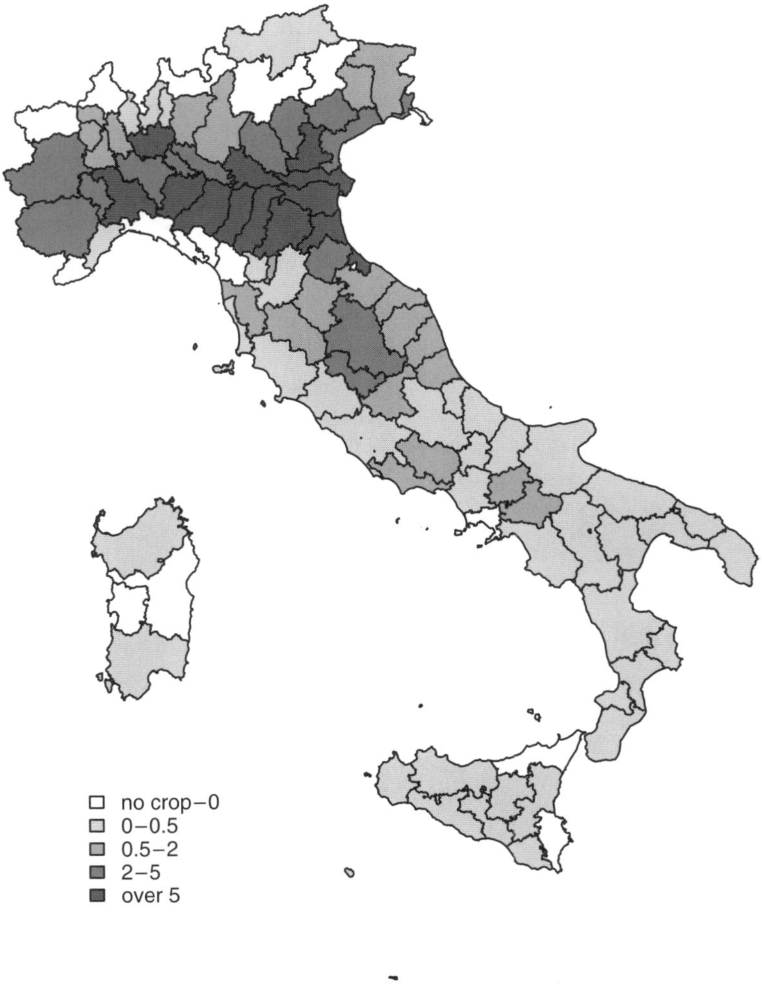

Figure 13.1 Results for soft wheat: small-area estimate under model 2 (% of area).

Table 13.4 shows the distribution of the likelihood ratio test (LRT) statistics calculated for the M different data sets, where the LRT is defined as $-2\log L \sim \chi_k^2$, with L being the ratio of the likelihoods under the two models and k is the difference between the number of parameters.

Here, the LRT compares models 2 and 3, and therefore $-2\log L \sim \chi_1^2$ under the null hypothesis of absence of spatial correlation. We find that the spatial correlation parameter θ is significant ($\chi_{1;0.05}^2 = 3.841$) for soft wheat and therefore for this crop the use of a spatial model improves the estimate of LCT. The small-area estimates for soft wheat at a provincial level are shown in Figure 13.1, and the coefficients of variation for models 1 and 2 are presented in Figures 13.2 and 13.3. A summary of the estimated standard errors (based on the D small areas) and the coefficients of variation under each model and for all crops are shown in Table 13.5. The direct survey estimates on small areas are always inefficient and can be considerably improved by using small-area models. The

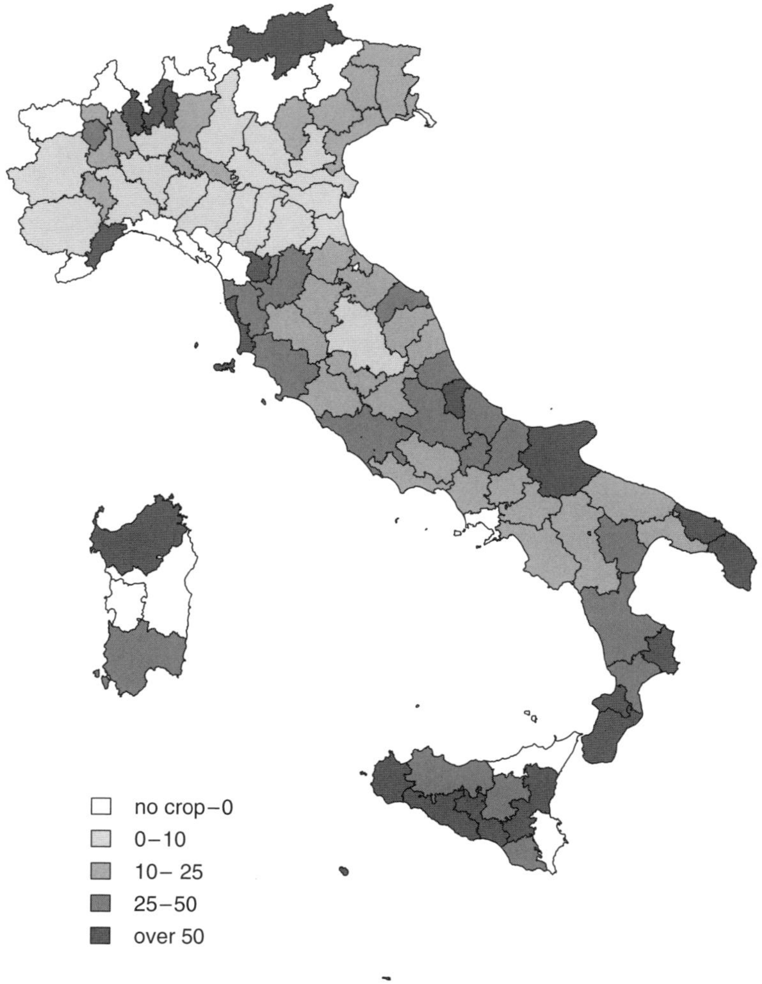

Figure 13.2 Results for soft wheat: CV of model 1.

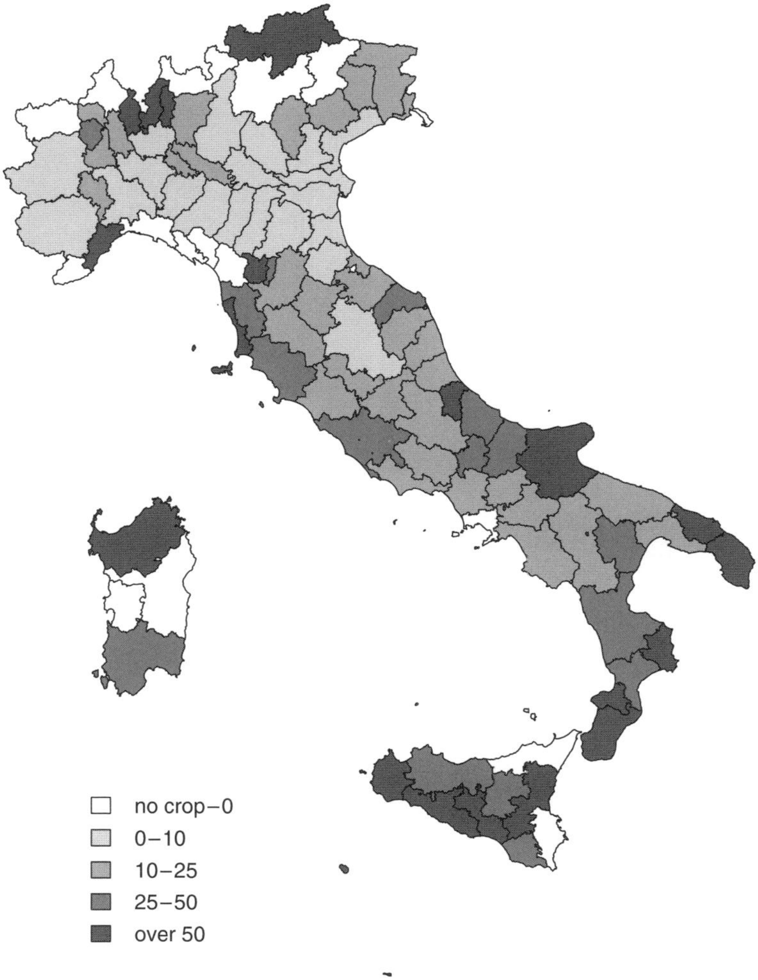

Figure 13.3 Results for soft wheat: CV of model 2.

spatial model improves the precision (SE) slightly in the case of soft wheat, where spatial autocorrelations have been found significant.

A measure of the gain in precision due to the small-area models is given by the relative efficiency defined as

$$RE_d = \left(\frac{MSE\,(\theta_d^{Mod_a})}{MSE\,(\theta_d^{Mod_b})} \right) \times 100, \tag{13.2}$$

that is, the MSE of the estimator obtained using one model against the MSE of the estimator under a different model. In Table 13.6 the minimum, maximum and average among the D small areas of this measure for the different crops are shown. There are differences in terms of RE between the crops: for durum wheat it ranges from 100 to 135, with an average of 108, whereas for tomatoes it ranges from 100 to 2783, with

Table 13.5 Average SE and CV of the estimates under models 1–3.

Crops	Model 1	Model 2	Model 3
	Average SE		
Durum wheat	848.37	799.04	799.04
Soft wheat	561.24	481.14	466.33
Barley	529.22	341.10	340.43
Maize	782.02	656.14	656.14
Sunflower	400.07	222.06	222.48
Soya	536.34	367.78	353.50
Sugar beet	370.68	204.67	204.67
Tomatoes	258.12	140.47	138.28
	Average CV		
Durum wheat	0.33	0.29	0.29
Soft wheat	0.27	0.27	0.27
Barley	0.34	0.33	0.33
Maize	0.27	0.21	0.21
Sunflower	0.22	0.21	0.21
Soya	0.41	0.34	0.34
Sugar beet	0.37	0.33	0.33
Tomatoes	0.42	0.34	0.34

Table 13.6 Percentage relative efficiency of the direct estimator in comparison with small-area models 2–3.

Crops	Min	Max	Average	Min	Max	Average
	RE model 2			RE model 3		
Durum wheat	100	135	108	100	136	108
Soft wheat	100	218	121	100	256	127
Barley	100	595	233	100	594	233
Maize	100	328	125	100	328	125
Sunflower	89	926	289	91	949	289
Soya	83	1084	210	100	1024	221
Sugar beet	83	1032	315	83	1033	315
Tomatoes	100	2783	492	100	3058	521

an average of 492. These differences can be explained by the differences in the spectral signatures between the different crops.

Figures 13.4–13.6 show the distribution of the provinces by RE for soft wheat, comparing models 1 and 2, models 1 and 3 and models 2 and 3, respectively. The scatter plots of the SE for the different comparisons of models are also given. Figures 13.4(a) and 13.5(a) indicate a large gain of efficiency by using remote sensing data in the estimator, whereas there is a small improvement by introducing spatial effects (Figure 13.6).

Finally, Figure 13.7 compares the distribution of the provinces by RE for soft wheat, for model 2 with a single imputation and with $M = 10$ different imputations to fill in

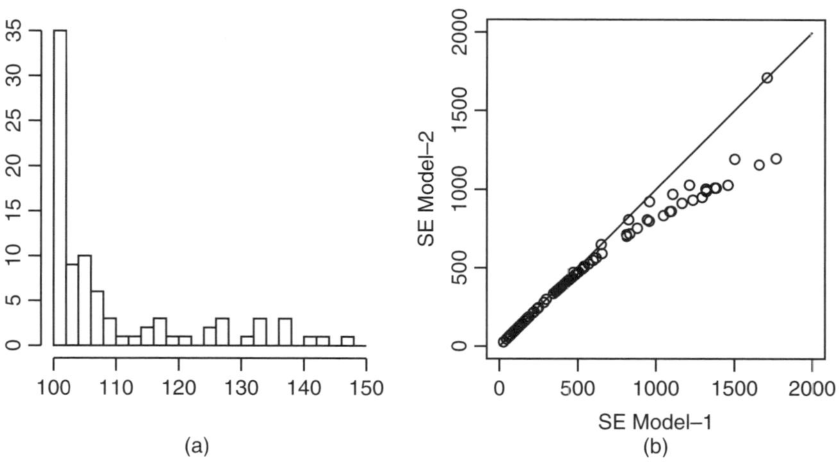

Figure 13.4 (a) Number of provinces by class of RE model 1 vs model 2; (b) SE model 1 vs SE model 2.

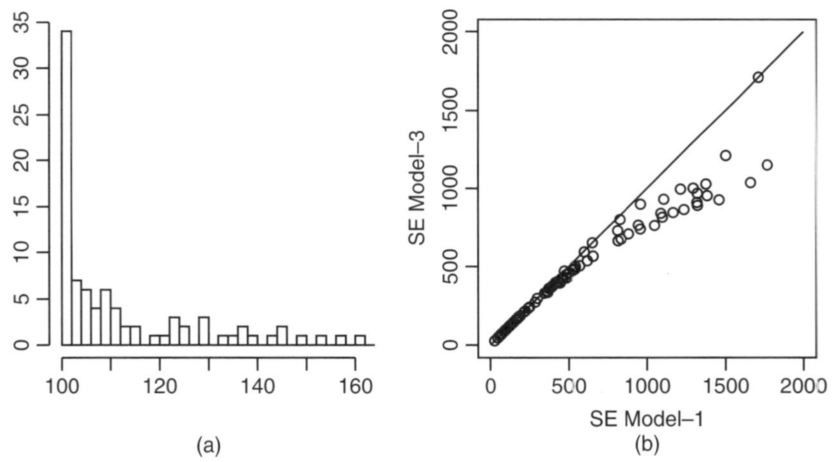

Figure 13.5 (a) Number of provinces by class of RE model 1 vs model 3; (b) SE model 1 vs SE model 3.

the missing information. The loss of efficiency underlined in Figure 13.7 reflects the uncertainty about the true value to impute (Rubin, 2004), as discussed in Section 13.3. This will allow valid confidence intervals from imputed data.

13.5 Conclusions

The results of this study confirm the superiority of the small-area models in comparison to the direct survey estimator, that is, the direct survey estimates can be definitely improved by defining basic area models. The introduction of a spatial autocorrelation among the small areas can be also used to increase the efficiency, but the model must be applied after an evaluation of the significance of the spatial correlation among the small areas.

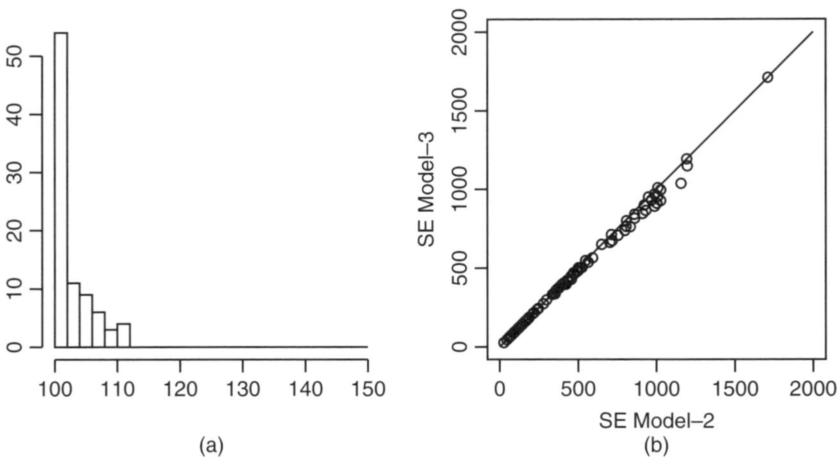

Figure 13.6 (a) Number of provinces by class of RE model 2 vs model 3; (b) SE model 2 vs SE model 3.

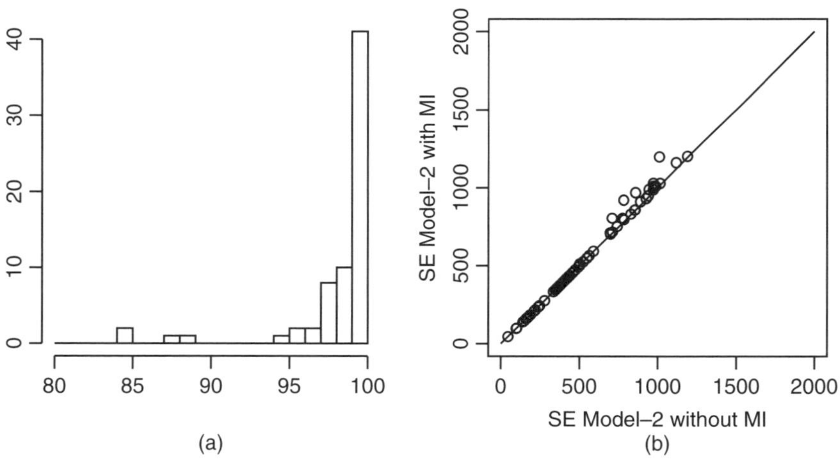

Figure 13.7 (a) Number of provinces by class of RE: model 2 without MI vs model 2 with MI; (b) SE model 2 without MI vs SE model 2 with MI.

Moreover, the choice of an appropriate proximity matrix W, respresenting the interaction between any pair of small areas, can be considered to strengthen the small-area estimates.

The introduction of a spatial autocorrelation among the small areas can be further exploited by using an alternative approach to small-area estimation. For instance, Salvati *et al.* (2007) incorporate spatial information into the of M-quantile model (Chambers and Tzavidis, 2006) by assuming that the regression coefficients vary spatially across the geography of interest (geographically weighted regression). The M-quantile regression model and the M-quantile geographically weighted regression model allow modelling of between-area variability without the need to specify the area-specific random components and therefore can be easily adapted to different needs.

Finally, it should be emphasized that the use of multiple imputation to fill in the missing information raises the possibility that the statistical model used to complete the data sets may be different or may be incompatible with that used in the analysis. In this context the behaviour of the small-area predictors can be analysed under a different imputer's model. Multiple-imputation inference when the imputer's model and the analyst's model are different has been exploited by Meng (1995) and Rubin (1996).

References

Allen, J.D. (1990) A look at the Remote Sensing Applications Program of the National Agricultural Statistics Service. *Journal of Official Statistics*, 6, 393–409.

Barnard, J. and Meng, X.L. (1999) Application of multiple imputation in medical studies: From AIDS to NHANES. *Statistical Methods in Medical Research*, 8, 17–36.

Brown, R.L. (1994) Efficacy of the indirect approach for estimating structural equation models with missing data: A comparison of five methods. *Structural Equation Modeling*, 1, 287–316.

Chambers, R. and Tzavidis, N. (2006) Models for small area estimation. *Biometrika*, 93, 255–268.

Cochran, W.G. (1977) *Sampling Techniques*, 3rd edition. New York: John Wiley & Sons. Inc.

Cressie, N. (1991) Small-area prediction of undercount using the general linear model. *Proceedings of Statistics Symposium 90: Measurement and Improvement of Data Quality*, pp. 93–105. Ottawa: Statistics Canada.

Flores, L.A. and Martnez, L.I. (2000) Land cover estimation in small areas using ground survey and remote sensing. *Remote Sensing of Environment*, 74(2), 240–248.

Graham, J.W., Hofer, S.M. and MacKinnon, D.P. (1996) Maximizing the usefulness of data obtained with planned missing value patterns: An application of maximum likelihood procedures. *Multivariate Behavioral Research*, 31, 197–218.

Hansen, M.H., Hurwitz, W.N. and Madow, W.G. (1953) *Sample Survey Methods and Theory*. New York: John Wiley & Sons, Inc.

Ibrahim, J.G. (1990) Incomplete data in generalized linear models. *Journal of the American Statistical Association*, 85, 765–769.

Laird, N.M. (1988) Missing data in longitudinal studies. *Statistics in Medicine*, 7, 305–315.

Little, R.J.A. and Rubin, D.B. (1987) *Statistical Analysis with Missing Data*. New York: John Wiley & Sons, Inc.

Madow, W.G., Nisselson, J. and Olkin, I. (1983) *Incomplete Data in Sample Surveys: Vol. 1. Report and Case Studies*. New York: Academic Press.

Meng, X.L. (1995), Multiple-imputation inferences with uncongenial sources of input (with discussion). *Statistical Science*, 10, 538–573.

Prasad, N. and Rao, J.N.K. (1990) The estimation of the mean squared error of small-area estimators. *Journal of the American Statistical Association*, 85, 163–171.

Rao, J.N.K. (2003) *Small Area Estimation*. Hoboken, NJ: John Wiley & Sons, Inc.

Robins, J.M., Rotnitzky, A. and Zhao, L.P. (1994) Estimation of regression coefficients when some regressors are not always observed. *Journal of the American Statistical Association*, 89, 846–866.

Rubin, D.B. (1976) Inference and missing data. *Biometrika*, 63, 581–592.

Rubin, D.B. (1996) Multiple imputation after 18+ years. *Journal of the American Statistical Association*, 91, 473–489.

Rubin, D.B. (2004) *Multiple Imputation for Nonresponse in Surveys*. Hoboken, NJ: Wiley-Interscience.

Salvati, N., Tzavidis, N., Pratesi, M. and Chambers, R. (2007) Small area estimation via M-quantile geographically weighted regression. CCSR Working Paper 2007-09, University of Manchester.

Schafer, J.L. (1997) *Analysis of Incomplete Multivariate Data*. London: Chapman & Hall.

Schafer, J.L. (1999) Multiple imputation: a primer. *Statistical Methods in Medical Research*, 8, 3–15.

Schafer, J.L. and Schenker, N. (2000) Inference with imputed conditional means. *Journal of the American Statistical Association*, 95, 144–154.

Sinharay, S., Stern, H.S. and Russell, D. (2001) The use of multiple imputation for the analysis of missing data. *Psychological Methods*, 6, 317–329.

Part IV

DATA EDITING AND QUALITY ASSURANCE

14

A generalized edit and analysis system for agricultural data

Dale Atkinson and Carol C. House
US Department of Agriculture, National Agricultural Statistics Service, USA

14.1 Introduction

In 1997 the responsibility for the quinquennial census of agriculture was transferred from the US Bureau of the Census (BOC) to the National Agricultural Statistics Service (NASS) in the US Department of Agriculture. This fulfilled a goal of NASS to become the national source of all essential statistics related to USA agriculture. It also provided an opportunity for the Agency to improve both the census and its ongoing survey and estimation programme through effective integration of the two.

The timing of the transfer, however, severely limited the changes the NASS could make for the 1997 Census of Agriculture. To complete this census the NASS formed a Census Division that had primary responsibility for managing the day-to-day operations of the census activities. This Division included former BOC employees who transferred to the NASS with the census. Much of the data collection, data capture and editing was contracted out to the BOC's National Processing Center (NPC) in Jeffersonville, Indiana, which had also assumed these functions in previous censuses.

The NASS was able to make significant changes in some of the census processes. Specifically, the Agency was able to utilize its 45 state statistical offices (SSOs) in coordinating census data collection with that of its ongoing survey programme. The SSOs also played a key role in the processes from macro-level editing through the final review of the data for publication. In previous censuses these processes had been centralized and the

Agricultural Survey Methods Edited by Roberto Benedetti, Marco Bee, Giuseppe Espa and Federica Piersimoni

states' data were reviewed sequentially, in a predetermined order. By decentralizing the review process, the states' data were reviewed concurrently – significantly reducing the time from initial data aggregation to publication. This allowed the publication of 1997 census data a year earlier than those of previous censuses.

However, some of the main benefits of the NASS acquiring the census of agriculture have yet to be realized – in particular, a proper integration of the census programme with the NASS's traditional programme figures to improve the quality and efficiency of each. These are benefits that Agency management has targeted for 2002 and beyond. To begin the process of integrating the programmes the NASS took two major steps. The first of these was the creation in late 1998 of the Project to Reengineer and Integrate Statistical Methods (PRISM). The team named to manage this project was charged with conducting a comprehensive review of all aspects of the NASS statistical programme and recommending any needed changes. The second step was a major structural reorganization of the Agency. This reorganization essentially absorbed the staff and functions of the Census Division, as formed for the 1997 census, into an enhanced survey/census functional structure. The reorganization was designed to increase efficiency and eliminate duplication of effort by integrating census responsibilities throughout the structure.

The census processing system needed to be re-engineered before 2002. With the transfer of census responsibility in 1997, the NASS had inherited an ageing system that had been used, largely unmodified, since 1982. It was out of date technology-wise and, to a lesser extent, methodology-wise. The system was relatively inflexible in that decision logic tables (DLTs) were 'hard coded' in Fortran. It was programmed to run on aging DEC VAX machines running the VMS operating system. While manual review and correction could be performed on standard PC screens, some functionality was lost when the system was used with display terminals other than the amber-screened DEC terminals for which it was designed. In general, the record review and correction process at both the micro and macro levels involved navigating an often frustrating combination of function and control keys. The system had served its purpose until the processing of the 1997 census, but it was time for a more up-to-date system.

In September 1999 the Processing Methodology Sub-Team of PRISM was chartered to specify a new edit, imputation and analysis system for the 2002 Census of Agriculture and subsequent large NASS surveys. This group reviewed the editing literature and processing systems used in NASS and other organizations (US Bureau of the Census, 1996; Weir, 1996) to synthesize the best of what was available into its recommendations for the new system. In February 2000 it published its findings and recommendations in an internal Agency research report. The report highlighted the team's guiding principles as follows:

1. *Automate as much as possible, minimizing required manual intervention.* Having dealt exclusively with much smaller sample surveys in the past, the NASS culture has been to touch every questionnaire and have statisticians manually specify needed data changes in response to automated edit flags. The sheer volume of data precludes this option for the census and necessitates a system that makes more editing/imputation decisions automatically, without manual intervention.

2. *Adopt a 'less is more' philosophy with regard to editing.* There is a tendency in many organizations to over-edit data automatically and/or manually. A leaner edit that focuses on critical data problems is less resource intensive and often more effective than a more complex one.

3. *Identify real data and edit problems as early as possible.* One of the concerns about the edit used for the 1997 census was that SSO analysts had nothing to review from the highly automated process for several months after editing started. Except for a few who were temporarily detailed to the NPC to correct edit failures, SSO statisticians were unable to see the data until they were weighted for non-response and aggregated. This was often six months after initial data collection. The delay caused problems that could have been more effectively handled earlier in the process and imposed additional stress on the SSOs by complicating and compressing their data review time.

4. *Design a system that works seamlessly.* While 'seamless' means different things to different people, what is needed is a system in which all the components interrelate smoothly such that the analyst can quickly and easily navigate to any screen and get any auxiliary data needed to identify and resolve a data problem. A system is definitely not seamless if the user has to log into various computer systems separately to obtain needed auxiliary data or run an ad hoc query. Lack of 'seamlessness' was a problem that reduced the effectiveness of the 1997 census processing system.

5. *Use the best features of existing products in developing the new system.* By the time the 1997 Census of Agriculture was completely put to rest, the 2002 Census of Agriculture was uncomfortably close at hand. The short developmental time would preclude 'reinventing the wheel.' It was imperative that the NASS incorporate the best aspects of what it and other organizations had already done research-wise and developmentally to expedite the process as much as possible.

In view of the above guiding principles the sub-team documented the features it felt the new system should include (Processing Methodology Sub-Team, 2000). Considerable emphasis was placed on minimizing unnecessary review and on the visual display of data. The sub-team discussed display attributes and methodologies that could be used to identify problematic data with high potential impact on published estimates. The 'features' section of their paper discussed the issue of refreshing the review screens as error corrections are made and stressed the need for the system to help manage the review process (i.e. to identify records that had already been reviewed, through colour and/or special characters). The sub-team concluded its paper with the following recommendations:

(i) As far as possible, use the Fellegi–Holt methodology in the new system.

(ii) Have the computer automatically correct everything with imputation at the micro level (i.e. eliminate the requirement for manual review).

(iii) Utilize the NASS data warehouse as the primary repository of historical data and ensure that it is directly accessible by all modules of the new system.

(iv) Design the system with tracking and diagnostic capabilities to enable the monitoring of the effect of editing and imputation. Develop analytics for a quality assurance programme to ensure edited/imputed data are trusted.

(v) Incorporate a score function to prioritize manual review.

(vi) Provide universal access to data and programme execution within the Agency.

(vii) Ensure that the system is integrated into the Agency's overall information technology architecture.

(viii) Make the system generalized enough, through modular design, to work over the entire scope of the Agency's survey and census programmes.

(ix) Enable users to enter and access comments anywhere in the system.

(x) Present as much pertinent information as possible on each screen of the system and provide on-screen help for system navigation.

(xi) Consider the use of browser and Java programming technology to assist in integrating parts of the system across software, hardware, and functions.

(xii) Designate a developmental team to take this report, develop detailed specifications and begin programming the system.

14.2 System development

In response to recommendation (xii), a number of working groups were formed to focus on various aspects of the processing system development. These included groups addressing check-in, data capture, edit specifications, interactive data review (IDR) screens, imputation, analysis, and census coverage evaluation. In order to ensure consistency of decisions across the working groups in assembling the system an oversight and technical decision-making body, the Processing Sub-Team, was formed of the leaders of the individual working groups. This sub-team was charged with considering the overall system flow and ensuring that the individual modules work together effectively. The sub-team members keep each other informed about the activities of their individual groups, thus ensuring consistency and that required connectivity is addressed. The sub-team also serves as the technical decision-making body for cross-cutting decisions that cannot be made by the individual working groups. The following sections describe plans for selected modules of the system, the progress made to date and some key issues that the working groups are grappling with.

14.2.1 Data capture

As was the case in 1997, the NASS will contract the printing, mailing and check-in of questionnaires and the data capture activities to the NPC. While all data capture for the 1997 Census of Agriculture was accomplished through key entry, the NASS's early discussions in preparing for 2002 indicated that scanning could be used to capture both an image of the questionnaire for interactive data review and the data themselves, through optical/intelligent character recognition (OCR/ICR). Preliminary testing done with the Agency's Retail Seed Survey supported the practicality of using scanning for data capture. Testing of the OCR/ICR process for this survey was conducted at three different confidence levels (65, 75 and 85%). The outcome of this small test was that at 65%, 4% of the characters were questionable; at 75%, 5–7%; and at 85%, 13%.

The NASS will utilize scanning with OCR/ICR as the primary mode of data capture for 2002. Current plans are to start with the industry standard confidence level of 85%, but this might be adjusted with further experience in using the system with agricultural census data. Results from the recently completed census of agriculture content test should help fine-tune the process. Questionable returns will be reviewed, with erroneous data re-entered by correct-from-image (CFI) key-entry operators. The scanning process will produce data and image files which will be sent to the Agency's leased mainframe computers at the National Information Technology Center (NITC) in Kansas City, Missouri, for further processing. The data will pass into the editing system and the images will be brought into the interactive data review screens that will be activated from the analysis system to review and correct problematic data.

14.2.2 Edit

As the edit groups began to meet on a regular basis, the magnitude of the task of developing a new editing system became obvious. The machine edit/imputation used for the 1997 census was enormous. It had consisted of 54 sequentially run modules of approximately 50 000 lines of Fortran code, and the sheer volume of the input DLTs was staggering. Up to 1997, the census questionnaires had changed very little from one census to the next, so the DLTs and Fortran code had required little modification. For 2002, however, an entirely new processing system would be built on a questionnaire that was also undergoing radical changes. Some of the questionnaire changes were necessitated by recent structural changes in agricultural production and marketing, while others were due to the planned use of OCR/ICR for data capture. In any case, the group members were saddled with the onerous task of working through the mountainous DLTs from 1997 to determine what routines were still applicable and, of these, which should be included in the new system specifications.

One of the key edit issues is reducing manual review without damaging data quality. In processing the 1997 census data, the complex edit corrected all critical errors and the staff at Jeffersonville manually reviewed all 'warning' errors. The approximate workload and time invested in this activity follows:

- Approximately 1.8 million records passed through the edit at least once. Of these, 470 000 (26%) were flagged with warning errors. About 200 000 (47%) of the flagged records required updates.
- About 4000 staff days were spent performing the review in Jeffersonville.

For 2002, the edit review (and analysis) will be performed in the NASS's SSOs. Considering the expected staff shortages in 2002 relative to 1997, the above figures would represent an intolerable commitment of staff resources. Furthermore, indications are that this amount of manual review is not altogether needed or (in some cases) desirable. Table 14.1 shows the relative impact of the automatic (computer) edit changes with no manual review; edit changes with/from manual review; and changes made during analytic review. Due to deficiencies in the edit coding, some changes made totally by computer could not be cleanly broken out from those with manual intervention, resulting in an overstatement of the manual edit effect. All changes made during analytic review resulted from human interaction and are considered part of the impact of manual review.

Table 14.1 Relative impact of the editing/imputation/analysis processing of the 1997 Census of Agriculture data (USA level).

Characteristic	Net effect of automated edit changes (%)	Net effect of edit manual review (%)	Net effect of analytic review (%)	Total effect (%)	Total manual effect (%)
Corn acres	(0.24)	(3.97)	0.26	(3.94)	(3.71)
Soybean acres	(0.20)	(2.33)	0.31	(2.22)	(2.02)
Wheat acres	(0.69)	(4.18)	(0.01)	(4.88)	(4.19)
Cotton acres	(0.10)	(0.29)	(0.27)	(0.66)	(0.56)
Cranberry acres	0.13	1.72	(4.04)	(2.18)	(2.32)
No. of cattle	0.74	4.75	(0.74)	4.74	4.01
No. of hogs	0.17	(4.23)	(3.92)	(7.98)	(8.15)

Table 14.1 shows that the overall effect of the edit/imputation/analysis process was relatively small for most items, especially crop acreages. Considerably larger adjustments are required for both non-response and undercoverage. While admittedly these numbers only reflect the impact on high-level aggregates (USA level) and the processing can often be more beneficial at lower levels (e.g. county totals), the size of the adjustments still raises questions about the efficacy of the extremely resource-intensive data editing and review process. Such considerations underpinned our two guiding principles of adopting a ‘less is more’ philosophy to editing and of automating as much as possible.

14.2.3 Imputation

Certainly one of the key considerations in moving to an automated system is determining how to impute for missing and erroneous data. The imputation group is currently working through the census questionnaire question by question and section by section to determine the most effective routines to use. Nearest-neighbour donor imputation will play a strong role in filling data gaps. The group is currently developing an SAS-based donor imputation module, which will provide near-optimal imputations in certain situations where high-quality matching variables are available. The group will be leveraging the Agency’s relatively new data warehouse capabilities of providing previously reported survey data. The data warehouse was populated with the 1997 census data and contains the data from most of the Agency’s surveys since 1997. As such, it serves as a valuable input into the imputation process, since many of the respondents in the current survey will have responded to one or more previous surveys. The warehouse data can provide direct imputations in some cases and identify items requiring imputation in many others.

A review of the imputation done for the 1997 Census of Agriculture and the plans for 2002 indicates the vast majority of the imputation performed will be deterministic (e.g. forcing subparts to equal a total). Deterministic imputation could amount to 70-80% of all imputation for the 2002 Census of Agriculture. Nearest-neighbour donor imputation will likely account for 10–20%, while direct imputation of historical data perhaps 5–10%.

14.3 Analysis

14.3.1 General description

The analysis system is perhaps the module of interest to the broadest audience in the NASS. This module will provide the tools and functionality through which analysts in Headquarters and our SSOs will interact with the data. All processes prior to this point are ones with no manual intervention or, in the case of data capture, one in which only a few will touch the data. As one of our senior executives aptly put it: 'All this other stuff – data capture, edit and imputation – will happen while I'm sleeping. I'm interested in what will go on when my eyes are open.' That's analysis!

Because of the broad interest in and the expected large number of users of the analysis system, the development team has made a special effort to solicit user input into its specification. The working group chartered to design and program this module circulated a hard-copy prototype of the proposed system to staff throughout the Agency early in 2001. This exercise resulted in very useful feedback from potential users. The feedback received was subsequently worked into the module specifications.

14.3.2 Micro-analysis

After the data have been processed through the edit and imputation steps, during which essentially all critical errors have been computer-corrected, they are ready for SSO review in the analysis system. The first of two analysis phases, micro-analysis, begins immediately. During micro-analysis SSOs will review (and update, if necessary) all records for which imputation was unsuccessful, all records failing consistency checks, and all those with specific items that were flagged for mandatory review. Such records are said to contain critical errors and must be corrected. This work will be done while data collection is ongoing, and will allow ample time for any follow-up deemed necessary. As review time permits, the system will also provide the capability to review records that have no critical errors, but may be nonetheless of concern. These would include those identified by the computer as influential or high scoring or with potential problems identified through graphical analytic views. Unlike the 1997 edit, warning errors will not be automatically corrected nor require manual intervention.

A score function is being developed for the 2002 Census of Agriculture to ensure that the records manually reviewed are those that are expected to have a substantial impact on aggregate totals. The quality of county aggregates is of particular concern with the census of agriculture. Therefore, the score function used for 2002 will be one that assigns high scores to records whose current report for selected characteristics represents a large percentage of the previous census's county total for that characteristic.

Micro-level graphics are simply a collection of record-level information shown together for all records for a specific item(s) of interest. The user will have the option of subsetting the graph by selecting a group of points or by specifying a subsetting condition. For some plots, the option of additional grouping and/or subgrouping of a variable(s) through the use of colours and symbols will be available (e.g. by size of farm, type of operation, race, total value of production, other size groups). Scatter plots, box-plots and frequency bar charts of various types will be provided. All graphics will provide drill-down capability to data values and the IDR screens to review and update problematic records.

Finally, the system will track IDs that have been previously reviewed, compare current values to historic data, allow for canned and ad hoc queries and have a comments feature to document actions. Micro-analysis will also include tables to review previously reported data for non-responding units. This will allow SSOs to focus non-response follow-up efforts on the most 'important' records.

14.3.3 Macro-analysis

The second phase of analysis, macro-analysis, begins immediately after preliminary weighting (adjusting for undercoverage and non-response). Macro-analysis uses tables and graphs to review data totals and farm counts by item, county and state. While the macro-analysis tools will retain the key objectives of the analytical review system used for the 1997 census, it will be much more interactive and user-friendly. The focal point of the macro-analysis will be a collection of graphics showing aggregate data at state and county levels. These graphics will include dot plots or bar charts of county rankings with historic comparisons, state maps with counties colour-coded by various statistics and scatter plots of current versus previous data.

The new macro-analysis tool will also be integrated more effectively with the Agency's data warehouse and its associated standard tools for user-defined ad hoc queries. Graphics or tables will be used to compare current census weighted totals and farm counts against previous census values and other published estimates. There will be a prepared library of database queries, in addition to the ability to build your own. Analysts will drill down to the IDR screens to verify/update records. If micro-analysis is done effectively, the number of issues to be dealt with in this phase will be fewer than in 1997, when no micro-analysis module was available.

The macro-edit can be run as soon as data collection is complete, the last records are run through edit and imputation, and preliminary weights are available. The objective of the macro-review will be the same as for 1997. That is, an analyst will be responsible for the complete review of all the state and county totals. According to a state's particular needs and characteristics the SSO's managers can elect to assign an analyst to a county for the review of all items, have a commodity specialist review items by state and county, or use a combination of both. In any case, every item in each county must be reviewed, and a check-off system will be provided in the analysis system to ensure this is achieved.

14.4 Development status

Timelines have been developed for specification and development of the various modules, and the groups are working hard to stick to them. Due to a number of factors beyond their control the developmental work started at least a year later than it should have, considering the magnitude of the system overhaul. In spite of the delays and overall staff shortages as compared to what was available for past censuses, the groups have done a fantastic job of moving ahead with the developmental work.

14.5 Conclusions

One of the key issues in edit development is determining what edits are essential to ensure the integrity of the data without over-editing. This is something that the edit group and Processing Sub-Team have struggled with. The team members represent an interesting blend of cultures. The longer-term, pre-census NASS staff developed within a culture of processing the returns from its sample surveys, where every questionnaire is hand-reviewed and corrected as necessary. While there is a need for some of this extra attention for sample surveys since survey weights can be high, this type of approach is nonetheless prone to manual over-editing. The NASS staff who came over with the census are much more comfortable with automatic editing/imputation and are perhaps overly interested in having the system account for every possible data anomaly. This approach can lead to an excessively complex system that automatically over-edits data.

The combination of these two cultures has resulted in some interesting discussions and decisions relative to the guiding principles of automating as much as possible and adopting a 'less is more' philosophy of editing. Everyone has his or her own pet anecdote indicating a 'crucial' situation that a reduced edit would not identify and correct. Such concerns have resulted in some compromises in the editing approach taken for 2002. The new processing system currently being developed will look more like an SAS version of the 1997 edit than the greatly reduced, predominantly error-localization driven system envisioned by the Processing Methodology Sub-Team. The bottom line for 2002 is that there will be 49 DLT edit modules, which will consist of much of the same type of intensive, sequential 'if–then' edit conditions that existed in 1997. There are some notable differences in the processes, however. There will be a presence of GEIS-type error localization (Statistics Canada, 1998) in the new system in addition to the 1997 style editing. Imputation has been moved out of the edit DLTs to a separate module of the system. This module will make strong use of nearest-neighbour donor imputation, enhanced by previously reported data from the Agency's data warehouse. The error-localization presence will help ensure that the imputations will pass all edits. The approach to be used in 2002 will serve as a foothold for the approach (Fellegi and Holt, 1976) initially endorsed by the Processing Methodology Sub-Team. For 2007 there will be a strong push to simplify the edit and increase the error-localization presence or move to the NIM-type approach of Statistics Canada (Bankier, 1999).

Another key issue in assembling the system lies in how much modularity/generality is possible. All efforts currently are, and need to be, directed at having a system in place and tested for the 2002 Census of Agriculture. Due to the tight time frame the developmental team is working with, some compromise on the goal of generality is inevitable. The evolving system is being developed modularly, however, so some retrofitting of generality should be possible.

One of the questions that have yet to be answered is whether runtimes and response times will be adequate. The current plans for the processing system are complex, requiring considerable cycling through the various sections of the questionnaire. Whether or not the runtimes on batch aspects of the system and response times on the interactive portions will be within workable tolerances will not be fully known until more of the system is built. If the answer in either case turns out to be negative, shortcuts will need to be taken to make the system workable.

Exactly what combination of processing platforms will be used in the final system is another issue that has yet to be fully decided. It will be comprised of some combination of the Agency's leased mainframe, its UNIX boxes and its Windows 98 machines on a Novell wide-area network. Since the system is being written in SAS, which will run on any of the three platforms, the processing platform decision has been delayed up to now. However, in the interest of seamless interoperability it will need to be made soon.

References

Bankier, M. (1999) Experience with the new imputation methodology used in the 1996 Canadian Census with extensions for future censuses. Paper presented to the Conference of European Statisticians, Rome, 2–4 June.

Fellegi, I.P., Holt, D. (1976) A systematic approach to automatic edit and imputation. *Journal of the American Statistical Association*, 71, 17–35.

Processing Methodology Sub-Team (2000) Developing a state of the art editing, imputation and analysis system for the 2002 Agricultural Census and Beyond. NASS Staff Report, February.

Statistics Canada (1998) Functional description of the Generalized Edit and Imputation System. Technical report.

US Census Bureau (1996) StEPS: Concepts and Overview. Technical report.

Todaro, T.A. (1999) Overview and evaluation of the AGGIES Automated Edit and Imputation System. Paper presented to the Conference of European Statisticians, Rome, 2–4 June.

Weir, P. (1996) Graphical Editing Analysis Query System (GEAQS). In *Data Editing Workshop and Exposition, Statistical Policy Working Paper 25*, pp. 126–136. Statistical Policy Office, Office of Management and Budget.

15

Statistical data editing for agricultural surveys

Ton De Waal and Jeroen Pannekoek
Department of Methodology, Statistics Netherlands, The Hague, The Netherlands

15.1 Introduction

In order to make well-informed decisions, business managers, politicians and other policy-makers need high-quality statistical information about the social, demographic, industrial, economic, financial, political, cultural, and agricultural aspects of society. National statistical institutes (NSIs) fulfil a very important role in providing such statistical information. The task of NSIs is complicated by the fact that present-day society is rapidly changing. Moreover, the power of modern computers enables end-users to process and analyse huge amounts of statistical information themselves. As a result, statistical information of increasingly great detail and high quality must be provided to end-users. To fulfil their role successfully NSIs also need to produce these high-quality data more and more quickly. Most NSIs have to face these tasks while their financial budgets are constantly diminishing.

Producing high-quality data within a short period of time is a difficult task. A major complicating factor is that data collected by statistical offices generally contain errors. In particular, the data collection stage is a potential source of errors. Many things can go wrong in the process of asking questions and recording the answers. For instance, a respondent may have given a wrong answer (either deliberately or by mistake) or an error may have been made at the statistical office while transferring the data from the questionnaire to the computer system. The presence of errors in the collected data makes

Agricultural Survey Methods Edited by Roberto Benedetti, Marco Bee, Giuseppe Espa and Federica Piersimoni

it necessary to carry out an extensive statistical data editing process of checking the collected data, and correcting it where necessary. For an introduction to statistical data editing in general we refer to Ferguson (1994).

Statistical agencies have always put a lot of effort and resources into data editing activities. They consider it a prerequisite for publishing accurate statistics. In traditional survey processing, data editing was mainly an interactive activity with the aim of correcting all data in every detail. Detected errors or inconsistencies were reported and explained on a computer screen. Clerks corrected the errors by consulting the form, or by recontacting the supplier of the information. This kind of editing is referred to as interactive or manual editing.

One way to address the challenge of providing huge amounts of high-quality data quickly is by improving the traditional editing and imputation process. Studies (see Granquist, 1995, 1997; Granquist and Kovar, 1997) have shown that generally not all errors have to be removed from each individual record (i.e. the data on a single respondent) in order to obtain reliable publication figures. It suffices to remove only the most influential errors. These studies have been confirmed by many years of practical experience at several NSIs. Well-known, but still relatively modern, techniques such as selective editing, macro-editing, and automatic editing can be applied instead of the traditional interactive approach. Selective (or significance) editing (Lawrence and McDavitt, 1994; Lawrence and McKenzie, 2000; Hedlin, 2003; see also Section 15.4 below) is applied to split the data into two streams: the critical stream and the non-critical stream. The critical stream consists of those records that are the most likely to contain influential errors; the non-critical stream consists of records that are unlikely to contain influential errors. The records in the critical stream are edited in a traditional, manual manner. The records in the non-critical stream are either not edited or are edited automatically. In practice, the number of records edited manually often depends directly on the available time and resources. Macro-editing (Granquist, 1990; De Waal *et al.*, 2000), that is, verifying whether figures to be published seem plausible, is often an important final step in the editing process. Macro-editing can reveal errors that would go unnoticed with selective editing or automatic editing. When automatic editing is applied the records are entirely edited by a computer instead of by a clerk. Automatic editing can lead to a substantial reduction in costs and time required to edit the data.

In this chapter we discuss selective and automatic editing techniques that can be used for agricultural surveys and censuses. We focus on automatic editing, for which we describe several algorithms. Agricultural surveys and censuses often contain many variables and many records. Especially for such surveys and censuses, automatic editing is an important editing approach. Provided that automatic editing has been implemented correctly in a general editing strategy (see Section 15.3) it is a very efficient approach, in terms of costs, required resources and processing time, to editing records. Our aim is to edit as many records as possible automatically, under the condition that the final edited data are of sufficiently high quality. In our opinion selective editing is an important companion to automatic editing as selective editing can be used to select the records that need interactive editing, thereby at the same time determining the records on which automatic editing can be applied successfully.

To automate the statistical data editing process one often divides this process into two steps. In the first step, the error localization step, the errors in the data are detected. During this step certain rules, so-called edits, are often used to determine whether a record is consistent or not. An example of such an edit is that the value of sales of milk and the value of sales of other agricultural products of a dairy farm must sum up to the total value of sales. These edits are specified by specialists based on statistical analyses and their subject-matter knowledge. Inconsistent records are considered to contain errors, while consistent records are considered error-free. If a record contains errors, the erroneous fields in this record are also identified in the error localization step. In the second step, the imputation step, erroneous data are replaced by more accurate data and missing data are imputed. The error localization step only determines which fields are considered erroneous; the imputation step actually determines values for these fields. To automate statistical data editing both the error localization step and the imputation step need to be automated. In this chapter we restrict ourselves to discussing the former step.

The remainder of this chapter is organized as follows. Section 15.2 describes the edits we consider in this chapter. Section 15.3 discusses the role of automatic editing in the entire data editing process. Selective editing is examined in Section 15.4. Sections 15.5–15.8 discuss various aspects of automatic editing. In particular, Section 15.6 describes some algorithms for automatic editing of so-called systematic errors and Section 15.8 of random errors. Section 15.9 concludes the chapter with a brief discussion.

15.2 Edit rules

At statistical agencies edit rules, or edits for short, are often used to determine whether a record is consistent or not. This is especially the case when automatic editing is applied. In this section we describe the edits we consider in this chapter.

We denote the continuous variables in a certain record by x_i $(i = 1, \ldots, n)$. The record itself is denoted by the vector $(x_1, \ldots, x_n)$. We assume that edit j $(j = 1, \ldots, J)$ is written in either of the two following forms:

$$a_{1j}x_1 + \cdots + a_{nj}x_n + b_j = 0 \tag{15.1}$$

or

$$a_{1j}x_1 + \cdots + a_{nj}x_n + b_j \geq 0. \tag{15.2}$$

An example of an edit of type (15.1) is

$$M + P = T, \tag{15.3}$$

where T is the total value of sales of a dairy farm, M its value of sales of milk, and P its value of sales of other agricultural products. Edit (15.3) says that the value of sales of milk and the value of sales of other agricultural products of a dairy farm should sum up to its total value of sales. A record not satisfying this edit is incorrect, and has to be corrected. Edits of type (15.1) are often called balance edits. Balance edits are usually 'hard' edits – edits that hold for all correctly observed records.

An example of an edit of type (15.2) is

$$P \leq 0.5T, \tag{15.4}$$

which says that the value of sales of other agricultural products of a dairy farm should be at most 50% of its total value of sales. Inequalities such as (15.4) are often 'soft' edits – edits that hold for a high fraction of correctly observed records but not necessarily for all of them. Moreover, 'soft' edits often only hold approximately true for a certain class of units in the population.

Edit j ($j = 1, \ldots, J$) is satisfied by a record $(x_1, \ldots, x_n)$ if (15.1) or (15.2) holds. A variable x_i is said to enter, or to be involved in, edit j given by (15.1) or (15.2) if $a_{ij} \neq 0$. That edit is then said to be involved in this variable. All edits given by (15.1) or (15.2) have to be satisfied simultaneously. We assume that the edits can indeed be satisfied simultaneously. Any field for which the value is missing is considered to be erroneous. Edits in which a variable with a missing value is involved are considered to be violated.

15.3 The role of automatic editing in the editing process

As mentioned in Section 15.1, we aim to edit as many records as possible automatically, while ensuring that the final data are of sufficiently high quality. Automatic editing alone is generally not enough to obtain data of sufficiently high statistical quality. We believe that the combined use of modern editing techniques leads to a reduction in processing time and a decrease in required resources in comparison to the traditional interactive method, while preserving or often even increasing quality. At Statistics Netherlands we apply such a combination of editing techniques to edit our structural annual business surveys.

For our structural annual business surveys we apply an edit and imputation approach that consists of the following main steps (see De Jong, 2002):

1. correction of 'obvious' (systematic) errors (see Section 15.6), such as thousand errors, sign errors, interchanged returns and costs, and rounding errors;
2. application of selective editing to split the records into a critical stream and a non-critical stream (see Lawrence and McKenzie, 2000; Hedlin, 2003);
3. editing of the data – the records in the critical stream are edited interactively, those in the non-critical stream are edited and imputed automatically;
4. validation of the publication figures by means of macro-editing.

During the selective editing step so-called plausibility indicators are used to split the records into a critical stream and a non-critical stream (see Section 15.4). Very suspicious or highly influential records are edited interactively. The remaining records are edited automatically (see Sections 15.7 and 15.8). The final validation step consists of detecting the remaining outliers in the data, and comparing the publication figures based on the edited and imputed data to publication figures from a previous year. Influential errors that were not corrected during automatic (or interactive) editing can be detected during this final step.

One could argue that with selective editing the automatic editing step is superfluous. Personally, we advocate the use of automatic editing, even when selective editing is used. We mention three reasons. First, the sum of the errors of the records in the non-critical stream may have an influential effect on the publication figures, even though each error itself may be non-influential. This may in particular be the case if the data contain systematic errors, since a substantial part of the data may be biased in the same direction. The automatic correction of 'obvious' systematic errors and random errors generally leads to data of higher statistical quality. Second, many non-critical records will be internally inconsistent (i.e. they will fail specified edits) if they are not edited, which may lead to problems when publication figures are calculated or when micro-data are released to external researchers. Finally, automatic editing provides a mechanism to check the quality of the selective editing procedures. If selective editing is well designed and well implemented, the records that are not selected for interactive editing need no or only slight adjustments. Records that are substantially changed during the automatic editing step therefore possibly point to an incorrect design or implementation of the selective editing step.

15.4 Selective editing

Manual or interactive editing is time-consuming and therefore expensive, and adversely influences the timeliness of publications. Moreover, when manual editing involves recontacting the respondents, it also increases the response burden. Therefore, most statistical institutes have adopted selective editing strategies. This means that only records that potentially contain influential errors are edited manually, whereas the remaining records are edited automatically. In this way, manual editing is limited to those errors where the editing has substantial influence on publications figures (see Hidiroglou and Berthelot, 1986; Granquist, 1995; Hoogland, 2002).

The main instrument in this selection process is the score function (see Farwell and Raine, 2000). This function assigns to each record a score that measures the expected influence of editing the record on the most important target parameters. Records with high scores are the first to be considered for interactive editing.

If interactive editing can wait until the data collection has been completed, the editing can proceed according to the order of priority implied by the scores until the change in estimates of the principle output parameters becomes unimportant or time or human resources are exhausted. In many cases, however, responses are received over a considerable period of time. Starting the time-consuming editing process after the data collection period will in such cases lead to an unacceptable delay of the survey process. Therefore, a threshold or cut-off value is determined in advance such that records with scores above the threshold are designated to be not plausible. These records are assigned to the so called 'critical stream', and are edited interactively, whereas the other records, with less important errors, are edited automatically. In this way the decision to edit a record is made without the need to compare scores across records.

A score for a record (record or global score) is usually a combination of scores for each of a number of important target parameters (the local scores). For instance, for the estimates of totals, local scores can be defined that measure the influence of editing each of the corresponding variables. The local scores are generally constructed so that they reflect the following two elements that together constitute an influential error: the size and likelihood of a potential error (the 'risk' component) and the contribution or influence of that record on the estimated target parameter (the 'influence' component). Scores can be defined as the product of these two components. The risk component can be measured by comparing the raw value with an approximation to the true value, called the 'anticipated value', which is often based on information from previous cycles of the same survey. Large deviations from the anticipated value are taken as an indication that the raw value may be in error and, if indeed so, that the error is substantial. Small deviations indicate that there is no reason to suspect that the raw value is in error and, even if it were, the error would be unimportant. The influence component can often be measured as the (relative) contribution of the anticipated value to the estimated total.

In defining a selective editing strategy we can distinguish three steps that will be described in more detail in the remainder of this section. These steps can be summarized as follows:

- defining local scores for parameters of interest such as (subgroup) totals (Section 15.4.1) and quarterly or yearly changes (Section 15.4.2) using reference values that approximate the true values as well as possible;
- defining a function that combines the local score to form a global or record score (Section 15.4.3);
- determining a threshold value for the global scores (Section 15.4.4).

15.4.1 Score functions for totals

For many agricultural surveys, the principal outputs are totals of a large number of variables such as acres and yield of crops, number of cattle and other livestock, value of sales for specific agricultural or animal products and operation expenses. Usually, these totals are published at the state level as well as for smaller regions such as counties or provinces.

A score function for totals should quantify the effect of editing a record on the estimated total. Let x_{ri} denote the value of a variable x_i in record r. The usual estimator of the corresponding total can then be defined as

$$\hat{X}_i = \sum_{r \in D} w_r \hat{x}_{ri}, \tag{15.5}$$

with D the data set (sample or census) and r denoting the records or units. The weights w_r correct for unequal inclusion probabilities and/or non-response. In the case of a census, non-response is of course the only reason to use weights in the estimator. The $\hat{x}_{ri}$ in (15.5) are edited values – that is, they have been subjected to an editing process in which some of the raw values (x_{ri}, say) have been corrected either by an automated process or by human intervention. For most records x_{ri} will be (considered) correct and the same as $\hat{x}_{ri}$.

The (additive) effect on the total of editing a single record can be expressed as the difference

$$d_{ri} = w_r(x_{ri} - \hat{x}_{ri}). \tag{15.6}$$

The difference d_{ri} depends on the (as yet) unknown corrected value $\hat{x}_{ri}$ and cannot therefore be calculated. A score function is based on an approximation to $\hat{x}_{ri}$ which is referred to as the 'anticipated value'. The anticipated value serves as a reference for judging the quality of the raw value. The following sources are often used for anticipated values:

- Edited data for the same farm from a previous version of the same survey, possibly multiplied by an estimate of the development between the previous and the current time point.
- The value of a similar variable for the same farm from a different source, in particular an administration such as a tax register.

- The mean or median of the target variable in a homogeneous subgroup of similar farms for a previous period. This subgroup of similar farms may be farms in the same size class and with the same main products.

Except for the unknown corrected value, the difference (15.6) depends also on an unknown weight w_r. Because the weights correct not only for unequal inclusion probabilities but also for non-response, they can only be calculated after the data collection is completed and the non-response is known. A score function that can be used during the data collection period cannot use these weights and will need an approximation. An obvious solution is to use, as an approximation, 'design weights' that only correct for unequal inclusion probabilities and which are defined by the sampling design (the inverse of the inclusion probabilities).

Using these approximations, the effect of editing a record r can be quantified by the score function

$$s_{ri} = v_r \left| x_{ri} - \tilde{x}_{ri} \right| = v_r \tilde{x}_{ri} \times \left| x_{ri} - \tilde{x}_{ri} \right| / \tilde{x}_{ri} = F_{ri} \times R_{ri}, \qquad (15.7)$$

say, with $\tilde{x}_{ri}$ the anticipated value and v_r the design weight. As (15.7) shows, this score function can be written as the product of an 'influence' factor ($F_{ri} = v_r \tilde{x}_{ri}$) and a 'risk' factor ($R_{ri} = \left| x_{ri} - \tilde{x}_{ri} \right| / \tilde{x}_{ri}$). The risk factor is a measure of the relative deviation of the raw value from the anticipated value. Large deviations are an indication that the raw value may be in error. If the anticipated value is the true value and the editor is capable of retrieving the true value, it is also the effect of correcting the error in this record. The influence factor is the contribution of the record to the estimated total. Multiplying the risk factor by the influence factor results in a measure of the effect of editing the record on the estimated total. Large values of the score indicate that the record may contain an influential error and that it is worthwhile to spend time and resources on correcting the record. Smaller values of the score indicate that the record does not contain very influential errors and that it can be entrusted to automatic procedures that use approximate solutions for the error detection and correction problems.

Often a scaled version of the score function (15.7) is used that can be obtained by replacing the influence component F_{ri} by the relative influence $F_{ri} / \sum_r F_{ri}$. Because

$$\sum_r F_{ri} = \sum_r v_r \tilde{x}_{ri} = \tilde{X}_i,$$

the resulting scaled score is $s'_{ri} = s_{ri} / \tilde{X}_i$, the original score divided by an estimate (based on the anticipated values) of the total. This scaling makes the score independent of the unit of measurement, which is an advantage when local scores are combined to form a record or global score (see Section 15.4.3).

Another well-known score function can be obtained by using a risk factor based on the ratio of the raw value and the anticipated value instead of the absolute difference. This risk factor, proposed Hidiroglou and Berthelot (1986), is defined as

$$R_{ri} = \max \left(\frac{\tilde{x}_{ri}}{x_{ri}}, \frac{x_{ri}}{\tilde{x}_{ri}} \right) - 1. \qquad (15.8)$$

This definition ensures that upward and downward multiplicative deviations from the anticipated value of the same size will lead to the same scores and that the score has a minimum value of zero for $x_{ri} = \tilde{x}_{ri}$.

Score functions can often be improved if a variable is available that is strongly correlated with the target variable. In such cases the ratio of these two variables can be more sensitive for detecting anomalous values than the target variable itself. For instance, the value of sales of specific crops can show much variation between farms, even within size classes, but the value of sales divided by the acres of harvested cropland is much less variable across farms. Therefore, erroneous values of sales are better detectable by using the ratio 'sales per acre' than by using sales itself. In fact, by using such a ratio errors in either of the variables are detected except in the unlucky case where both variables deviate in the same direction. Score functions based on ratios can be obtained by replacing the x_{ri} and $\tilde{x}_{ri}$ in the risk factors in (15.7) and (15.8) by the raw value of the ratio and an anticipated value of the ratio, respectively.

For censuses, panel surveys and sample surveys with a sampling fraction of unity in some strata (e.g. strata corresponding to large size classes), a value of the target variable from a previous period may be available for most farms or the most important farms. This historical value can be a very effective variable when used in a ratio based risk factor. Score functions that use historical values are considered in more detail in the next subsection. As an anticipated value for this ratio one could use, for instance, the median of the ratios, for a subgroup of farms, for a previous period.

15.4.2 Score functions for changes

Agricultural surveys are generally repeated surveys, and interest lies not only in yearly estimated totals of variables related to, for instance, production, labour and costs of agricultural enterprises but also in the changes in these totals over time. If a measure of change is the parameter of interest, an editing strategy should be oriented towards finding suspect values with a large influence on the change. For repeated surveys where values for the same farm are available for a previous time point the raw unedited change from time $t-1$ to t in the target variable for a farm r is $\delta_{ri} = x_{ri,t}/\hat{x}_{ri,t-1}$, where we have assumed that the $t-1$ data have already been edited. This change now takes the role of the target variable and an anticipated value must be defined so that suspect values of the change can detected. Hidiroglou and Berthelot (1986) propose using the median of the changes δ_{ri} in a homogeneous subgroup of units as reference value. A drawback of this idea is that the scores cannot be calculated and selective editing cannot be started until the amount of response is large enough to estimate the medians reliably. As an alternative that is not hampered by this drawback, Latouche and Berthelot (1992) propose using the median of the changes at a previous cycle of the survey (i.e. the changes between $t-2$ and $t-1$), which seems reasonable only if the change between $t-2$ and $t-1$ is similar to the change between $t-1$ and t. This is the case for variables that are gradually and moderately increasing over time, such as hourly labour costs. Another way to obtain an anticipated value, especially for short-term statistics, is to estimate a time series model with a seasonal component for a series of $x_{ri,t}$ values and to use the prediction from this model for the current value as an anticipated value for $x_{ri,t}$. By dividing this anticipated value for $x_{ri,t}$ by the corresponding edited value for $t-1$, an anticipated value for the change is found that also does not rely on the current data except for the record for which the score is calculated.

With the anticipated value, a risk factor can be defined for which, in the case of year-to-year changes, it is more common the use the multiplicative form (15.8) than the additive form (15.6). A score function can then be defined by multiplying this risk factor by an influence factor. Hidiroglou and Berthelot (1986) propose as a measure of influence (the unweighted version of)

$$F_{ri} = \left[\max\left(v_{r,t} x_{ri,t}, w_{r,t-1}\hat{x}_{ri,t-1}\right)\right]^{c}, \tag{15.9}$$

with $0 \leq c \leq 1$. The parameter c can be used to control the importance of the influence: higher values give more weight to the influence factor. In empirical studies at Statistics Canada, Latouche and Berthelot (1992) found 0.5 to be a reasonable value for c. The maximum function in (15.9) has the effect that an error in the reported value $x_{ri,t}$ is more likely to lead to an overestimation of the influence than to lead to an underestimation. This is because a too low reported value $x_{ri,t}$ can never result in an influence value smaller than $\hat{x}_{ri,t-1}$, whereas a too high value can increase the influence in principle without limit. A scaled version of a score with influence factor (15.9) can be obtained by dividing $w_{r,t-1}\hat{x}_{ri,t-1}$ and $v_{r,t}x_{ri,t}$ by their respective totals. The total of $w_{r,t-1}\hat{x}_{ri,t-1}$ is simply the population estimate for the previous period, $\hat{X}_{i,t-1} = \sum_{r \in D_{t-1}} w_{r,t-1}\hat{x}_{ri,t-1}$. The actual total, however, must be approximated if we cannot assume that all the data are already collected. An obvious approximation is obtained by using the anticipated values resulting in $\tilde{X}_{i,t} = \sum_{r \in D_t} v_{r,t}\tilde{x}_{ri,t}$.

15.4.3 Combining local scores

In order to select a whole record for interactive editing, a score on the record level is needed. This 'record score' or 'global score' combines the information from the 'local scores' that are defined for a number of important target parameters. The global score should reflect the importance of editing the complete record. In order to combine scores it is important that the scores are measured on comparable scales. It is common, therefore, to scale local scores before combining them into a global score. In the previous subsection we have seen one option for scaling local scores – by dividing by the (approximated) total. Another method is to divide the scores by the standard deviation of the anticipated values (see Lawrence and McKenzie, 2000). This last approach has the advantage that deviations from anticipated values in variables with large variability will lead to less high scores and are therefore less likely to be designated as suspect values than deviations in variables with less natural variability.

Scaled or standardized local scores have been combined in a number of ways. The global score is often defined as the sum of the local scores (see Latouche and Berthelot, 1992). As a result, records with many deviating values will get high scores. This can be an advantage because editing many variables in the same record is relatively less time-consuming than editing the same number of variables in different records, especially if it involves recontacting the respondent. But a consequence of the sum score is also that records with many, but only moderately deviating values will have priority for interactive editing over records with only a few strongly deviating values. If it is deemed necessary for strongly deviating values in an otherwise plausible record to be treated by specialists, then the sum score is not the global score of choice.

An alternative for the sum of the local scaled scores, suggested by Lawrence and McKenzie (2000), is to take the maximum of these scores. The advantage of the maximum is that it guarantees that a large value of any one of the contributing local scores will lead to a large global score and hence manual review of the record. The drawback of this strategy is that it cannot discriminate between records with a single large local score and records with numerous equally large local scores. As a compromise between the sum and max functions, Farwell (2005) proposes the use of the Euclidian metric (the root of the sum of the squared local scores). These three proposals (sum, max, Euclidean metric) are all special cases of the Minkowski metric (see Hedlin, 2008) given by

$$S_r^{(\alpha)} = \left(\sum_{i=1}^{n} s_{ri} \right)^{1/\alpha}, \tag{15.10}$$

where $S_r^{(\alpha)}$ is the global score as a function of the parameter α, s_{ri} the ith local score and n the number of local scores. The parameter α determines the influence of large values of the local scores on the global score, the influence increasing with α. For $\alpha = 1$ equation (15.10) is the sum of the local scores, for $\alpha = 2$ it is the Euclidean metric, and for $\alpha \to \infty$ it approaches the maximum of the local scores.

For the extensive and detailed questionnaires that are often used in agricultural surveys, it may be more important for some variables to be subjected to an interactive editing process than for others. In such cases the local scores in the summation can be multiplied by weights that express their relative importance (see Latouche and Berthelot, 1992).

15.4.4 Determining a threshold value

Record scores are used to split the data into a critical stream of implausible records that will be edited interactively and a non-critical stream of plausible records that will be edited automatically. This selection process is equivalent to determining the value of the binary plausibility indicator variable defined by

$$PI_r = \begin{cases} 1 & \text{(plausible)} \quad \text{if } S_r \geq C, \\ 0 & \text{(implausible)} \quad \text{otherwise,} \end{cases} \tag{15.11}$$

with C a cut-off value or threshold value and S_r the record score or global score.

The most prominent method for determining a threshold value is to study, by simulation, the effect of a range of threshold values on the bias in the principal output parameters. Such a simulation study is based on a raw unedited data set and a corresponding fully interactively edited version of the same data set. These data must be comparable with the data to which the threshold values are applied. Data from a previous cycle of the same survey are often used for this purpose.

The simulation study now proceeds according to the following steps:

- Calculate the global scores according to the chosen methods for the records in the raw version of the data set.
- Simulate that only the first $100p\%$ of the records is designated for interactive editing. This is done by replacing the values of the $100p\%$ of the records with the highest scores in the raw data by the values in the edited data. The subset of the $100p\%$ edited records is denoted by E_p.
- Calculate the target parameters using both the $100p\%$ edited raw data set and the completely edited data set.

These steps are repeated for a range of values of p. The absolute value of the relative difference between the estimators calculated in the last step is called the absolute pseudo-bias (Latouche and Berthelot, 1992). For the estimation of the total of variable i, this absolute pseudo-bias is given by

$$B_i(p) = \frac{1}{\hat{X}_i}\left|\sum_{r \notin E_p} w_r(x_{ri} - \hat{x}_{ri})\right|. \tag{15.12}$$

As (15.12) shows, the absolute pseudo-bias is determined by the difference in totals of the edited and non-edited values for the records not selected (for interactive editing). If the editing results in correcting all errors (and only errors) then (15.12) equals the absolute value of the relative bias that remains because not all records have been edited. However, because it is uncertain that editing indeed reproduces the correct values, (15.12) is an approximation of this bias, hence the name 'pseudo-bias'.

The pseudo-bias at $100p\%$ editing can also be interpreted as an estimator of the gain in accuracy that can be attained by also editing the remaining $100(1 - p)\%$ of the records. By calculating the pseudo-bias for a range of values of p, we can trace the gain in accuracy as a function of p . If sorting the records by their scores has the desired effect, this gain will decline with increasing values of p. At a certain value of p one can decide that the remaining pseudo-bias is small enough and that it is not worthwhile to pursue interactive editing beyond that point. The record score corresponding to this value of p will be the threshold value.

The pseudo-bias as described above is based on a comparison between interactive editing and not editing at all. In most cases the alternative to interactive editing is automatic editing rather than not editing. Assuming that automatic editing does at least not lead to more bias than not editing at all, the value of the pseudo-bias according to (15.12) is an upper limit of the pseudo-bias in situations where automatic editing is applied.

The simulation approach can also be used to compare different score functions. One obvious approach for this is to determine, for different score functions, the number of records that need to be edited to obtain the same value of the pseudo-bias. The most effective score function is the one for which this number is lowest.

15.5 An overview of automatic editing

When automatic editing is applied, records are edited by computer without human intervention. Automatic editing is the opposite of the traditional interactive approach to the editing problem, where each record is edited manually.

We can distinguish two kinds of errors: systematic ones and random ones. A systematic error is an error reported consistently by (some of) the respondents. It can be caused by the consistent misunderstanding of a question by (some of) the respondents. Examples are when gross values are reported instead of net values, and particularly when values are reported in units instead of, for instance, the requested thousands of units (so-called 'thousand errors'). Random errors are not caused by a systematic deficiency, but by accident. An example is an observed value where a respondent mistakenly typed in a digit too many.

Systematic errors, such as thousand errors, can often be detected by comparing a respondent's present values with those from previous years, by comparing the responses to questionnaire variables with values of register variables, or by using subject-matter knowledge. Other systematic errors, such as redundant minus signs, can be detected and corrected by systematically exploring all possible inclusions/omissions of minus signs. Rounding errors – a class of systematic errors where balance edits are violated because the values of the involved variables have been rounded – can be detected by testing whether failed balance edits can be satisfied by slightly changing the values of the variables involved. Once detected, a systematic error is often simple to correct. Automatic editing of systematic errors is discussed in more detail in Section 15.6.

Generally speaking, we can subdivide the methods for automatic localization of random errors into methods based on statistical models, methods based on deterministic checking rules, and methods based on solving a mathematical optimization problem. Methods based on statistical models, such as outlier detection techniques, are extensively discussed in the literature. We therefore do not discuss these techniques in this chapter.

Deterministic checking rules state which variables are considered erroneous when the edits in a certain record are violated. An example of such a rule is: if component variables do not sum to the corresponding total variable, the total variable is considered to be erroneous. Advantages of this approach are its transparency and simplicity. A drawback is that many detailed checking rules have to be specified, which can be time- and resource-consuming to do. Another drawback is that maintaining and checking the validity of a large number of detailed checking rules can be complex. Moreover, in some cases it may be impossible to develop deterministic checking rules that are powerful enough to identify errors in a reliable manner. A final disadvantage is that bias may be introduced as one aims to correct random errors in a systematic manner.

To formulate the error localization problem as a mathematical optimization problem, a guiding principle (i.e. an objective function) for identifying the erroneous fields in a record is needed. Freund and Hartley (1967) were among the first to propose such a guiding principle. It is based on minimizing the sum of the distance between the observed data and the 'corrected' data and a measure for the violation of the edits. That paradigm never became popular, however, possibly because a 'corrected' record may still fail to satisfy the specified edits. A similar guiding principle, based on minimizing a quadratic function measuring the distance between the observed data and the 'corrected' data subject to the constraint that the 'corrected' data satisfy all edits, was later proposed by Casado Valero *et al.* (1996). A third guiding principle is based on first imputing missing data and potentially erroneous data for records failing edits by means of donor imputation, and then selecting an imputed record that satisfies all edits and that is 'closest' to the original record. This paradigm forms the basis of the nearest-neighbour imputation methodology (NIM). Thus far, NIM has mainly been used for demographic data. For details on NIM we refer to Bankier *et al.* (2000).

The best-known and most often used guiding principle is the (generalized) paradigm of Fellegi and Holt (1976), which says that the data in each record should be made to satisfy all edits by changing the fewest possible fields. Using this guiding principle, the error localization problem can be formulated as a mathematical optimization problem (see Sections 15.7 and 15.8). The (generalized) Fellegi–Holt paradigm can be applied to numerical data as well as to categorical (discrete) data. In this chapter we restrict ourselves to numerical data as these are the most common data in agricultural surveys and censuses.

15.6 Automatic editing of systematic errors

In this section we discuss several classes of systematic errors and ways to detect and correct them.

As already mentioned, a well-known class of systematic errors is the so-called thousand errors. These are cases where a respondent replies in units rather than in the requested thousands of units. The usual way to detect such errors is – similar to selective editing – by considering 'anticipated' values, which could, for instance, be values of the same variable from a previous period or values available from a register. One then calculates the ratio of the observed value to the anticipated value. If this ratio is higher than a certain threshold value, say 300, it is assumed that the observed value is 1000 times too large. The observed value is then corrected by dividing it by 1000. A minor practical problem occurs when the anticipated value equals zero. Usually this problem can easily be solved in practice.

A more complex approach for detecting and correcting thousand errors, or more generally unity measure errors, i.e. any error due to the erroneous choice by some respondents of the unity measure in reporting the amount of a certain variable, has been proposed by Di Zio *et al.* (2005). That approach uses model-based cluster analysis to pinpoint various kinds of unity measure errors. The model applied consists of a finite mixture of multivariate normal distributions.

Balance edits are often violated by the smallest possible difference. That is, the absolute difference between the total and the sum of its components is equal to 1 or 2. Such inconsistencies are often caused by rounding. An example is when the terms of the balance edit $x_1 + x_2 = x_3$ with $x_1 = 2.7$, $x_2 = 7.6$ and $x_3 = 10.3$ are rounded to integers. If conventional rounding is used, x_1 is rounded to 3, x_2 to 8 and x_3 to 10, and the balance edit becomes violated.

From a purely statistical point of view, rounding errors are rather unimportant as by their nature they have virtually no influence on publication figures. Rounding errors may be important, however, when we look at them from the point of view of the data editing *process*. Some statistical offices apply automatic editing procedures for random errors, such as automatic editing procedures based on the Fellegi–Holt paradigm (see Section 15.7). Such automatic editing procedures are computationally very demanding. The complexity of the automatic error localization problem increases rapidly as the number of violated edit rules increases, irrespective of the magnitude of these violations. A record containing many rounding errors may hence be too complicated to solve for an automatic editing procedure for random errors, even if the number of random errors is actually low. From the point of view of the data editing process it may therefore be advantageous to resolve rounding errors at the beginning of the editing process.

Scholtus (2008a, 2008b) describes a heuristic method for resolving rounding errors. The method does not lead to solutions that are 'optimal' according to some criterion, such as that the number of changed variables or the total change in value is minimized. Instead the method just leads to a good solution. Given that the statistical impact of resolving rounding errors is small, a time-consuming and complex algorithm aimed at optimizing some target function is not necessary anyway. The heuristic method is referred to as the 'scapegoat algorithm', because for each record assumed to contain rounding errors a number of variables, the 'scapegoats', are selected beforehand and the rounding errors are resolved by changing only the values of the selected variables. Under certain very mild conditions, the algorithm guarantees that exactly one choice of values exists for the selected variables such that the balance edits become satisfied. Different variables may be selected for each record to minimize the effect of the adaptations on published aggregates.

In general, the solution obtained might contain fractional values, whereas most business survey variables are restricted to integer values. If this is the case, a controlled rounding algorithm could be applied to the values to obtain an integer-valued solution (see Salazar-González *et al.*, 2004). Under certain additional mild conditions, which appear to be satisfied by most data sets arising in practice, the problem of fractional values does not occur, however. For details we refer to Scholtus (2008a, 2008b).

Rounding errors often occur in combination with other 'obvious' systematic errors. For instance, a sign error might be obscured by the presence of a rounding error. Scholtus (2008a, 2008b) provides a single mathematical model for detecting sign errors and rounding errors simultaneously.

15.7 The Fellegi–Holt paradigm

In this section we describe the error localization problem for random errors as a mathematical optimization problem, using the (generalized) Fellegi–Holt paradigm. This mathematical optimization problem is solved for each record separately.

For each record $(x_1, \ldots, x_n)$ in the data set that is to be edited automatically, we wish to determine – or, more precisely, to ensure the existence of – a synthetic record $(x_1^*, \ldots, x_n^*)$ such that $(x_1^*, \ldots, x_n^*)$ satisfies all edits j $(j = 1, \ldots, J)$ given by (15.1) or (15.2), none of the x_i^* $(i = 1, \ldots, n)$ is missing, and

$$\sum_{i=1}^{n} u_i y_i \tag{15.13}$$

is minimized, where the variables y_i $(i = 1, \ldots, n)$ are defined by $y_i = 1$ if $x_i^* \neq x_i$ or x_i is missing, and $y_i = 0$ otherwise. Here u_i is the so-called reliability weight of variable x_i $(i = 1, \ldots, n)$. A reliability weight of a variable expresses how reliable one considers the values of this variable to be. A high reliability weight corresponds to a variable whose values are considered trustworthy, a low reliability weight to a variable whose values are considered not so trustworthy. The variables whose values in the synthetic record differ from the original values, together with the variables whose values were originally missing, form an optimal solution to the error localization problem. The above formulation is a mathematical formulation of the generalized Fellegi–Holt paradigm. In the original Fellegi–Holt paradigm all reliability weights were set to 1 in (15.13). A solution to the error localization problem is basically just a list of all variables that need to be changed. There may be several optimal solutions to a specific instance of the error localization problem. Preferably, we wish to find all optimal solutions. Later, one of these optimal solutions may then be selected, using a secondary criterion. The variables involved in the selected solution are set to missing and are subsequently imputed during the imputation step by a statistical imputation method of one's choice, such as regression imputation or donor imputation (for an overview of imputation methods see Kalton and Kasprzyk, 1982; Kovar and Whitridge, 1995; Schafer, 1997; Little and Rubin, 2002).

In the present chapter we will not explore the process of selecting one optimal solution from several optimal solutions, nor will we explore the imputation step.

Assuming that only few errors are made, the Fellegi–Holt paradigm is obviously a sensible one. Provided that the set of edits used is sufficiently powerful, application of this paradigm generally results in data of higher statistical quality, especially when used in combination with other editing techniques, such as selective editing.

A drawback of using the Fellegi–Holt paradigm is that the class of errors that can safely be treated is limited to random errors. A second drawback is that the class of edits that can be handled is restricted to 'hard' edits. 'Soft' edits cannot be handled as such, and – if specified – are treated as hard edits.

15.8 Algorithms for automatic localization of random errors

An overview of algorithms for solving the localization problem for random errors in numerical data automatically based on the Fellegi–Holt paradigm is given in De Waal and Coutinho (2005). In this section we briefly discuss a number of these algorithms.

15.8.1 The Fellegi–Holt method

Fellegi and Holt (1976) describe a method for solving the error localization problem automatically. In this section we sketch their method; for details we refer to the original article by Fellegi and Holt (1976). The method is based on generating so-called *implicit*, or *implied*, edits. Such implicit edits are logically implied by the explicit edits – the edits specified by the subject-matter specialists. Implicit edits are redundant. They can, however, reveal important information about the feasible region defined by the explicit edits. This information is already contained in the explicitly defined edits, but there that information may be rather hidden.

The method proposed by Fellegi and Holt starts by generating a well-defined set of implicit and explicit edits that is referred to as the *complete set of edits* (see Fellegi and Holt, 1976, for a precise definition of this set). It is called the complete set of edits not because all possible implicit edits are generated, but because this set of (implicit and explicit) edits suffices to translate the error localization problem into a so-called set-covering problem (see Nemhauser and Wolsey, 1988, for more information about the set-covering problem in general). The complete set of edits comprises the explicit edits and the so-called essentially new implicit ones (Fellegi and Holt, 1976). Once the complete set of edits has been generated one only needs to find a set of variables S that covers the violated (explicit and implicit) edits of the complete set of edits – that is, in each violated edit of the complete set of edits at least one variable of S should be involved, with a minimum sum of reliability weights in order to solve the error localization problem.

The complete set of edits is generated by repeatedly selecting a variable, which Fellegi and Holt (1976) refer to as the generating field. Subsequently, all pairs of edits (explicit or implicit) are considered, and it is checked whether (essentially new) implicit edits can be obtained by eliminating the selected generating field from these pairs of edits by means of Fourier–Motzkin elimination. For inequalities Fourier–Motzkin elimination basically consists of using the variable to be eliminated to combine these inequalities pairwise (if possible). For instance, if we use Fourier–Motzkin elimination to eliminate x_2 from the edits $x_1 \leq x_2$ and $x_2 \leq x_3$, we obtain the new edit $x_1 \leq x_3$ that is logically implied by the two original ones. If the variable to be eliminated is involved in a balance edit, we use this equation to express this variable in terms of the other variables and then use this expression to eliminate the variable from the other edits (for more on Fourier–Motzkin elimination, we refer to Duffin, 1974, and De Waal and Coutinho, 2005). This process continues until no (essentially new) implicit edits can be generated, whatever generating field is selected. The complete set of edits has then been determined.

Example 1 illustrates the Fellegi–Holt method. This example is basically an example provided by Fellegi and Holt themselves, except for the fact that in their article the edits indicate a condition of edit failure (i.e. if a condition holds true, the edit is violated), whereas here the edits indicate the opposite condition of edit consistency (i.e. if a condition holds true, the edit is satisfied).

Example 1. Suppose we have four variables x_i $(i = 1, \ldots, 4)$. The explicit edits are given by

$$x_1 - x_2 + x_3 + x_4 \geq 0, \tag{15.14}$$

$$-x_1 + 2x_2 - 3x_3 \geq 0. \tag{15.15}$$

The (essentially new) implicit edits are then given by

$$x_2 - 2x_3 + x_4 \geq 0, \tag{15.16}$$

$$x_1 - x_3 + 2x_4 \geq 0, \tag{15.17}$$

$$2x_1 - x_2 + 3x_4 \geq 0. \tag{15.18}$$

For instance, edit (15.16) is obtained by selecting x_1 as the generating field, and eliminating this variable from edits (15.14) and (15.15). The above five edits form the complete set of edits as no further (essentially new) implicit edit can be generated.

An example of an implicit edit that is not an essentially new one is

$$-x_1 + 3x_2 - 5x_3 + x_4 \geq 0,$$

which is obtained by multiplying edit (15.15) by 2 and adding the result to edit (15.14). This is not an essentially new implicit edit, because no variable has been eliminated from edits (15.14) and (15.15).

Now, suppose we are editing a record with values $x_1 = 3$, $x_2 = 4$, $x_3 = 6$, and $x_4 = 1$, and suppose that the reliability weights are all equal to 1. Examining the explicit edits, we see that edit (15.14) is satisfied, whereas edit (15.15) is violated. From the explicit edits it is not clear which of the fields should be changed. If we also examine the implicit edits, however, we see that edits (15.15), (15.16) and (15.17) fail. Variable x_3 occurs in all three violated edits. So, we can conclude that we can satisfy all edits by changing x_3. For example, x_3 could be made equal to 1. Changing x_3 is the only optimal solution to the error localization problem in this example.

An important practical problem with the Fellegi–Holt method for numerical data is that the number of required implicit edits may be very high, and in particular that the generation of all these implicit edits may be very time-consuming. For most real-life problems the computing time to generate all required implicit edits becomes extremely high even for a small to moderate number of explicit edits. An exception to the rule that the number of (essentially new) implicit edits becomes extremely high is formed by ratio edits – that is, ratios of two variables that are bounded by constant lower and upper bounds. For ratio- edits the number of (essentially new) implicit edits is low, and the Fellegi–Holt method is exceedingly fast in practice (see Winkler and Draper, 1997).

15.8.2 Using standard solvers for integer programming problems

In this section we describe how standard solvers for integer programming (IP) problems can be used to solve the error localization problem. To apply such solvers we make the assumption that the values of the variables x_i $(i = 1, \ldots, n)$ are bounded. That is, we assume that for variable x_i $(i = 1, \ldots, n)$ constants α_i and β_i exist such that

$$\alpha_i \leq x_i \leq \beta_i \tag{15.19}$$

for all consistent records. In practice, such values α_i and β_i always exist although they may be very large, because numerical variables that occur in data of statistical offices are by nature bounded.

The problem of minimizing (15.13) so that all edits (15.1) and (15.2) and all bounds (15.19) become satisfied is an IP problem. It can be solved by applying commercially available solvers for IP problems. McKeown (1984), Riera-Ledesma and Salazar-González (2003) also formulate the error localization problem for continuous data as a standard IP problem. Schaffer (1987) and De Waal (2003a) give extended IP formulations that include categorical data.

15.8.3 The vertex generation approach

In this section we briefly examine a well-known and popular approach for solving the error localization problem that is based on generating the vertices of a certain polyhedron. If the set of edits (15.1) and (15.2) is not satisfied by a record $(x_1^0, \ldots, x_n^0)$, where x_i^0 $(i = 1, \ldots, n)$ denotes the observed value of variable x_i, then we seek values $x_i^+ \geq 0$ and $x_i^- \geq 0$ $(i = 1, \ldots, n)$ corresponding to positive and negative changes, respectively, to x_i^0 $(i = 1, \ldots, n)$ such that all edits (15.1) and (15.2) become satisfied. The objective function (15.13) is to be minimized subject to the constraint that the new, synthetic record $(x_1^0 + x_1^+ - x_1^-, \ldots, x_n^0 + x_n^+ - x_n^-)$ satisfies all edits (15.1) and (15.2). That is, the x_i^+ and x_i^- $(i = 1, \ldots, n)$ have to satisfy

$$a_{1j}(x_1^0 + x_1^+ - x_1^-) + \cdots + a_{nj}(x_n^0 + x_n^+ - x_n^-) + b_j = 0 \tag{15.20}$$

and

$$a_{1j}(x_1^0 + x_1^+ - x_1^-) + \cdots + a_{nj}(x_n^0 + x_n^+ - x_n^-) + b_j \geq 0 \tag{15.21}$$

for each edit j $(j = 1, \ldots, J)$ of type (15.1) or (15.2), respectively . For convenience, we assume that all variables x_i $(i = 1, \ldots, n)$ are bounded from above and below – an assumption we also made in Section 15.8.2. For convenience we also assume that the constraints for the x_i^+ and the x_i^- $(i = 1, \ldots, n)$ resulting from these lower and upper bounds are incorporated in the system given by the constraints (15.20) and (15.21). If the value of x_i $(i = 1, \ldots, n)$ is missing, we fill in a value larger than the upper bound on x_i. As a consequence, the value of x_i will necessarily be modified, and hence be considered erroneous.

The set of constraints given by (15.20) and (15.21) defines a polyhedron for the unknowns x_i^+ and x_i^- $(i = 1, \ldots, n)$. This polyhedron is bounded, because we have assumed that all variables x_i $(i = 1, \ldots, n)$ are bounded from above and below. The vertex generation approach for solving the error localization problem is based on the observation that an optimal solution to the error localization problem corresponds to a vertex of the polyhedron defined by the set of constraints (15.20) and (15.21) (for a simple proof of this observation we refer to De Waal and Coutinho, 2005).

This observation implies that one can find all optimal solutions to the error localization problem by generating the vertices of the polyhedron defined by the constraints (15.20) and (15.21), and then selecting the ones with the lowest objective value (15.13). Chernikova (1964, 1965) proposed an algorithm that can be used to find the vertices of a system of linear inequalities given by

$$\mathbf{Ax} \leq \mathbf{b}, \tag{15.22}$$

where $\mathbf{x}$ and $\mathbf{b}$ are vectors, and $\mathbf{A}$ is a matrix. It can be extended to systems including equations besides inequalities. The original algorithm of Chernikova has been modified in order to make it more suitable for the error localization problem (see Rubin, 1975, 1977; Sande, 1978; Schiopu-Kratina and Kovar, 1989; Fillion and Schiopu-Kratina, 1993). The modified algorithm is much faster than the original one. A detailed discussion on how to accelerate Chernikova's algorithm for the error localization problem is beyond the scope of the present chapter, but the main idea is to avoid the generation of suboptimal solutions as much as possible. For a summary of several papers on Chernikova's algorithm and an extension to categorical data we refer to De Waal (2003a, 2003b).

Modified versions of Chernikova's algorithm have been implemented in several computer systems, such as the Generalized Edit and Imputation System (GEIS; see Kovar and Whitridge, 1990) by Statistics Canada, an SAS program developed by the Central Statistical Office of Ireland (Central Statistical Office, 2000), CherryPi (De Waal, 1996) by Statistics Netherlands, and Agricultural Generalized Imputation and Edit System (AGGIES; see Todaro, 1999) by the National Agricultural Statistics Service. AGGIES has been developed especially for automatic editing of data from agricultural surveys and censuses.

15.8.4 A branch-and-bound algorithm

De Waal and Quere (2003) propose a branch-and-bound algorithm for solving the error localization problem. The basic idea of the algorithm we describe in this section is that for each record a binary tree is constructed. In our case, we use a binary tree to split up the process of searching for solutions to the error localization problem. We need some terminology with respect to binary trees before we can explain our algorithm. Following Cormen *et al.* (1990), we recursively define a binary tree as a structure on a finite set of nodes that either contains no nodes, or comprises three disjoint sets of nodes: a root node, a left (binary) subtree and a right (binary) subtree. If the left subtree is non-empty, its root node is called the left child node of the root node of the entire tree, which is then called the parent node of the left child node. Similarly, if the right subtree is non-empty, its root node is called the right child node of the root node of the entire tree, which is then called the parent node of the right child node. All nodes except the root node in a binary tree have exactly one parent node. Each node in a binary tree can have at most two (non-empty) child nodes. A node in a binary tree that has only empty subtrees as its child nodes is called a terminal node, or also a leaf. A non-leaf node is called an internal node. In each internal node of the binary tree generated by our algorithm a variable is selected that has not yet been selected in any predecessor node. If all variables have already been selected in a predecessor node, we have reached a terminal node of the tree.

We first assume that no values are missing. After the selection of a variable two branches (i.e. subtrees) are constructed: in one branch we assume that the observed value of the selected variable is correct, in the other branch we assume that the observed value is incorrect. By constructing a binary tree we can, in principle, examine all possible error patterns and search for the best solution to the error localization problem.

In the branch in which we assume that the observed value is correct, the variable is fixed to its original value in the set of edits. In the branch in which we assume that the observed value is incorrect, the selected variable is eliminated from the set of edits. A variable that has either been fixed or eliminated is said to have been treated (for the corresponding branch of the tree). For each node in the tree we have an associated set of edits for the variables that have not yet been treated in that node. The set of edits corresponding to the root node of our tree is the original set of edits.

Eliminating a variable is non-trivial as removing a variable from a set of edits may imply additional edits for the remaining variables. To illustrate why edits may need to be generated, we give a very simple example. Suppose we have three variables (x_1, x_2 and x_3) and two edits ($x_1 \leq x_2$ and $x_2 \leq x_3$). If we want to eliminate variable x_2 from these edits, we cannot simply delete this variable and the two edits, but have to generate the new edit $x_1 \leq x_3$ implied by the two old ones, for otherwise we could have that $x_1 > x_3$ and the original set of edits cannot be satisfied. To ensure that the original set of edits can be satisfied Fourier–Motzkin elimination is used.

In each branch of the tree the set of current edits is updated. Updating the set of current edits is the most important aspect of the algorithm. How the set of edits is updated depends on whether the selected variable is fixed or eliminated. Fixing a variable to its original value is done by substituting this value in all current edits, failing as well as non-failing ones. Conditional on fixing the selected variable to its original value, the new set of current edits is a set of implied edits for the remaining variables in the tree. That is, conditional on the fact that the selected variable has been fixed to its original value, the remaining variables have to satisfy the new set of edits. As a result of fixing the selected variable to its original value some edits may become tautologies, that is, satisfied by definition. An example of a tautology is '$1 \geq 0$'. Such a tautology may, for instance, arise if a variable x has to satisfy the edit $x \geq 0$, the original value of x equals 1, and x is fixed to its original value. These tautologies may be discarded from the new set of edits. Conversely, some edits may become self-contradicting relations. An example of a self-contradicting relation is '$0 \geq 1$'. If self-contradicting relations are generated, this particular branch of the binary tree cannot result in a solution to the error localization problem. Eliminating a variable by means of Fourier–Motzkin elimination amounts to generating a set of implied edits that do not involve this variable. This set of implied edits has to be satisfied by the remaining variables. In the generation process we need to consider all edits, both the failing edits and the non-failing ones, in the set of current edits pairwise. The generated set of implied edits plus the edits not involving the eliminated variable become the set of edits corresponding to the new node of the tree.

If values are missing in the original record, the corresponding variables only have to be eliminated from the set of edits (and not fixed).

After all variables have been treated we are left with a set of relations involving no unknowns. If and only if this set of relations contains no self-contradicting relations, the variables that have been eliminated in order to reach the corresponding terminal node of the tree can be imputed consistently, that is, such that all original edits can be satisfied (see Theorems 1 and 2 in De Waal and Quere, 2003). The set of relations involving no unknowns may be the empty set, in which case it obviously does not contain any self-contradicting relations. In the algorithm we check for each terminal node of the tree whether the variables that have been eliminated in order to reach this node can be imputed consistently. Of all the sets of variables that can be imputed consistently we select the ones with the lowest sum of reliability weights. In this way we find all optimal solutions to the error localization problem (see Theorem 3 in De Waal and Quere, 2003).

We illustrate the branch-and-bound approach by means of an example.

Example 2. Suppose the explicit edits are given by

$$T = M + P, \tag{15.23}$$

$$P \leq 0.5T, \tag{15.24}$$

$$0.1T \leq P, \tag{15.25}$$

$$T \geq 0, \tag{15.26}$$

$$T \leq 550N, \tag{15.27}$$

where T again denotes the total value of sales of a dairy farm, M its value of sales of milk, P its value of sales of other agricultural products, and N the number of milk cows. Let us consider a specific erroneous record with values $T = 100$, $G = 60\,000$, $P = 40\,000$ and $N = 5$. The reliability weights of the variables T, G and P equal 1, and the reliability weight of variable N equals 2. Edits (15.25)–(15.27) are satisfied, whereas edits (15.23) and (15.24) are violated. As edits (15.23) and (15.24) are violated, the record contains errors.

We select a variable, say T, and construct two branches: one where T is eliminated and one where T is fixed to its original value. We consider the first branch, and eliminate T from the set of edits. We obtain the following edits:

$$P \leq 0.5(M + P), \tag{15.28}$$

$$0.1(M + P) \leq P, \tag{15.29}$$

$$M + P \geq 0, \tag{15.30}$$

$$M + P \leq 550N. \tag{15.31}$$

Edits (15.28)–(15.30) are satisfied, edit (15.31) is violated. Because edit (15.31) is violated, changing T is not a solution to the error localization problem. If we were to continue examining the branch where T is eliminated by eliminating and fixing more variables, we would find that the best solution in this branch has an objective value

(15.13) equal to 3. We now consider the other branch where T is fixed to its original value. We fill in the original value of T in edits (15.23)–(15.27), and obtain (after removing any tautology that might arise) the following edits:

$$100 = M + P, \tag{15.32}$$

$$P \leq 50, \tag{15.33}$$

$$10 \leq P, \tag{15.34}$$

$$100 \leq 550N. \tag{15.35}$$

Edits (15.34) and (15.35) are satisfied, edits (15.32) and (15.33) are violated. We select another variable, say P, and again construct two branches: one where P is eliminated and one where P is fixed to its original value. Here, we only examine the former branch, and obtain the following edits (again after removing any tautology that might arise):

$$100 - M \leq 50,$$

$$10 \leq 100 - M, \tag{15.36}$$

$$100 \leq 550N. \tag{15.37}$$

Only edit (15.36) is violated. We select variable M and again construct two branches: one where M is eliminated and another one where M is fixed to its original value. We only examine the former branch, and obtain edit (15.37) as the only implied edit. As this edit is satisfied by the original value of N, changing P and M is a solution to the error localization problem. By examining all branches of the tree, including the ones that we have skipped here, we find that this is the only optimal solution to this record.

15.9 Conclusions

To a large extent data editing of agricultural surveys and censuses is similar to editing of economic data in general. The same kinds of methods and approaches can be used. For agricultural surveys and censuses, for which often extensive and detailed questionnaires are sent out to many units in the population, it is very important that the actually implemented algorithms and methods can handle a large number of variables, edits and records quickly and accurately. A combination of selective and automatic editing provides a means to achieve this.

In our discussion of data editing we have focused on identifying errors in the data as this has traditionally been considered as the most important aim of data editing in practice. In fact, however, this is only one of the goals of data editing. Granquist (1995) identifies the following main goals of data editing:

1. to identify error sources in order to provide feedback on the entire survey process;
2. to provide information about the quality of the incoming and outgoing data;
3. to identify and treat influential errors and outliers in individual data;
4. when needed, to provide complete and consistent individual data.

In recent years, goals 1 and 2 have gained in importance. The feedback on other survey phases can be used to improve those phases and reduce the amount of errors arising in these phases. Data editing forms part of the entire statistical process at NSIs. A direction for potential future research is hence the relation between data editing and other steps of the statistical process, such as data collection (see, e.g., Børke, 2008). In the next few years goals 1 and 2 are likely to become even more important.

From our discussion of data editing the reader may have got the feeling that the basic problems of data editing are fixed, and will never change. This is not the case. The world is rapidly changing, and this certainly holds true for data editing. The traditional way of producing data, by sending out questionnaires to selected respondents or interviewing selected respondents, and subsequently processing and analysing the observed data, is in substantial part being replaced by making use of already available register data. This presents us with new problems related to data editing. First, differences in definitions of the variables and units between the available register data and the desired information have to be resolved before register data can be used. This can be seen as a special form of data editing. Second, the external register data may have to be edited themselves. Major differences between editing self-collected survey data and external register data are that in the former case one knows, in principle, all the details regarding the data collection process whereas in the latter case one does not, and that in the former case one can recontact respondents as a last resort whereas in the latter case this is generally impossible. Another difference is that the use of register data requires co-operation with other agencies, for instance tax offices. An increased use of register data seems to be the way of the future for most NSIs. The main challenge for the near future for data editing is to adapt itself so we can handle these data efficiently and effectively.

References

Bankier, M., Poirier, P., Lachance, M. and Mason, P. (2000) A generic implementation of the nearest-neighbour imputation methodology (NIM). In *Proceedings of the Second International Conference on Establishment Surveys*, pp. 571–578. Alexandria, VA: American Statistical Association.

Børke, S. (2008) Using 'Traditional' Control (Editing) Systems to Reveal Changes when Introducing New Data Collection Instruments. UN/ECE Work Session on Statistical Data Editing, Vienna.

Casado Valero, C., Del Castillo Cuervo-Arango, F., Mateo Ayerra, J. and De Santos Ballesteros, A. (1996) Quantitative data editing: Quadratic programming method. Presented at the COMPSTAT 1996 Conference, Barcelona.

Central Statistical Office (2000) Editing and calibration in survey processing. Report SMD-37, Central Statistical Office, Ireland.

Chernikova, N.V. (1964) Algorithm for finding a general formula for the non-negative solutions of a system of linear equations. *USSR Computational Mathematics and Mathematical Physics*, 4, 151–158.

Chernikova, N.V. (1965) Algorithm for finding a general formula for the non-negative solutions of a system of linear inequalities. *USSR Computational Mathematics and Mathematical Physics*, 5, 228–233.

Cormen, T.H., Leiserson, C.E. and Rivest, R.L. (1990) *Introduction to Algorithms*. Cambridge, MA: MIT Press; New York: McGraw-Hill.

De Jong, A. (2002) Uni-Edit: Standardized processing of structural business statistics in the Netherlands. UN/ECE Work Session on Statistical Data Editing, Helsinki.

De Waal, T. (1996) CherryPi: A computer program for automatic edit and imputation. UN/ECE Work Session on Statistical Data Editing, Voorburg.

De Waal, T. (2003a) Processing of erroneous and unsafe data. PhD thesis, Erasmus University, Rotterdam.

De Waal, T. (2003b) Solving the error localization problem by means of vertex generation. *Survey Methodology*, 29, 71–79.

De Waal, T. and Coutinho, W. (2005) Automatic editing for business surveys: An assessment of selected algorithms. *International Statistical Review*, 73, 73–102.

De Waal, T. and Quere, R. (2003) A fast and simple algorithm for automatic editing of mixed data. *Journal of Official Statistics*, 19, 383–402.

De Waal, T., Renssen, R. and Van de Pol, F. (2000), Graphical macro-editing: Possibilities and pitfalls. In *Proceedings of the Second International Conference on Establishment Surveys*, pp. 579–588. Alexandria, VA: American Statistical Association.

Di Zio, M., Guarnera, U. and Luzi, O. (2005), Editing Systematic unity measure errors through mixture modelling. *Survey Methodology*, 31, 53–63.

Duffin, R.J. (1974) On Fourier's analysis of linear inequality systems. *Mathematical Programming Studies*, 1, 71–95.

Farwell, K. (2005) Significance editing for a variety of survey situations. Paper presented at the 55th Session of the International Statistical Institute, Sydney.

Farwell, K. and Rain, M. (2000) Some current approaches to editing in the ABS. In *Proceedings of the Second International Conference on Establishment Surveys*, pp. 529–538. Alexandria, VA: American Statistical Association.

Fellegi. I.P. and Holt. D. (1976) A systematic approach to automatic edit and imputation. *Journal of the American Statistical Association*, 71, 17–35.

Ferguson, D.P. (1994) An introduction to the data editing process. In *Statistical Data Editing (Volume 1): Methods and Techniques*. Geneva: United Nations.

Fillion, J.M. and Schiopu-Kratina, I. (1993) On the Use of Chernikova's Algorithm for Error Localization. Report, Statistics Canada.

Freund, R.J. and Hartley, H.O. (1967) A procedure for automatic data editing. *Journal of the American Statistical Association*, 62, 341–352.

Granquist, L. (1990) A review of some macro-editing methods for rationalizing the editing process. In *Proceedings of the Statistics Canada Symposium*, pp. 225–234.

Granquist, L. (1995) Improving the traditional editing process. In B.G. Cox, D.A. Binder, B.N. Chinnappa, A. Christianson, M.J. Colledge and P.S. Kott (eds), *Business Survey Methods*, pp. 385–401. New York: John Wiley & Sons, Inc.

Granquist, L. (1997) The new view on editing. *International Statistical Review*, 65, 381–387.

Granquist, L. and Kovar, J. (1997) Editing of survey data: How much is enough? In L. Lyberg, P. Biemer, M. Collins, E. De Leeuw, C. Dippo, N. Schwartz and D. Trewin (eds), *Survey Measurement and Process Quality*, pp. 415–435. New York: John Wiley & Sons, Inc.

Hedlin, D. (2003) Score functions to reduce business survey editing at the U.K. Office for National Statistics. *Journal of Official Statistics*, 19, 177–199.

Hedlin, D. (2008) Local and global score functions in selective editing. UN/ECE Work Session on Statistical Data Editing, Vienna.

Hidiroglou, M.A. and Berthelot, J.M (1986) Statistical editing and imputation for periodic business surveys. *Survey Methodology*, 12, 73–78.

Hoogland, J. (2002) Selective editing by means of plausibility indicators. UN/ECE Work Session on Statistical Data Editing, Helsinki.

Kalton, G. and Kasprzyk, D. (1982) Imputing for missing survey responses. In *Proceedings of the Section on Survey Research Methods*, 22–31. Alexandria, VA: American Statistical Association.

Kovar, J. and Whitridge, P. (1990) Generalized Edit and Imputation System: Overview and applications/ *Revista Brasileira de Estadťstica*, 51, 85–100.

Kovar, J. and Whitridge, P. (1995) Imputation of business survey data. In B.G. Cox, D.A. Binder, B.N. Chinnappa, A. Christianson, M.J. Colledge and P.S. Kott (eds), *Business Survey Methods*, pp. 403–423. New York: John Wiley & Sons, Inc.

Latouche, M. and Berthelot, J.M. (1992) Use of a score function to prioritize and limit recontacts in editing business surveys. *Journal of Official Statistics*, 8, 389–400.

Lawrence, D. and McDavitt, C. (1994) Significance editing in the Australian Survey of Average Weekly Earning. *Journal of Official Statistics*, 10, 437–447.

Lawrence, D. and McKenzie, R. (2000) The general application of significance editing. *Journal of Official Statistics*, 16, 243–253.

Little, R.J.A. and Rubin, D.B. (2002) *Statistical Analysis with Missing Data*, 2nd edition. Hoboken, NJ: John Wiley & Sons, Inc.

McKeown, P.G. (1984) A mathematical programming approach to editing of continuous survey data. *SIAM Journal on Scientific and Statistical Computing*, 5, 784–797.

Nemhauser, G.L. and Wolsey, L.A. (1988) *Integer and Combinatorial Optimization*. New York: John Wiley & Sons, Inc.

Riera-Ledesma, J. and Salazar-Gonzaález, J.J. (2003) New algorithms for the editing and imputation problem. UN/ECE Work Session on Statistical Data Editing, Madrid.

Rubin, D.S. (1975) Vertex generation and cardinality constrained linear programs. *Operations Research*, 23, 555–565.

Rubin, D.S. (1977) Vertex generation methods for problems with logical constraints. *Annals of Discrete Mathematics*, 1, 457–466.

Sande, G. (1978) An algorithm for the fields to impute problems of numerical and coded data. Technical report, Statistics Canada.

Schafer, J.L. (1997) *Analysis of Incomplete Multivariate Data*. London: Chapman & Hall.

Schaffer, J. (1987) Procedure for solving the data-editing problem with both continuous and discrete data types. *Naval Research Logistics*, 34, 879–890.

Salazar-González, J.J., Lowthian, P., Young, C., Merola, G., Bond, S. and Brown, D. (2004), Getting the best results in controlled rounding with the least effort. In J. Domingo-Ferrer and V. Torra (eds), *Privacy in Statistical Databases*, pp. 58–71. Berlin: Springer.

Schiopu-Kratina I., Kovar J.G. (1989) *Use of Chernikova's Algorithm in the Generalized Edit and Imputation System*, Methodology Branch Working Paper BSMD 89-001E, Statistics Canada.

Scholtus, S. (2008a) Algorithms for detecting and resolving obvious inconsistencies in business survey data. UN/ECE Work Session on Statistical Data Editing, Vienna.

Scholtus, S. (2008b) Algorithms for correcting some obvious inconsistencies and rounding errors in business survey data. Discussion paper, Statistics Netherlands, Voorburg.

Todaro, T.A. (1999) Overview and evaluation of the AGGIES automated edit and imputation system. UN/ECE Work Session on Statistical Data Editing, Rome.

Winkler, W.E. and Draper, L.A. (1997) The SPEER edit system. In *Statistical Data Editing (Volume 2): Methods and Techniques*. Geneva: United Nations.

16

Quality in agricultural statistics

Ulf Jorner[1] and Frederic A. Vogel[2]

[1]*Statistics Sweden, Stockholm, Sweden*

[2]*The World Bank, Washington DC, USA*

16.1 Introduction

There is a long history of discussions about quality issues related to the collection of data and the preparation of statistics. Deming (1944) was one of the first to point out issues concerning errors in surveys. The Hansen *et al.* (1961) paper on measurement errors in censuses and surveys was one of the first to make the point that errors affecting surveys were also a census issue.

Governmental organizations that produce official statistics have always been concerned with the quality of the statistics they produce and have developed measurements such as accuracy, timeliness and relevancy. Accuracy was generally defined in terms of sampling errors, or total survey error. The main measure of timeliness was the period from the reference point to data dissemination. Relevance was more difficult to define, but it mainly meant whether the data helped answer the questions of the day. The main point is that the first measures of quality were defined by the producing organization. Some examples of quality measures are described in Vogel and Tortora (1987).

There is no generally accepted definition of the concept of quality, but a typical modern definition would be something like 'the ability of a product or service to fulfil the expectations of the user'. From this starting point, Statistics Sweden in 2000 adopted the following definition: quality of statistics refers to all aspects of statistics that are of relevance to how well it meets users' needs and expectations of statistical information. Obviously, this and similar definitions need to be operationalized in order to be useful; we will return below with examples of how this is done in Sweden and in USA.

Agricultural Survey Methods Edited by Roberto Benedetti, Marco Bee, Giuseppe Espa and Federica Piersimoni

One main point is that the modern approach is user-oriented, while the earliest measures of statistical quality were producer-oriented. Another point is that it is relative rather than absolute and even to some extent subjective. At least, different users tend to value a given component in different ways. As an example, a research worker would rank accuracy high, while a decision-maker would put more emphasis on timeliness. A final point is what is not included (i.e. costs). While costs are sometimes included in the quality concept, thus making a low cost contribute towards a higher quality, this has distinct disadvantages. The most obvious is perhaps that in general usage, quality is often weighed against costs. It is good to keep this possibility, so that for statistics we may also speak of value for money.

The purpose of this chapter is to describe how measures of quality need to change to become more customer focused. The rapid development of the Internet, for example, means that an organization's products are available around the world and used for purposes far beyond the original intent. The rapid access to data on the web is changing the meaning of timeliness and the increased sophistication of data users is also raising the need for new measures of quality. The authors will describe how the National Agricultural Statistics Service (NASS) in the USA and Statistics Sweden are attempting to redefine how they monitor and report the quality of their products. At the same time, it should be stressed that it is an evolution, not a revolution. Original measures of quality, such as accuracy, are not superseded by new components, but rather supplemented.

A unique characteristic of agricultural statistics is that many of the items being measured are perishable or have a relatively short storage time. Many of the products are seasonable rather than being available on a continuous flow. It is also generally the case that the prices are highly volatile. Small changes in quantities translate into much larger changes in price. This places a premium on accurate forecasts and estimates. Because of the perishable and seasonable nature of agricultural products, the most important agricultural statistics are those that can be used to forecast future supplies. Knowledge of crop acres planted, for example, if available shortly after planting, provides an early season measure of the future production. Inventories of animal breeding stock can be used to forecast future meat supplies.

16.2 Changing concepts of quality

The most significant event is that the concepts of quality are being customer or externally driven rather than determined internally by the statistical organization. The availability of official statistics on the Internet is increasing the audience of data users. In previous times, the data users were fewer in number and usually very familiar from long experience with the data they were being provided. The new audience of data users is increasingly becoming more sophisticated and also more vocal about their needs.

16.2.1 The American example

The following sections describe the six components of quality used by the National Agricultural Statistics Service and how they are measured:

1. Comprehensiveness.
2. Accuracy.

3. Timeliness.
4. Comparability.
5. Availability.
6. Confidentiality and security.

Comprehensiveness

The measure of quality is whether the subject matter is covered. In agriculture, one example is that crop production is provided by including estimates of acres planted and harvested, yields per acre, production, quantities in storage at the end of harvest, measures of utilization such as millings, exports, feed use, and the average prices to measure value of production. A comprehensive view of crop production includes the land use, the yield that is a measure of productivity, total production and stocks that measure the total supply, and the utilization that measures the disappearance. One way the NASS evaluates the comprehensiveness of its statistics program is through annual meetings with organizations and people using the data. For example, every October, the NASS sponsors a widely advertised open forum and invites all interested people to meet with subject-matter specialists to seek their input about the content, scope, coverage, accuracy, and timeliness of USDA statistics. Input received at recent meetings was that because of the emergence of biotech crop varieties, information was needed about the proportion of the crop acreages planted to those varieties. As a result, the data on acres planted and for harvest for several crops are now provided by biotech class.

The NASS also has a formal Statistics Advisory Committee comprised of producers, agribusinesses, academia, and other governmental users. This advisory committee meets once a year with the NASS to review the content of the statistics programme and to recommend where changes in agriculture are requiring a change in the NASS programme. As a result of these meetings, the NASS will be including questions about contract production in agriculture in the next census of agriculture.

Accuracy

Traditional measures include sampling errors and measures of coverage. These are basic measures that need to be used by the statistical organization in its internal review of its data. Certainly, the organization needs to define standards for sampling errors and coverage that determine whether or not to publish data not meeting those standards. The NASS also publishes a measure of error for major data series and uses a method called the root mean squared error (RMSE) which is the difference between an estimator and the 'true' value. The RMSE as defined by the NASS for this usage is the average difference between the first estimate or forecast and the final estimate,

$$RMSE = \frac{\sqrt{\sum_i (\tilde{X}_i - \hat{X}_i)^2}}{n},$$

where $\tilde{X}_i$ is the first estimate or forecast and $\hat{X}_i$ is the final revised estimate.

Using corn as an example, the NASS publishes a forecast of production in August each year. This forecast is updated each month during the growing season. After harvest, a preliminary final estimate of production is published. A year later, the preliminary final estimate may be revised based upon utilization and stocks data. Five years after the census of agriculture, the estimates of production for the census year and years between census periods are again reviewed and revised if suggested by the census results. The RMSE is published each month during the growing season and each time a revision is published. The reliability table also shows the number of times the forecast is above or below the final and the maximum and minimum differences. The primary gauge of the accuracy of reports is whether they meet the needs of the data users. If revisions exceed their expectations, the NASS will be pressed to either improve the accuracy or discontinue the report.

Timeliness

One measure of timeliness is the time between the survey reference period and the date of publication. There is a trade-off between accuracy and timeliness. Timeliness needs vary depending upon the frequency the data are made available. Those using data to make current marketing decisions need the data to be current. It is not helpful to not learn until long after harvest and after much of the crop has already been sold that the current year's corn crop was at a record high level. On the other hand, someone making long-term investment decisions may be more concerned with whether the census results are consistent with history. The NASS has the following standards for timeliness. Monthly reports with a first of month reference date should be published during the same month. The report should be published before the next data collection period begins. Quarterly and semi-annual reports should also be published within one month of the reference dates. The census data published every 5 years should be published a year and a month after the census reference period.

Another issue of timeliness is that the data user should know when the data are to become available. Each October, the NASS publishes a calendar showing the date and hour that over 400 reports will be published during the coming calendar year. The final issue related to timeliness is the need to be punctual and meet the dates and hours as published in the calendar.

Market-sensitive reports should not be released while the markets are open. There should be a policy to only release such reports in the morning before markets open or at the end of the day when markets have closed. This may be a moot point when electronic trading allows continuous marketing, but until then, the timing of the release of the data needs to be considered.

Comparability

Data should be comparable over time and space. It is necessary to carefully define for the data users what is being measured by defining the basic characteristics and ensure they remain the same. If it becomes necessary to change a concept or definition, then there should be a bridge showing how the data relationship changed. This may require that a parallel survey be conducted for a period to measure the affect of the change in definition. Comparability is essential for measures of change. Even subtle changes such as the way a question is asked on a survey form can change the final result.

Data should be comparable to other related information or there should be an explanation of the reasons for the departure. An example would be when livestock inventories are declining, but meat supplies are increasing. The reason could be that animals are being fattened to much heavier weights. The statistical organization needs to 'mine' its data to ensure it is comparable to internal and external sources, or to explain the difference.

Data should be easily understood or detected. The NASS recently published some data showing the percentage of the corn acreage planted to bio-tech herbicide resistant and the percentage planted to insect resistant varieties. The percentages were not additive because some seeds contained both characteristics. However, data users added them anyway because the arrangement of the data made them think they could. The point is that statistical organizations need to ensure that appropriate inferences can be made about published data even if this requires additional tables or explanation. If pieces are not additive, or totals are not the same as the sum of the parts, then they must be presented so that there is clear understanding.

Availability

Official statistics need to be available to everyone and at the same time. Data collected using public funds should be made available to the public. The integrity of the statistical agency should not be at stake by granting special access to some people or organizations before others. Special care needs to be taken to ensure that there is no premature release of information – all users need to be treated equally and fairly. The NASS operates under laws and regulations that require strict security measures ensuring that data are released only at the appropriate time and place. In addition, the data should be made available immediately to everyone via printed copies, press releases, and the website.

Confidentiality and security

The quality of official statistics is dependent upon maintaining the trust of the farms and agribusinesses reporting on their operations. A basic tenet should be that their data are strictly confidential and used only for statistical purposes. The statistical agency should seek laws and regulations that provide protection from the data being used for taxation and/or regulatory purposes. As the data are tabulated and preliminary estimates being derived, there need to be precautions to prevent premature disclosure of the results. First, a premature disclosure may be misleading if it differs from the final results. It may also give some data users an unfair advantage in the use of the data. The key to an organization maintaining its integrity and public trust is to first being totally committed to maintaining confidentiality of individually reported data and ensuring the security of the estimates until they are released to the public. None of the data in the 400+ reports published by the NASS each year are subject to political approval. In fact, they are not even made available to cabinet officials including the Secretary of Agriculture until they are released to the public. In some instances, the Secretary will enter NASS secure work areas minutes before data release to obtain a briefing about the report.

16.2.2 The Swedish example

The Swedish agricultural statistical system is, apart from scale, rather similar to the American system. The same can be said about the trends in quality concepts. The Swedish experiences will thus not be given as duplicates, or for a comparison. Rather, they will be used to expand the discussion of changing concepts of quality.

The Swedish quality concept for official statistics has five components:

1. Contents.
2. Accuracy.
3. Timeliness.
4. Coherence, especially comparability.
5. Availability and clarity.

For some background to this, as well as some useful references, see Elvers and Rosén (1999).

Contents

This component is very similar to the comprehensiveness component of the NASS. It is interesting to note that this type of component was earlier often called 'relevance'. The reason behind the change is that what is relevant for one user may be irrelevant to another. A good example is given by the use of lower thresholds. In Swedish agricultural statistics, this threshold is 2 hectares of arable land (or large numbers of animals or significant horticultural production). This reduction of the population of farms will rule out less than 1% of crop production and would thus be acceptable to most users. On the other hand, it may be very disturbing for persons studying rural development in less favoured areas. Thus, with a user-oriented quality approach it seems more natural to talk about contents or comprehensiveness.

An interesting aspect of Swedish statistics is the great use made of administrative data. This source of data is also prominent in agricultural statistics. As a prime example, the Common Agricultural Policy in the European Union makes a wide range of subsidies available to farmers. For most crops and animals, the applications from farmers give a very good coverage of the population in question. This forms a very cheap source of data, and moreover the data have been checked extensively, without cost to the statisticians. On the other hand, the statisticians have no control over the definitions used, the thresholds applied, the crops included or excluded, etc. Much work is thus required to convert administrative data into statistical data. In Sweden, the definitive way to do this has not yet been decided upon, but Wallgren and Wallgren (1998) give a good overview of possibilities and problems.

With administrative data, another problem is that the source may run dry due to political or administrative decisions. As an example, tax returns were for many years used in Sweden to produce statistics on farmers' income (with due regard to possible systematic errors). A simplification of Swedish tax legislation made farm incomes virtually indistinguishable from other forms of income in tax returns, and thus this statistical series was ended.

Accuracy

When measures of accuracy became available in the mid-1950s, this measure tended to become almost synonymous with quality. Lately, it has fallen in importance compared with, for example, timeliness and availability.

Accuracy of course involves both precision and bias. While the former is easily measured (e.g. as standard error), the latter is both difficult and expensive to measure.

As an example, reporting errors in crop areas were measured in the field for a random sample of farms in Sweden up to 1995, but this was the discontinued for cost reasons. Even when measurements were made, the number of measurements did not allow error estimates for individual years, but rather estimates for a time period. In a similar way, nation-wide coverage checks were never made for the Swedish Farm Register, although spot checks indicated that the undercoverage of crop areas and livestock was negligible while the number of farms may be underestimated by 1–2%.

Timeliness

Timeliness is perhaps the component of quality that has gained most in importance by the transition from producer-oriented to user-oriented quality. It is also naturally in conflict with components as accuracy, as shorter production times mean less time to follow up possible sources of errors. One attempt to both eat the cake and have it is to use preliminary statistics. However, there is a price to pay with regard to clarity, as in the American example above.

And, while timeliness is one of the best examples of components that have lately had high priority, it is interesting to note that as early as 1877 it was one of three components (expedition, completeness and reliability) that defined quality in Swedish agricultural statistics.

Coherence, especially comparability

The more explicit role of users in the process of choosing appropriate levels of quality also highlights the eternal dilemma between the need for changes in definitions, etc. and the need for long time series. Users such as research workers will give priority to long, unbroken time series and thus good comparability over time, while policy-makers are more interested in statistics that give a good and up-to-date picture of current agriculture. Official statisticians often have to devise compromises; a good but often expensive solution is to provide parallel series for a few years at the point of change. As an example, when the population and domains of study were changed for the Swedish Farm Accountancy Survey on Sweden's accession to the European Union, series for both the old and new classification were published for both 1997 and 1998.

Availability and clarity

Statistics Sweden makes an annual assessment of changes in the five quality components, and the one with the highest level of improvement over the last years is availability and clarity. This also holds for agricultural statistics.

The increased availability of agricultural statistics makes metadata more and more essential. Metadata are data about data; the term was coined as early as 1973 (see Sundgren, 1975). As a simple example, taken from Sundgren, the figure 427 621 is quite meaningless as such; the information – metadata – that it is the number of milk cows in Sweden in 2000 gives it some meaning. However, serious users of statistics would need more background information: what date it refers to (2 June 2000), whether it is based on a sample (yes, $n = 10\,009$), the data collection method (mail questionnaire), etc. Other users would be interested in non-response rate (6.3%), standard error (0.6%), or in the properties of the frame. Still other would like to know how this figure could be compared to earlier figures for Sweden (based on different techniques) or to figures for other countries, or whether there is a regional breakdown available (and if so, the corresponding quality measures).

Two aspects of metadata should be stressed here. First, different users have different needs for metadata, thus the system for retrieving metadata must be flexible, and of course preferably easily accessible via the Internet, for example. Second, metadata should not only act as some kind of tag on the actual figures, but also provide a means to search for information. A user should be able to search for, say, 'milk cows' and find the appropriate statistical series and its properties. Good metadata are thus both an end and a means as regards good availability.

16.3 Assuring quality

Measuring quality, however difficult it might be in specific cases, is only a lower level of quality ambition. Setting and meeting quality goals are, both from the user's and the producer's perspective, a higher ambition. The term 'quality assurance' is often used for the process of ensuring that quality goals are consistently met.

16.3.1 Quality assurance as an agency undertaking

Traditionally, we tend to think of quality as a property of a single variable. We measure this quality and try to control and hopefully improve it. Of course, we realize that the quality properties of different variables are interconnected, as they are often collected in the same survey, or that they share common processes. As an example, the same frame is normally used for different surveys and the same field staff, etc. Thus, we must consider a whole system when considering quality assurance. Figure 16.1 illustrates this point.

The figure illustrates not only the fact that the variables are interconnected, but also that we generally have problems measuring the effects of individual efforts to improve quality; measurements are made in one dimension whereas effects appear in another. To assure quality at each intersection of the grid is impractical, thus quality assurance should apply to the whole system producing the agricultural statistics in question, or more appropriately even to the whole agency that produces it.

It is useful to separate the quality of the statistics from the quality of the process that produces it. Thus, a process that assures a predetermined level of quality in a cost-efficient

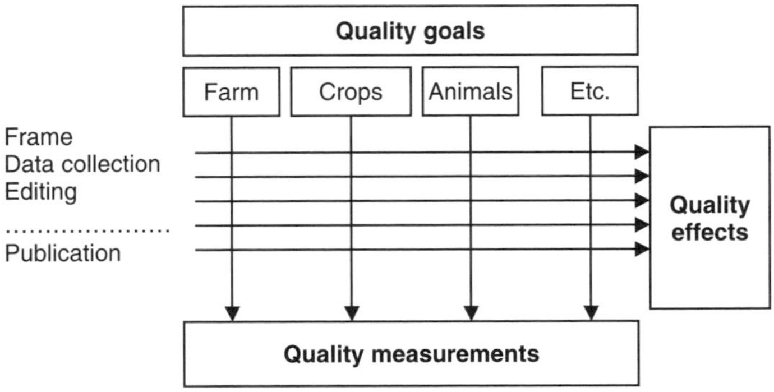

Figure 16.1 The system of quality assurance.

way is in itself 'good', even if the quality level is low. In a way, quality assurance can be looked upon as the link between quality in the process and quality in the product.

Most agencies producing agricultural statistics have been subject to budget cuts, but have been able to reduce the adverse effects on the quality of agricultural statistics through improvements in processes. The application of new techniques, such as computers and the Internet, as well as new data sources, such as administrative registers and remote sensing, have been part of this process, but also new management ideas, such as total quality management (TQM) and business process reengineering. It should be remembered that an approach such as TQM involves both gradual improvement in existing procedures and development of totally new procedures.

Finally, it should be stressed that maintaining the integrity of an agency, for example by having and demonstrating independence from the political sphere, is of paramount importance in assuring quality in the eyes of users.

16.3.2 Examples of quality assurance efforts

The statistical agency should adhere to all of the traditional principles of quality control that have been well documented elsewhere. The entire statistical process from frame development needs controls embedded to control quality. Basic things such as data edits and analysis are crucial and should be a normal part of doing business. The following paragraphs describe some non-traditional things that the NASS has been doing to enhance data quality.

Technical review teams. Teams have been formed for two purposes. The NASS has an office in 45 states in addition to the headquarters units. The headquarters units develop the specifications for the survey, analysis, and estimation processes that are carried out by the state offices. To ensure that the states are properly interpreting and carrying out the correct procedures, teams of subject-matter specialists are periodically sent to visit state offices on a rotating basis to do a complete review of the office. The team is responsible for documenting their findings, and the state is responsible for correcting areas that did not meet the standards. It is sometimes found that one reason why the states were not following correct procedures was that the headquarters' instructions were not clear or complete. The other purpose of technical teams is to review a specific estimating program. The most recent effort involved a review of the survey and estimating processes used to generate estimates of grain in storage. A team of experts reviewed the statistical and data collection processes from the state offices through the preparation of the final estimates. As a result of this review, the sample design is being revamped, and a new edit and analysis system is being developed.

Data warehouse. The NASS has developed a data warehouse system that provides the capability to store all data for every farm in the USA. It currently contains individual farm data for the 1997 Census of Agriculture and all survey data collected for each farm since that time. Edit and analysis systems are being developed that will incorporate the use of historic information on an individual farm basis.

Statistical research. The NASS's Statistical Research Division operates outside the operational programme. It does periodic quality check re-interviews, special analysis to identify non-sampling errors, develops new sample designs and estimators, and provides in-house consulting services to monitor and improve the quality of statistics.

Hotline. A staff is assigned to responding to a toll-free telephone number or an email address. Any farmer or data user in the country can make a toll-free call to this number or send an email message to ask any question or raise any concern about anything the NASS does. A log is maintained and used to determine areas needing improvement.

Outreach: As described above, the NASS relies upon an advisory board for input in addition to the formal programme of data user meetings. NASS staff also meet regularly with commodity organizations and other governmental organizations to keep current with their needs for quality.

16.4 Conclusions

As data users become more sophisticated in their use of public data, their demands for quality will be a major force to determine the quality of official statistics. Data producers must take into account that data users will attempt to use information for purposes that go beyond the capabilities of the data. As an example, user-friendly quality declarations/metadata will be very important.

Quality assurance of agricultural statistics will also be more important, in order to prevent large accidental errors that might lead users to wrong and perhaps costly decisions. Such assurance will apply to the agency as such rather than to individual statistical series. Use of unconventional approaches and new techniques will be ever more important in this work.

The trends in USA and Sweden are probably representative for other countries as well. Of course, there are deviations from country to country, for example in how far these trends have progressed, and the relative importance of different quality components. Hopefully, the guiding principles defining quality in Sweden and the USA could provide examples and standards that will be useful to other countries.

The final measure of quality is whether the statistics being produced are believable. The ultimate test is whether the public accepts the data as being accurate assessments of the current situation. This public perception is ultimately based on its trust in the integrity of the statistical organization. The foundation of any statistical organization's integrity is to have the public's trust that it protects the confidentiality of individual reported data, ensures statistics are secure before release, and releases results to everyone at the same time.

References

Deming, W.E. (1944) On errors in surveys. *American Sociological Review*, 9, 359–369.

Elvers, E. and Rosén, B. (1999) Quality concepts for official statistics. In S. Kotz, B. Campell and D. Banks (eds), *Encyclopedia of Statistical Sciences*, Update Vol. 3, pp. 621–629. New York: John Wiley & Sons, Inc.

Hansen, M., Hurwitz, W. and Bershad, M.A. (1961) Measurements errors in censuses and surveys. *Bulletin of the International Statistical Institute*, 38, 359–374.

Sundgren, B. (1975) *Theory of Data Bases*. New York: Petrocelli/Charter.

Vogel, F. and Tortora, R.D. (1987) Statistical standards for agricultural surveys. In *Proceedings of the 46th Session of the International Statistics Institute*, Tokyo. Voorburg, Netherlands: ISI.

Wallgren, A. and Wallgren, B. (1998) How should we use IACS data? Statistics Sweden, Örebro.

17

Statistics Canada's Quality Assurance Framework applied to agricultural statistics

Marcelle Dion[1], Denis Chartrand[2] and Paul Murray[2]

[1]*Agriculture, Technology and Transportation Statistics Branch, Statistics Canada, Canada*

[2]*Agriculture Division, Statistics Canada, Canada*

17.1 Introduction

Statistics Canada's product is information. The management of quality therefore plays a central role in the overall management of the Agency, and in the strategic decisions and day-to-day operations of a large and diverse work group such as its Agriculture Division. The Agency's Quality Assurance Framework[1] defines what is meant by data quality. This Framework is based on six dimensions: relevance, accuracy, timeliness, accessibility, interpretability and coherence.

The first part of this chapter summarizes the recent evolution of the structure of the agriculture industry in Canada and users' needs. This is followed by a description of Agriculture Division's framework for collecting, compiling, analysing and disseminating quality agriculture data.

The third part of the chapter briefly describes the Agency's Quality Assurance Framework. The remainder of the chapter examines the framework for managing data quality

[1] See `http://stdweb/standards/IMDB/IMDB-nutshell.htm`

Agricultural Survey Methods Edited by Roberto Benedetti, Marco Bee, Giuseppe Espa and Federica Piersimoni

in Statistics Canada under each of the six dimensions noted above. Specific emphasis is placed on how the Agriculture Division manages data quality for its agricultural statistics programme.

17.2 Evolution of agriculture industry structure and user needs

The Canadian agriculture and agri-food system is a complex integrated production, transformation and distribution chain of industries supplying food, beverages, feed, tobacco, biofuels and biomass to domestic and international consumers (Chartrand, 2007). It is an integral part of the global economy, with trade occurring at each stage of the chain. However, the relative contribution of primary agriculture to Gross Domestic Product (GDP) and employment has been declining significantly. Although the value of agricultural production has tripled since 1961, the Canadian economy as a whole has grown at a faster rate (by six times), driven mainly by growth in the high-tech and service sectors (Agriculture and Agri-Food Canada, 2006). The result has been a drop in the share of primary agriculture to about 1.3% of GDP. On the other hand, the agriculture and agri-food industry as a whole remains a significant contributor to the Canadian economy, accounting for 8.0% of total GDP and 12.8% of employment in 2006 (Agriculture and Agri-Food Canada, 2008).

The rapid pace of the evolution in the structure of the agriculture industry can be partially illustrated by examining the changing number and size of farm operations in Canada. After peaking in 1941, farm numbers have fallen steadily while total agricultural land has been relatively stable, resulting in a continual increase in average farm size in Canada during this time (see Figure 17.1). Moreover, agriculture production has become much more concentrated on larger farms. Figure 17.2 shows that an increasingly smaller proportion of farms accounts for the majority of sales over time.

As farms become larger, with more complex legal and operating structures, the tasks of collecting, processing and analysing data for the industry, and, ultimately, measuring its

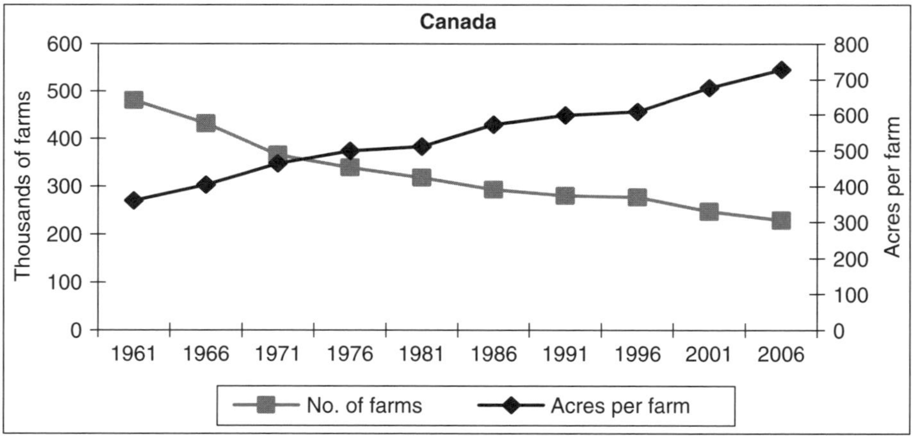

Figure 17.1 Farm numbers fall while average farm size increases. The trend towards fewer but larger farms continues in Canada. *Source*: Census of Agriculture.

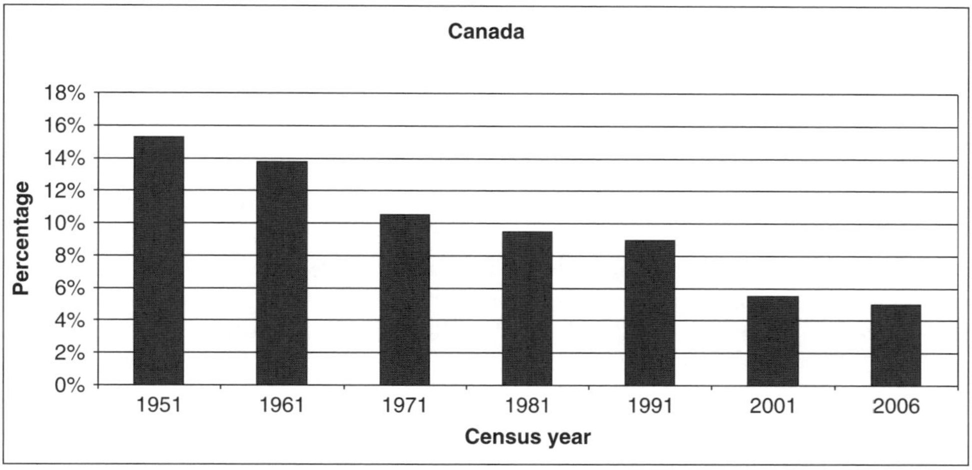

Figure 17.2 A smaller percentage of farms is needed to account for half of the agricultural sales. Concentration has been increasing – sales becoming more concentrated in larger operations. *Source*: Census of Agriculture.

performance, have also become more complicated and difficult. The growing importance of vertically integrated operations, increased contractual arrangements and more varied marketing opportunities such as direct marketing, dual markets and numerous payment options have added to this complexity.

The structure of Canadian agricultural production and marketing began changing more rapidly in the latter part of the twentieth century with the increasing adoption of new technologies on farms (including more substitution of capital for labour and realization of economies of scale), changes in domestic and foreign agriculture policy, and rising globalization, among others. For example, *Growing Forward*, Canada's agricultural policy framework implemented in 2008, 'is a new commitment to Canada's agriculture sector that's focused on achieving results, reflects input from across the sector, and will deliver programs that are simpler, more effective and tailored to local needs'.[2]

At the same time as policy agendas of governments were changing, users of agriculture production, financial, trade, environmental and other related data identified needs for more detailed and integrated statistical information that would illuminate current issues. They also put more emphasis on the quality of the statistics, as will be discussed in the following paragraphs.

17.3 Agriculture statistics: a centralized approach

The production and dissemination of national and provincial estimates on the agriculture sector is part of Statistics Canada's mandate. The statistical agency carries out monthly, quarterly, annual and/or seasonal data collection activities related to crop and livestock surveys and farm finances as needed. It also conducts the quinquennial Census of Agriculture in conjunction with the Census of Population (to enhance the availability of

[2] http://www4.agr.gc.ca/AAFC-AAC/display-afficher.do?id=1200339470715&lang=e

socio-economic data for this important industry, while reducing overall collection and public communication costs and strengthening the coverage of the Census of Agriculture), and produces and publishes economic series on the agriculture sector that flow to the System of National Accounts (SNA) to form the agriculture component of the GDP (Dion, 2007; Statistics Canada, Agriculture Division, 2005)

Administrative data supplement limited survey taking in supply-managed agriculture sectors such as dairy and poultry. Taxation data are mined extensively to tabulate annual disaggregated financial characteristics by major farm types, revenue classes and regions for Canadian farm operations, as well as farm family income.

The extensive cost-recovery programme constantly evolves to meet the changing needs of clients, and provides valuable auxiliary statistical information to complement the core survey programme on topics such as agri-environment, farm management, specialty crop production, farm assets and liabilities, etc. The Farm Register currently provides the key agriculture survey frames.

Joint collection agreements are in force with most provincial and territorial departments of agriculture to minimize burden by reducing or eliminating duplicative local surveying. These agreements cover production surveys and, increasingly, provincially administered stabilization programme data. This latter source is important for verifying survey data and has the potential for replacing some survey data in the future to reduce respondent burden while ensuring data quality is maintained or improved.

As can be seen in Figure 17.3, all of the above collection activities are an integral part of the Agriculture Statistics Framework that feed into the SNA for the calculation of the GDP for the agriculture sector. An integrated statistical framework plays two important roles: a quality assurance role and a 'fitness for use' role facilitating the interpretability of the information.

The *quality assurance* process of the agriculture statistics programme is built into each program activity and is thus being carried out continually. There are, though, two significant occasions occurring in the Division that are worth special mention in the assurance of high-quality data: on an annual basis with the provision of information to the SNA, and on a quinquennial basis with the release of the Census of Agriculture (CEAG) data.

As can be seen in Figure 17.3, the Farm Register provides the frame for most survey activities; the crop, livestock and financial data series serve as inputs to the derivation of the farm income series. In turn, the production of the farm income estimates is a coherence exercise that allows the validation and revision (if necessary) of the input data.

Every five years, the same holds true for the CEAG. It allows for an update of the Farm Register used to draw samples for the various crop, livestock and financial surveys and provides a base to revise estimates (including farm income estimates) produced between censuses. In turn, sample surveys and farm income series are useful tools to validate census data.

Metadata, 'the information behind the information', are stored on the corporate repository of information, the Integrated Metadata Base (IMDB),[3] which provides an effective vehicle for communicating metadata to data users in their determination of 'fitness for use' (Johanis, 2001). These metadata, organized around the hierarchical structure of

[3] Metadata for all agriculture surveys are available at `http://www.statcan.gc.ca/english/sdds-/indext.htm`

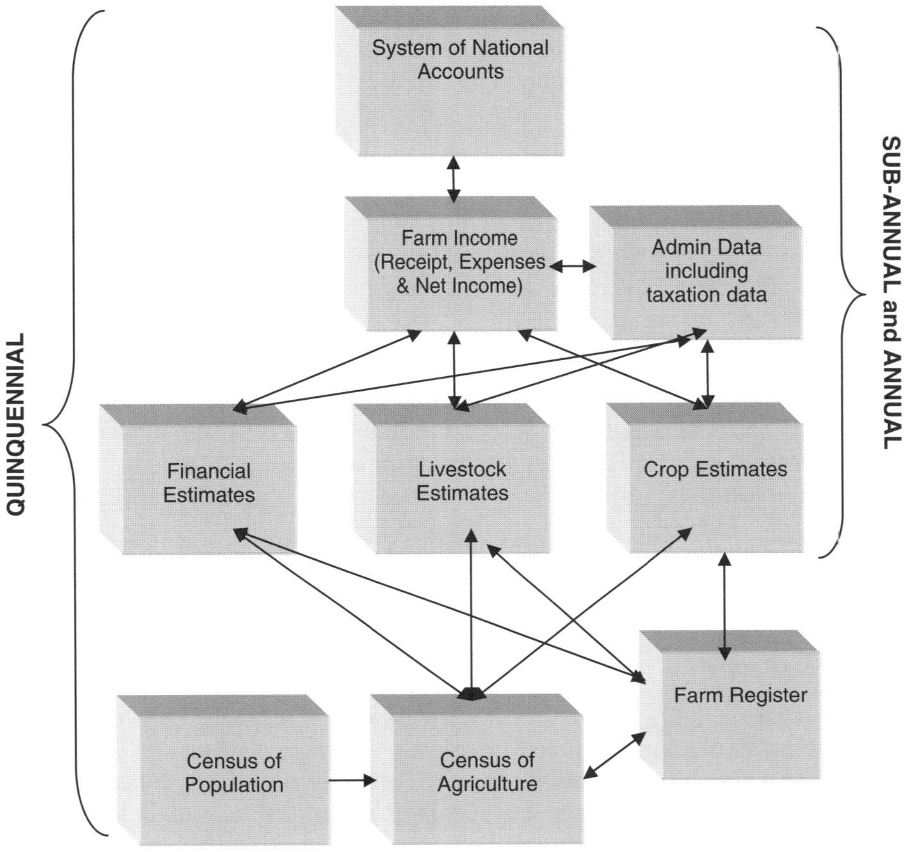

Figure 17.3 Agriculture Statistics Framework.

the farm income series shown in Figure 17.3, facilitate users' understanding of the interrelationships between the various statistical activities. They also increase users' awareness that the 'fitness for use' test of the farm income series must take into account the metadata information of all the farm income series' inputs. Additional information on the role of metadata in managing interpretability is presented in Section 17.5.5.

17.4 Quality Assurance Framework

The Quality Assurance Framework describes the approaches that Statistics Canada takes to the management of quality for all its programmes including those related to agriculture (Statistics Canada, 2002). Statistics Canada has defined data quality in terms of 'fitness for use'. Quality is important, but it is a matter of degree. One needs very high standards of accuracy, timeliness, etc. for some statistical applications, but one can 'make do' with much less accuracy, timeliness, etc. for some other purposes. This is what the 'fitness for use' concept is all about. Six dimensions of quality have been identified within the concept of 'fitness for use'.

1. The *relevance* of statistical information reflects the degree to which it meets the real needs of users. It is concerned with whether the available information sheds light on the issues of most importance to users.
2. The *accuracy* of statistical information is the degree to which the information correctly describes the phenomena it was designed to measure. It is usually characterized in terms of error in statistical estimates and is traditionally decomposed into bias (systematic error) and variance (random error) components. It may also be described in terms of the major sources of error that potentially cause inaccuracy (coverage, sampling, non-response, response (reporting and/or recording errors), etc.).
3. The *timeliness* of statistical information refers to the delay between the reference point (or the end of the reference period) to which the information pertains, and the date on which the information becomes available. It is typically involved in a trade-off against *accuracy*. The timeliness of information will influence its *relevance* (as will the other dimensions).
4. The *accessibility* of statistical information refers to the ease with which it can be obtained by users. This includes the ease with which the existence of information can be ascertained, as well as the suitability of the form or medium through which the information can be accessed. The cost of the information may also be an aspect of accessibility for some users.
5. The *interpretability* of statistical information reflects the availability of the supplementary information and metadata necessary to interpret and utilize it appropriately. This information normally covers the underlying concepts, variables and classifications used, the methodology of collection, and indicators of the accuracy of the statistical information.
6. The *coherence* of statistical information reflects the degree to which it can be successfully brought together with other statistical information within a broad analytic framework and over time. The use of standard concepts, classifications and target populations promotes coherence, as does the use of common methodology across surveys. Coherence does not necessarily imply full numerical consistency.

Management of the six dimensions of quality at Statistics Canada occurs within a matrix management framework – project management operating within a functional organization. The design or redesign of a statistical programme normally takes place within an interdisciplinary project management structure in which the sponsoring programme area and the involved service areas – particularly for collection and processing operations, for informatics support, for statistical methodology, and for marketing and dissemination support – all participate. It is within such a project team that the many decisions and trade-offs necessary to ensure an appropriate balance between concern for quality and considerations of cost and response burden are made. It is the responsibility of functional organizations (divisions) to ensure that project teams are adequately staffed with people able to speak with expertise and authority for their functional area while being sensitive to the need to weigh competing pressures in order to reach a project team consensus. The Agriculture Division relies very heavily on the use of such teams to ensure the overall quality of its products.

17.5 Managing quality

17.5.1 Managing relevance

The management of relevance embraces those processes that lead to the determination of what information the Agency produces. It deals essentially with the translation of user needs into programme approval and budgetary decisions within the Agency. The processes that are used to ensure relevance also permit basic monitoring of other elements of quality and assessment of user requirements in these other dimensions. Since these needs evolve over time, a process for continuously reviewing programmes in the light of client needs and making necessary adjustments is essential.

For the Agriculture Division, these processes can be described under three broad headings: agriculture client and stakeholder feedback mechanisms; programme review exercises; and data analysis activities. A priority-setting process translates the information from the three previous processes into program and budget decisions.

Agriculture client and stakeholder feedback mechanisms

The Agriculture Division maintains its relevance with adjustments, as required, to its regular programme. Specific guidance and feedback mechanisms to ensure the relevance of agriculture data are provided regularly through several key committees and constant contact with its broad and diverse user community and through its strong cost recovery initiatives.

Feedback mechanisms serve to maintain awareness of the issues of interest to each major client and stakeholder group, and the information needs likely to flow from these issues. The Agriculture Division also uses these feedback mechanisms to obtain information from current users of its products and services on their level of satisfaction, and to identify potential new markets for information.

The Agriculture Division works directly with the provinces, Agriculture and Agri-Food Canada (AAFC) and various other users on an ongoing basis. In addition, the Division profits from interactions with a number of key committees, all of which meet annually:

- The Advisory Committee on Agriculture Statistics, with membership composed of respected academic and industry leaders, assists in determining the strategic direction of the Division.
- The Federal-Provincial-Territorial Committee on Agriculture Statistics includes representation of all provincial and territorial governments, and has the goal of exchanging information about the shared agricultural statistics programmes to ensure data quality.
- The North American Tripartite Committee on Agricultural Statistics, with representatives from Canada, Mexico and the United States, deals with agricultural statistics matters in a North American context.

Other programme-related consultations have increased for several reasons recently. Intensive consultations related to the Census of Agriculture take place each cycle for questionnaire content and output development. This broadly-based process includes representatives from all levels of government, industry, academia and consultants. Users are asked to provide feedback and propose changes to the CEAG questionnaire based on their requirements. It is important to keep in mind that, for example, the data from the 2011 CEAG need to meet the users' projected requirements until 2017.

Regarding outputs, users are asked to comment on accessibility of CEAG products, the suite of standard products, reference products, geographic products, analytic products and the services provided by CEAG staff. The consultations on products proposed for the CEAG allow users to provide indications of their satisfaction and their comments, permitting the planning of new products and services.

Feedback on content and outputs gathered during these consultations can also be relevant for the ongoing survey programme.

Central to the excellent working relationship between the Agriculture Division and AAFC – its major external client – is an Interdepartmental Letter of Agreement (ILOA) between the two organizations. A new agreement – beginning in April 2009 – is the third such five-year accord. The ILOA permits longer-term planning and better prioritization of data needs for AAFC's *Growing Forward* policy and other departmental priorities; allows flexibility for new projects to be added and other projects to be deleted as priorities change; and permits Statistics Canada to smooth out staff peaks and troughs through stable, planned cost-recovery funding. The agreement contributes significantly to improved relevance by funding particular surveys to fill important data gaps. By contributing to larger sample sizes for other surveys, accuracy is increased and the Division's ability to disseminate more disaggregated data is enhanced. The agreement specifies an annual review of projects and funding by AAFC and Agriculture Division senior management, thus ensuring relevance by allowing for the reallocation of funds to emerging issues, such as the collection of data that assisted in the understanding of the 2003 bovine spongiform encephalopathy (BSE) crisis in the cattle sector.

In other circumstances, specialized workshops are organized to address specific relevance issues, resulting in key recommendations that lead to strategies to guide related work in the Agriculture Division. For example, the farm income accounts have been scrutinized in recent years as numerous users search for ways to better depict the sector's performance and the well-being of producers and their families. In response, a specialized workshop – the 2007 Workshop on Farm Income Measures – was organized jointly by AAFC and Statistics Canada. The technical, international workshop provided an opportunity for experts in the farm income area to discuss issues related to current farm income measures and identify options to better describe the economic situation in agriculture. At the forefront was the strong sense that the Agency should develop and produce an integrated broadly-based set of farm income and other performance measures that are released at the same time, ideally based on the same time period, and that provide data by the various types of agriculture production and farm revenue size.

Programme review exercises

In addition to the regular flow of information coming from liaison mechanisms or specialized workshops both at the national and international levels, programmes are periodically reviewed to assess whether they are meeting user needs. Every four years, each Statistics Canada programme is required to produce a report that is a strategic review of its relevance and direction, including the results of consultation with its clients and stakeholders. These Quadrennial Program Reviews (QPRs) consolidate and analyse the feedback obtained from clients and may present proposals for addressing any identified programme weaknesses, data gaps or data adjustments. The strategic direction presented in the Agriculture Division's QPR in December 2009 reflects issues and paves the way for its programme, including data quality management, for the next four years.

Every second year, each programme must supplement its QPR with a mid-term review – known as a Biennial Program Report (BPR) – documenting its performance and laying out changes to its strategic directions described in the QPR submitted two years prior to the BPR.

Data analysis activities

Data analysis serves several purposes in the management of quality, including an important role in maintaining relevance. While its primary purposes may be to advance understanding and to discover further insights from existing outputs, it also provides a valuable source of feedback on the adequacy and completeness of the collected data. By identifying questions the data cannot answer, it pinpoints gaps and weaknesses in data holdings. The development of certain longitudinal surveys, several record linkage activities, the creation of a metadata base, harmonized calibration, and attention to data integration and standardization of concepts are among the initiatives that can be attributed, at least in part, in response to obstacles faced in undertaking analysis.

The use of analytic frameworks such as the SNA or the Agriculture Statistics Framework, explained in Section 17.3, to integrate and reconcile data coming from different sources is an important element in identifying gaps and weaknesses in our data. While research based on aggregate-level data has been and still is useful, there is a shift to the use of longitudinal databases and the use of microdata in support of research. Longitudinal information and micro-data analysis are better suited to the understanding of impact and outcome that in turn increase the relevance of our data and also identify gaps that would not have been recognized otherwise.

The Agriculture Division has a six-year longitudinal tax database of farm taxfilers as well as an AGPOP linkage database. The taxfiler base is very useful in analysing factors affecting the financial performance and profitability of different subsets of farms over time. The AGPOP linkage database (a rich source of socio-economic data resulting from linking the censuses of agriculture and population, and produced every five years since 1971) supports analysis related to Canadian farm operators and their families, contributing to a better understanding of this segment of the Canadian population.

The Agriculture Division has been working with the AAFC Research Networks and the Rural Secretariat at AAFC for several years to discuss topics of importance and establish priorities for research to be carried out. The Division also has a vibrant University Research Partnership Program targeted at graduate student research in Canada on topics addressing issues in agriculture, food, the rural economy, and the environment as they affect agriculture and associated industries.

Priority setting and funding

In some cases, adjustments to the programme can be made to maintain or improve relevance with negligible cost. However, in many instances, maintaining relevance will imply the need for significant resources, either provided by clients (or more rarely, from the Statistics Canada corporate reserve) or freed up within the programme by reallocating from other parts of the programme. Efficiencies are another potential source of resources for maintaining relevance, but at Statistics Canada most of these are appropriated for broader corporate priorities and so are not available to individual divisions. In the case of agriculture statistics, there are a few major clients (SNA, AAFC, the provinces) that account for most of the usage of the product, so reallocating involves negotiating with them, bilaterally and multilaterally, as noted earlier. They are the ones for whom the product has to be kept relevant, so it is logical that they also be the ones to face the trade-offs on where reallocations can be made. In many cases, the major client, AAFC, provides funding to address data gaps affecting policy development and programme monitoring.

In summary, the Agency uses a variety of mechanisms to keep abreast of clients' information requirements, to obtain users' views on its existing products and services, to seek professional advice on issues and priorities, and to identify weaknesses and gaps in its data. This information provides a basis for the management judgements that have to be made on revising the Agency's (including the Agriculture Division's) programme. In addition to user needs, costs, respondent burden and public sensitivities, the Agency's capacity and expertise have to be taken into account. Judgements must be made in the light of current public policy priorities as to which statistical programmes are in most need of redevelopment or further investment, which can be eliminated or which new programmes need to be funded.

17.5.2 Managing accuracy

Processes described previously under 'relevance' determine which programmes are going to be carried out, their broad objectives, and the resource parameters within which they must operate. Within those 'programme parameters', the management of accuracy requires particular attention during three key stages of a survey process: survey design; survey implementation; and assessment of survey accuracy. These stages typically take place in a project management environment, outlined in Section 17.4, which is crucial to a proper balancing of conflicting considerations in the development of a statistical programme.

Survey design

The collection, processing and compilation of data require the use of sound statistical and analytical methods and models; effective designs, instruments, operational methods and procedures; and efficient systems and algorithms. A number of primary aspects of design need to be addressed in every programme to ensure that accuracy considerations requiring attention during the design stage have been identified. Consideration of overall trade-offs between accuracy, cost, timeliness and respondent burden are all normally made at this stage.

In order to ensure that the design of statistical programmes efficiently meets the objectives set for them, a variety of means have been put in place to provide guidance and information. The use of specialized staff for subject-matter, methodology, operations and systems to participate in programme design greatly aids the process, as do specialized resource and support centres for certain functions (e.g., questionnaire design and testing, data analysis) and a standard survey frame (the Farm Register) for agriculture surveys. In addition, the development and implementation of a number of internal policies – on the review and testing of questionnaires; standard definitions of concepts, variables and classifications for common subject-matter areas; peer and institutional reviews; and a policy on use of software products and applications that identifies recommended and supported software for key functions – all clarify requirements for divisional and other project team staff.

Good design contains built-in protection against implementation errors (through quality assurance processes, for example). The results of implementation depend not only on the specific design, but also on the instruments of implementation, including the plans for human and other resources, the supervisory structure, the schedules, the operations, the procedures and checks, the training, the publicity, etc., developed and specified during the design phase.

Survey implementation

Mechanisms for monitoring implementation of quality assurance processes are built into Agriculture Division survey processes as part of design. Two types of information are required. The first is a system to produce timely information to monitor and correct, in real time, any problems arising while the survey is in progress. The second need is for information to assess, after the event, whether the design was carried out as planned, whether some aspects of the design were problematic in operation, and what lessons were learned from the operational standpoint to aid design in the future. Using Statistics Canada's management information systems as a framework, the Division manages and monitors implementation through:

- regular reporting and analysis of response rates and completion rates during the collection phase;
- monitoring refusal and conversion rates of non-respondents into respondents;
- monitoring interviewer and respondent feedback;
- monitoring of edit failure rates and the progress of corrective actions;
- monitoring the results of quality control procedures during collection and processing;
- monitoring of expenditures against progress and budget;
- development, implementation and monitoring of contingency plans.

To ensure quality in light of the high technical content of many design issues, the Agriculture Division's programmes incorporate independent technical review into their design, implementation and accuracy assessment plans, where appropriate. In addition to those highlighted above, the Division profits from the guidance of internal technical review committees for major systems development; referral of issues of technical standards, or general methods or approaches to senior statistical methods committees; and the review of the practices of other national statistical agencies and the exchange of experiences with them. The use of work-in-progress reviews (validation consultations) with selected provincial specialists for key crop, livestock and farm income data series – subject to the procedures laid out in Statistics Canada's Policy on The Daily and Official Release – also aids in ensuring the accuracy of divisional data.

Assessment of survey accuracy

The third key stage of the survey process is the assessment of accuracy – what level of accuracy has actually been achieved? Though described last, it needs to be a consideration at the design stage since the measurement of accuracy often requires information to be recorded as the survey is taking place.

As with design, the extent and sophistication of accuracy assessment measures will depend on the size of the survey, and on the significance of the uses of the estimates. Statistics Canada's Policy on Informing Users of Data Quality and Methodology requires at least the following four primary areas of accuracy assessment to be considered in all surveys: coverage of the survey in comparison to a target population; sampling error where sampling was used; non-response rates or percentages of estimates imputed; and any other serious accuracy or consistency problems with the survey results.

While the Agriculture Division actively participates in Agency quality assurance review processes seeking to further reduce the risk of data errors, the Division employs several other specific processes and activities that also contribute significantly. They are described below.

Farm Register and the Large Agricultural Operations Statistics Group With the increasingly complex structure of Canadian primary agriculture, specialized procedures are required to ensure that all farms are covered and represented accurately by the CEAG and regular-programme survey activity. As noted in the description of the Agriculture Statistics Framework in Section 17.3, the Farm Register – a registry of all Canadian farming operations – plays an important role in ensuring the overall quality of survey activities in the Agriculture Division.

The Farm Register is kept up-to-date in various ways:

- the Farm Update Survey, an ongoing survey that is designed to verify births, deaths and mergers of farming operations based on signals received from various lists such as taxation files, industry associations' producer lists, etc.;
- feedback from ongoing surveys and Census of Agriculture collection operations;
- Large Agricultural Operations Statistics processes (see below).

The Agriculture Division's Large Agricultural Operations Statistics (LAOS) group is responsible for all aspects of the Division's interaction with very large and complex farming operations, ensuring the collection of relevant and accurate data. These farming operations are only contacted by a single employee of the Agriculture Division using an agreed reporting schedule with the farm representative. During these contacts, the organizations are profiled to gain a better understanding of their legal and operating structure in order to ensure complete and accurate coverage. Specialized collection and processing procedures have been implemented for the past several censuses to ensure that coverage was maintained and to offer these key respondents the ability to customize the manner in which they provide information to the CEAG and surveys, thus also reducing respondent burden. There is an increasing number of large, corporate enterprises involved in agricultural production. At the time of the last CEAG in 2006, the number of operations handled this way (765) was 23.0% higher than in 2001.

In addition, the relationships among another group of complex operations – large operations which are connected through contracts rather than through ownership – must be regularly validated by the LAOS group. Arrangements between enterprises and their contract farms can vary widely from one reporting period to the next. It is important to profile all participants in these arrangements so that the data can be collected from the appropriate parties to ensure that neither undercoverage nor double-counting occurs.

Having an up-to-date Farm Register benefits the coverage of all surveys as they draw on the Register for their sample selection. The Census of Agriculture, while also benefiting in the past, will have an increased reliance on the Farm Register with the 2011 cycle. In order to reduce the dependence on a large decentralized workforce for the 2011 CEAG, the target for mail-out (as opposed to direct delivery by Census field staff) to farms has been increased from the 6% achieved in 2006 to 90% in 2011. To achieve this target, the Farm Register will be used to support the mail-out and missing farm follow-up activities.[4]

Survey redesign It is fair to say that Agriculture Division surveys impose a significant burden on farmers which may in turn reduce response rates and quality of data. This has the potential to increase as the Division strives to deliver new outputs to meet user demands in sensitive areas such as agri-environmental statistics. The Agriculture Division uses the opportunity provided by survey redesigns to minimize burden by reducing, for example, sample overlap among surveys. This opportunity arises every five years once the CEAG results have been released.

Within this context, goals of the 2007 divisional survey redesign included reducing response burden while not degrading data quality, improving the robustness of the survey designs so the surveys will continue to perform well for five years until the next redesign, redrawing samples using a Farm Register that has been updated with the latest CEAG results, and standardizing methods and concepts as much as possible. An inter-divisional working group composed of Business Survey Method Division (BSMD) and Agriculture Division representatives was formed and analysis began in early 2007 with all divisional survey redesigns implemented by early 2008.

The redesign also incorporated methodological improvements. For example, the overall number of strata in the Crops Survey population was reduced, which will result in more stable estimates over time. The stratification of the Atlantic Survey was modified to improve the data quality of the crops variables, a requirement identified before the redesign.

Administrative data The Agriculture Division continues to actively pursue the use of administrative data sources to reduce response burden and collection costs, but also to ensure coverage and to confront survey estimates as a means of improving overall data quality.

The use of administrative data requires the same level of scrutiny as survey data to understand their strengths and limitations. It is important to continually monitor changes

[4] The Missing Farms Follow-up Survey was a follow-up procedure where the Farm Register and external validated lists were matched to CEAG questionnaires at a particular point in processing. Non-matches were contacted, within resource constraints, to obtain a CEAG questionnaire. Step D on the Census of Population questionnaire asked whether anyone in the household operated a farm. Again, if a CEAG questionnaire was not received for a positive Step D response, follow-up action was initiated. Both activities improved coverage of the CEAG.

incorporated into the concepts, methodology and collection methods of administrative data to evaluate the impact of such changes on the production of estimates.

The Division relies heavily on administrative data, only surveying producers when the required data are not available elsewhere. While it uses taxation data extensively, other types of administrative data provided by industry and government have been an integral part of the Division's programme for many years. In the latter case, since the production of administrative data is related to a particular government programme, changes in the type of data available, their disappearance altogether or the emergence of new data impact the quality of Agriculture Division statistical programmes. Divisional analysts have well-established processes to monitor the quality of the wide variety of administrative data used in their programmes. A critical component has been developing excellent relationships with the owners of these data sources.

Financial estimates The production process used to generate financial estimates for the agriculture industry is critical in the assessment of the quality of both survey and administrative data. Operating much like a smaller-scale version of the SNA for the agriculture industry, the Farm Income and Prices Section (FIPS) of the Agriculture Division integrates a vast amount of survey data and administrative data from federal and provincial governments, marketing boards, industry organizations and others in preparing economic estimates of the agriculture industry for the SNA and external users. In order to play such a role, analysts involved in the development of these estimates must rely on tools that are dependable and easy to understand. A new standardized system, developed in 2008/09, is currently being implemented. It will lead to a more convenient flow of information and give staff more time for coherence analysis, reducing the risk of poor data quality and errors affecting data integrity. The new system uses Agency-approved system architecture and software and includes increased uniformity of naming conventions, improved revision, edit and analysis features, and enhanced audit capabilities, while reducing manual data entry and transcription errors.

Checks and balances: Census of Agriculture As noted earlier, a key component for ensuring data quality in the Agriculture Division is the quinquennial CEAG. Significant changes are planned for the 2011 CEAG to address changing industry and data collection environments. This plan will still retain integration with the Census of Population on key critical activities such as collection, communications, delivery and registration, an internet collection option, help lines and capture. These partnerships will continue to provide significant efficiencies to the overall census programme and several of these joint processes have in the past proven valuable in increasing response rates and coverage for the CEAG.

The decision to aim for a 90% mail-out/mail-back over the whole collection period in 2011 will make the missing farms follow-up process central to the success of the next census. The centralized Missing Farms Follow-up Survey, instituted in 2006, was a powerful test of the strength of the enhanced content of the Farm Register and Step D (identification of the presence of a farm operator in the household) on the Census of Population questionnaire in pursuing non-response and coverage follow-up for the Census of Agriculture.

The quality of ongoing survey estimates depends partly on the quality of CEAG results, as a coherence exercise – known as intercensal revisions – between Census and survey estimates, follows each cycle of the CEAG. This is a quality assurance process where survey and other estimates are confronted and revised, if necessary, based on CEAG data (see Section 17.5.6 for a more complete explanation).

Error reduction in subject-matter sections Despite the serious efforts of many groups throughout the processes mentioned in this Section, errors remain in all estimates to some limited extent. The Agriculture Division conducts many surveys based on samples of agricultural operations on a monthly, quarterly, occasional and annual basis to prepare estimates for release to the public. As such, these estimates are subject to sampling and non-sampling errors. The overall quality of the estimates depends on the combined effect of these two types of errors. Quality problems also occur in the administrative data and the associated models and assumptions used in estimation.

Sampling errors arise because estimates are derived from sample data and not the entire population. These errors depend on factors such as sample size, sampling design and the method of estimation. Inevitably, there are frame errors and omissions, both at census time (despite the Missing Farms Follow-up Survey) and with the intercensal divisional surveys, and non-response bias (perhaps reduced by making surveys mandatory). The Division's efforts to keep the Farm Register current, as described earlier in this section, are crucial in ensuring accuracy. In addition, by funding larger sample sizes for some surveys, such as the special crop surveys, AAFC contributes to increasing accuracy.

Non-sampling errors can occur whether a sample is used or a complete census of the population is taken, and can be introduced at various stages of data processing as well as inadvertently by respondents. Population coverage, differences in the interpretation of questions, incorrect or missing information from respondents, and mistakes in recording, coding and processing of data are examples of non-sampling errors (Murray and Culver, 2007).

Each of the subject-matter sections in the Agriculture Division has its own rigorous verification and validation procedures to ensure high data quality. The IMDB notes that each survey, such as the Field Crop Reporting Series and the Livestock Survey,[5] employs detailed systematic procedures to minimize errors, some of which are briefly described below.

The computer-assisted telephone interviewing applications used for collection contain range and consistency edits and 'help' text. A set of reports is run to identify problem items early in collection for remedial action (e.g., variables with a significant number of edit failures or missing information). The data processing phase includes checking interviewer notes, manually reviewing significant inconsistencies and reviewing the top contributors to the unweighted and weighted estimates for each variable in each province.

Total non-response (e.g. refusals and no contacts) is accounted for by weighting adjustments to each stratum. Some item non-response is estimated deterministically (using other information in the respondent's questionnaire). Some missing information is imputed manually during the edit process, and some using a donor imputation method. The automated imputation system looks for donors within the stratum and then verifies that the donor record and the record to be imputed are acceptable. A final review of the imputed data is then performed. Finally, analysis of survey estimates and administrative data, as described elsewhere in the chapter, is a critical step in ensuring overall data quality.

[5] Statistics Canada's Integrated Metadata Base is available at http://www.statcan.gc.ca/english/-sdds/indext.htm

The potential error introduced by sampling can be estimated from the sample itself by using a statistical measure called the coefficient of variation (CV).[6] Measures of accuracy are an important input for Program Review (Section 17.5.1) for assessing whether user requirements are being met, and for allowing appropriate analytic use of the data. They are also a crucial input to the management of interpretability as elaborated in Section 17.5.5. Tables 17.1 and 17.2 show CVs and response rates for recent Agriculture Division surveys, while Table 17.3 displays the response burden imposed on Canadian farm operators by the Division's surveys.[7]

In addition to specific actions noted above, with Statistics Canada's current strong desire to avoid errors, the Agency has pursued corporately-driven quality assurance processes. For example, as described in Section 17.6, Statistics Canada has introduced a Quality Secretariat to promote and support the use of sound quality management practices. This group has undertaken an analysis of issues that arise during the last

Table 17.1 Coefficients of variation for agriculture surveys, Canada, 2003/04 to 2008/09 (error levels may be higher for commodities or provinces where production is less commercially significant).

	Coefficient of variation (%)					
Survey area	2003/04	2004/05	2005/06	2006/07	2007/08	2008/09
Major crops	2.29	2.54	2.82	2.78	3.70	3.90
Special crops	3.77	4.41	4.67	5.63	5.40	5.30
Cattle	0.98	1.07	0.88	1.26	0.89	0.98
Hogs	1.22	1.18	1.26	1.30	1.13	1.27

Weighted average of two surveys per year for cattle and special crops, four surveys for hogs, and six surveys for major crops.

Table 17.2 Response rates for agriculture census and surveys, Canada, 2003/04 to 2008/09.

	Response rate (%)					
Survey area	2003/04	2004/05	2005/06	2006/07	2007/08	2008/09
Census				95.7		
Cattle	92.5	92.9	91.9	92.6	91.5	91.1
Hogs	94.2	93.0	92.6	91.0	92.5	93.9
Major and special crops	84.0	78.0	81.0	82.0	78.0	79.9
Farm financial	85.3	83.3	83.3	84.2	82.2	81.2
Environmental				80.0		

Weighted average of two surveys per year for cattle, four surveys for hogs, and six surveys for major crops and special crops.

[6] An important property of probability sampling is that sampling error can be computed from the sample itself by using a statistical measure called the coefficient of variation. Over repeated surveys, 95 times out of 100, the relative difference between a sample estimate and what should have been obtained from an enumeration of all farming operations would be less than twice the coefficient of variation. This range of values is referred to as the confidence interval.

[7] Data in these tables are taken from the Division's BPR of December 2007 for the earlier years and from other internal sources for the most recent years.

Table 17.3 Survey and response burden for agriculture, Canada, 2004 to 2008.

	Year				
Response burden	2004	2005	2006	2007	2008
Total no. of surveys	29	29	30	29	30
Total no. of survey occasions	130	129	130	122	122
Burden (potential cost, hours)	72320	68692	68034	79717	52502
Regular programme	56227 78%	55043 80%	53570 79%	51474 65%	38508 73%
Cost recovery programme	16093 22%	13649 20%	14464 21%	28243 35%	13994 27%

Year 2006 excludes Census of Agriculture and related follow-up activity. Includes a Farm Register administrative clean-up survey in preparation for the Census of Agriculture; year 2007 includes the Farm Environmental Management Survey, conducted every five years.
Source: Ombudsman Small Business Response Burden, Statistics Canada

step of the survey process, the release of data in *The Daily* (Statistics Canada's official release bulletin), providing as a result a set of best practices that can be incorporated into an individual survey's quality assurance processes. The Agriculture Division has also implemented its own quality review team to ensure that Agency directives and best practices relating to data quality are applied uniformly throughout the Division.

17.5.3 Managing timeliness

The desired timeliness of information derives from considerations of relevance – for what period does the information remain useful for its main purposes? The answer to this question varies with the rate of change of the phenomena being measured, with the frequency of measurement, and with how quickly users must respond using the latest data. Specific types of agriculture data require different levels of timeliness. Data on crop area, stocks and production, for example, must be available soon after the reference date in order to provide useful market information, while data from the Census of Agriculture, which provide a broad and integrated picture of the industry, have a longer 'shelf life' and are not so time-sensitive. Agriculture economic data must be provided to the SNA branch in accordance with a predetermined time frame and revision schedule to integrate with all other economic data used in producing Canadian GDP figures. In addition, as noted in Section 17.5.1, the ILOA with AAFC gives the Agriculture Division the capacity to rapidly shift priorities to meet data needs on emerging issues, such as the collection of data that assisted in understanding the 2003 BSE situation in the cattle sector.

Planned timeliness is a design decision, often based on trade-offs with accuracy: are later but more accurate data preferable to earlier and less accurate data? Improved timeliness is not, therefore, an unconditional objective. A statistical agency could hold back its data for a long time until they were perfected to the maximum, but would not realistically do so because it recognizes the value of timeliness. On the other hand, if it rushed out the release of estimates by reducing quality assurance, there could be greater accuracy problems.

Cost is, of course, an important factor, no matter what decision is taken. Programme managers must find an optimal balance, and that optimum can shift over time depending on the requirements of users. For example, in order to achieve an improved balance, the Census of Agriculture carefully modified its processes and released its 'farm' and 'farm operator' data – over 300 variables – on 16 May 2007, exactly one year after the Census reference date. This was the first time that operator data were released this early, and also the first time that all data were available free online down to the census consolidated subdivision level, both done to meet users' demands for increased timeliness of these data.

Timeliness is an important characteristic that should be monitored over time to warn of deterioration. User expectations of timeliness are likely to heighten as they become accustomed to immediacy in all forms of service delivery thanks to the pervasive impact of technology.

Major Agriculture Division information releases all have release dates announced well in advance. This not only helps users plan, but it also provides internal discipline and, importantly, undermines any potential effort by interested parties to influence or delay any particular release for their benefit. The achievement of planned release dates should be monitored as a timeliness performance measure. Changes in planned release dates should also be monitored over longer periods.

For some divisional programmes, the release of preliminary data followed by revised and final figures is used as a strategy for making data more timely. In such cases, the tracking of the size and direction of revisions can serve to assess the appropriateness of the chosen timeliness–accuracy trade-off. It also provides a basis for recognizing any persistent or predictable biases in preliminary data that could be removed through estimation. To assist users in this determination, the Agriculture Division has published a measure, such as Thiel's root mean square prediction error,[8] for major variables for many years.

To be able to gauge timeliness, there are informal practices to release:

- monthly survey results about 60 days after the reference month;
- quarterly survey results during the second quarter following the reference quarter;
- annual survey results 15 months after the reference year.

As can be seen from Table 17.4,[9] release dates for major series have been stable in recent years and are within acceptable ranges.

17.5.4 Managing accessibility

Statistical information that users do not know about, cannot locate, cannot access or cannot afford, is of no value to them. Accessibility of information refers to the ease with which users can learn of its existence, locate it, and import it into their own working environment. Agency-wide dissemination policies and delivery systems determine most aspects of accessibility. Programme managers are responsible for designing statistical

[8] Theil's root mean square prediction error provides an indication of the expected size of revisions to an estimate. It represents the average percentage difference between the initial and current estimates during the period in question.

[9] Data in this table are taken from the Division's BPR of December 2007 for the earlier years and from other internal sources for the most recent year (Statistics Canada, Agricultural Division, 2007).

Table 17.4 Timeliness of selected agriculture statistics releases, Canada, 2001/02, 2004/05 and 2008/09.

			Elapsed time to release (Actual number of days)		
Publication	Frequency	Source(s)	2001/02	2004/05	2008/09
Livestock					
Cattle and sheep statistics	semi-annual	Producer survey/ administrative data	48–49	48–49	47–49
Hog statistics	quarterly	Producer survey/ administrative data	23–49	23–49	23–49
Red meat stocks	quarterly	Industry survey	24–29	24–29	24–30
Dairy statistics	quarterly	Administrative data/industry survey	44–46	44–46	43–46
Aquaculture	annual	Unified enterprise survey	250–260	292	303
Field crop reports	seasonal	Producer survey/ administrative data	20–25	19–25	21–22
Fruit and vegetable production	seasonal	Producer Survey	na	42	51
Farm cash receipts	quarterly	Administrative data	56–58	55–57	54–56
Net farm income	annual	Administrative data	148	148	147
Food available for consumption	annual	Administrative data/model	150	150	148

products, choosing appropriate delivery systems and ensuring that statistical products are properly included within corporate catalogue systems.

Statistics Canada's dissemination objective is to maximize the use of the information it produces while ensuring that dissemination costs do not reduce the Agency's ability to collect and process data in the first place. Its strategy is to make 'public good' information of broad interest available free of charge through several media (including the press, the internet, research data centres and libraries) while charging for products and services that go beyond satisfying a broad public demand for basic information.

The Daily is Statistics Canada's first line of communication with the media and the public (including respondents), released online (`www.statcan.gc.ca`) at 8:30 a.m. Eastern time each working day. It provides a comprehensive one-stop overview of new information available from Statistics Canada. Each monthly, quarterly, annual or occasional Agriculture Division release of data to the public is announced in *The Daily* with a short analytic note, charts and tables, as appropriate for the data series.

In recent years, not only are users' requirements for agriculture data changing, but their preferred ways of accessing data and information are also evolving. Long gone are the days when printed publications were the main product demanded by the Agriculture Division's clients. To aid users in finding what they need, the Agriculture Division produces and releases free on Statistics Canada's website, an electronic publication

Table 17.5 CANSIM downloads, agriculture statistics, 2003/04 to 2008/09.

	Year					
Programme area	2003/04	2004/05	2005/06	2006/07	2007/08	2008/09
CANSIM downloads, total	351 038	383 058	637 454	764 859	680 086	796 037
Crops	123 361	131 887	208 139	261 521	219 046	201 060
Livestock	103 769	104 668	119 719	151 178	124 111	154 009
Farm income and prices	123 908	146 503	309 596	352 160	336 929	440 968

Includes the number of vectors (time series) downloaded by customers and employees, excluding those by employees for publishing and maintenance purposes.
Source: `http://cansim2.statcan.gc.ca:81/mis`

entitled *People, Products and Services*,[10] which provides extensive information about all divisional data series.

The Division's publications have been available electronically in PDF format at no charge on Statistics Canada's website for several years. More recently, to conform to the Government of Canada's 'common look and feel' requirements for electronic information, the Division has continued to work on making its publications available in HTML format as well as in PDF to allow users more flexibility in accessing and analysing the data. Some specialized divisional products, such as its Crop Condition Assessment Program, Canada Food Stats, and the Extraction System of Agricultural Statistics (ESAS), are also directly available on the Agency's website or on CD-ROM. In addition, the role of the Division's Client Service group was broadened to include more specialized requests and strengthened to permit better service to users. Growing CANSIM (Statistics Canada's on-line database) downloads, as indicated in Table 17.5, provide further evidence that the Division's data are accessible and used.

17.5.5 Managing interpretability

Statistical information that users cannot understand – or can easily misunderstand – has no value and may be a liability. Providing sufficient information to allow users to properly interpret statistical information is therefore a responsibility of the Agency. 'Information about information' has come to be known as meta-information or metadata.

Metadata are at the heart of the management of the interpretability indicator, by informing users of the features that affect the quality of all data published by Statistics Canada. The information provides a better understanding of the strengths and limitations of data, and how they can be effectively used and analysed. Metadata may be of particular importance when making comparisons with data across surveys or sources of information, and in drawing conclusions regarding change over time, differences between geographic areas and differences among subgroups of the target populations of surveys.[11]

The type of meta-information provided covers the data sources and methods used to produce the data published from statistical programmes, indicators of the quality of the data as well as the names and definitions of the variables, and their related classifications. Statistics Canada's IMDB also provides direct access to questionnaires.

[10] Cat. no. 21F0003GIE.
[11] See `http://dissemination.statcan.gc.ca/english/concepts/background.htm`

Statistics Canada's Policy on Informing Users of Data Quality and Methodology ensures that divisions comply with the Agency's requirements for ensuring accurate and complete metadata as it requires that all statistical products include or refer to documentation on data quality and methodology. The underlying principle behind the policy is that data users first must be able to verify that the conceptual framework and definitions that would satisfy their particular data needs are the same as or sufficiently close to those employed in collecting and processing the data. Users also need to be able to assess the degree to which the accuracy of the data and other quality factors are consistent with their intended use or interpretation. Individual IMDB records can be accessed on the Agency's website through hyperlinks from CANSIM, the on-line catalogue of products, summary tables and *The Daily*, as well as through a button found on its home page (Born, 2004).

Essentially, the information in the IMDB covers what has been measured, how it was measured, and how well it was measured. Users clearly need to know what has been measured (to assess its relevance to their needs), how it was measured (to allow appropriate analytical methods to be used), and how well it was measured (to have confidence in the results). Since we can rarely provide a profile of all dimensions of accuracy, the description of methodology also serves as a surrogate indicator of accuracy: it allows the user the option of assessing whether the methods used were scientific, objective and carefully implemented. It is important for statistical agencies to publish good metadata because by doing so they show openness and transparency, thereby increasing the confidence of users in the information they produce.

A further aid to Statistics Canada's clients is interpretation of data as they are released. Commentary in *The Daily* and in associated materials focuses on the primary messages that the new information contains. Directed particularly at the media and the public, such commentary increases the chance that at least the first level of interpretation to the public will be clear and correct. Statistics Canada attaches very high priority to the interpretability of its releases, especially those in *The Daily*, which get close and repeated advance scrutiny by senior management and by colleagues in other divisions such as the SNA branch and the Communications and Library Services Division.

Moreover, Statistics Canada's Policy on Highlights of Publications requires that all statistical publications contain a section that highlights the principal findings in the publication. In addition to the descriptive analysis in *The Daily* and in publications, major contributions to the interpretability of the Division's and other related data are found in its specialized analytic publications. Interpretive divisional publications, such as *Understanding Measurements of Farm Income*,[12] also aid in users' basic understanding of our data.

The metadata support all of the Agency's dissemination activities including its online data tables, CANSIM and summary tables, publications, analytical studies and *The Daily*. The metadata also support data collection activities. The IMDB is the source for the survey information displayed on the Information for Survey Participants module on Statistics Canada's website.

17.5.6 Managing coherence

Coherence of statistical data includes coherence between different data items pertaining to the same point in time, coherence between the same data items for different points in time,

[12] Cat. no. 21-525-X.

and international coherence. Three complementary approaches are used for managing coherence in Statistics Canada.

The first approach to the first element is the development and use of standard frameworks, concepts, variables and classifications for all the subject-matter topics that are measured. This aims to ensure that the target of measurement is consistent across programmes, that consistent terminology is used across programmes, and that the quantities being estimated bear known relationships to each other. The Agriculture Division implements this element through the adoption and use of frameworks such as the SNA, the Agriculture Statistics Framework (described in Section 17.3) and by employing standard classification systems for all major variables. For example, the Agriculture Division employs the North American Industry Classification System for industry determination and the Standard Geographical Classification (SGC) for subnational statistics. The important issue of international comparability is addressed by adhering to international standards where these exist.

The second approach aims to ensure that the process of measurement does not introduce inconsistency between data sources even when the quantities being measured are defined in a consistent way. The development and use of common frames, methodologies and systems for data collection and processing contribute to this aim. Examples of ways in which this approach is implemented in the Agriculture Division include the following:

- the use of a common Farm Register as the frame for all agricultural surveys;
- the use of commonly formulated questions when the same variables are being collected in different surveys;
- the application of 'harmonized' methodologies and the embodiment of common methodology for survey functions (e.g., sampling, editing, estimation);
- the application of the Quality Guidelines document to encourage consistent consideration of design issues across surveys;
- the use of Agency centres of expertise in certain methodologies and technologies to exchange experience, identify best practice, develop standards, and provide training.

The third approach analyses the data themselves and focuses on the comparison and integration of data from different sources or over time. Conceptual frameworks play an important role by providing a basis for establishing coherence or recognizing incoherence. Examples of this approach in the Agriculture Division are the use of supply/disposition balance sheets in the analysis of crop, livestock and food data; the data integration approach to producing farm income data; and the validation consultation exercises (known as 'work-in-progress' consultations) conducted with provincial governments.

Checks and Balances: Census of Agriculture Survey Coherence Staff from the subject-matter sections in the Agriculture Division actively participated in the validation and certification of data from the CEAG, preparing them well for the major divisional intercensal revision (ICR) project, a quality assurance process where survey and other estimates are confronted and revised, if necessary, based on CEAG data. The ICR process is an excellent opportunity to review the accuracy and coherence of data, to search for and integrate new data sources and to evaluate the best methods to represent the myriad of data series for Canadian farms. Consultation with provincial agricultural representatives is a critical part of the review process for livestock, crop and financial data.

Checks and Balances: Processing Farm Income and Prices Data The use of analytic frameworks such as the SNA and the Agriculture Statistics Framework to integrate and reconcile data coming from different sources is an important element in identifying coherence issues, gaps and weaknesses in our data. As the major Agriculture Division integrator of data from a wide array of survey and administrative sources, the Farm Income and Prices Section collects, analyses, integrates and formats major economic series for the SNA branch and outside users.

The Division is currently involved in an informatics project to develop an integrated data system to streamline and facilitate the processing of regular-programme crop, livestock and financial survey and administrative aggregate data, leaving more time for data coherence and other analysis, and thus reducing the risk of errors.

17.6 Quality management assessment

Quality management assessment at Statistics Canada encompasses key elements of the Quality Assurance Framework (QAF), a framework for reporting on data quality and the Integrated Metadata Base (Julien and Born, 2006). Within this structure, a systematic assessment of surveys, focusing on the standard set of processes used to carry them out, is crucial. It has several objectives:

- to raise and maintain awareness of quality at the survey design and implementation levels;
- to provide input into programme reports and corporate planning;
- to respond to future requests from the Office of the Auditor General and Treasury Board;
- to reveal gaps in the QAF.

In order to promote and support the use of sound quality management practices across the Agency, Statistics Canada has created a Quality Secretariat. Its activities are overseen by the Agency's Methods and Standards Committee.

In addition to overseeing a systematic assessment of the processes of the many surveys carried out by Statistics Canada, the Quality Secretariat has undertaken an analysis of issues that arise during the last step of the survey process, the release of data in *The Daily*. Through a series of tables on the types of errors that occur at this last stage, a list of best practices that can be incorporated into an individual survey's quality assurance processes has been developed. The awareness, the continuous monitoring of what is happening at release time and the implementation of the best practices have contributed to an increase in the quality of data releases.

The Agriculture Division has also implemented a quality review team under the direction of one of its assistant directors to ensure that Agency directives and best practices relating to data quality are applied uniformly throughout the Division.

17.7 Conclusions

The management of the six dimensions of quality takes place within the organizational environment of the Agency. While all aspects of that environment influence how effectively quality management can be carried out, some are critical to its success and deserve explicit mention in this Quality Assurance Framework. Programme managers are helped in fulfilling their programme objectives through a series of measures aimed at creating an environment and culture within the Agency that recognizes the importance of quality to the Agency's effectiveness. These measures include the recruitment of talented staff and their development to appreciate quality issues, l and an open and effective network of internal communications. They include explicit measures to develop partnerships and understandings with the Agency's suppliers (especially respondents). Finally, they also include programmes of data analysis and methodological research that encourage a search for improvement.

Statistics Canada's Quality Assurance Framework consists of a wide variety of mechanisms and processes, acting at various levels throughout the Agency's programmes and across its organization. The effectiveness of this framework depends not on any one mechanism or process but on the collective effect of many interdependent measures. These build on the professional interests and motivation of the staff. They reinforce each other as means to serve client needs. They emphasize the Agency's objective professionalism, and reflect a concern for data quality. While the overall framework is presented in this chapter as a set of separate components, the important feature of the regime is the synergy resulting from the many players in the Agency's programmes, operating within a framework of coherent processes and consistent messages.

Statistics Canada has spent a great deal of effort in developing, implementing, managing and documenting rigorous guidelines and processes to ensure that it produces high-quality data. Within the environment of a rapidly changing agriculture industry and increasingly demanding user requirements, the Agriculture Division follows these processes and has adapted them to fit its own particular situation in meeting the needs of its data suppliers, its data users and its employees.

Acknowledgements

The opinions expressed in this chapter are those of the authors and do not necessarily reflect the official position of Statistics Canada. The authors wish to acknowledge the contributions of Jeffrey Smith and Philip Smith in the preparation of this chapter.

References

Agriculture and Agri-Food Canada (2006) *An Overview of the Canadian Agriculture and Agri-Food System 2006*, Publication 10013E.

Agriculture and Agri-Food Canada (2008) *An Overview of the Canadian Agriculture and Agri-Food System 2008*, Publication 10770E.

Born, A. (2004) Metadata as a tool for enhancing data quality in a statistical agency. Paper presented to the European Conference on Quality and Methodology in Official Statistics.

Chartrand, D. (2007) How to build an integrated database in agriculture and food: The Farm Income and Prices Section database – a case study. Paper presented to the Fourth International Conference on Agriculture Statistics, Beijing, 22–24 October.

Dion, M. (2007) Metadata: An integral part of Statistics Canada's Data Quality Framework. Paper presented to the Fourth International Conference on Agriculture Statistics, Beijing, 22–24 October.

Johanis, P. (2001) Role of the Integrated Metadata Base at Statistics Canada. In *Statistics Canada International Symposium Series – Proceedings*.

Julien, C. and Born, A. (2006) Quality Management Assessment at Statistics Canada. Paper preseneted to the European Conference on Quality in Survey Statistics.

Murray, P. and Culver, D. (2007) Current farm income measures and tools: Gaps and issues. Paper presented to the Agriculture and Agri-Food Canada/Statistics Canada Workshop on Farm Income Measures, Ottawa.

Statistics Canada (2002) *Statistics Canada's Quality Assurance Framework*, Cat. no. 12-586-XIE.

Statistics Canada, Agriculture Division (2005) Quadrennial Program Review 2001/02-2004/05.

Statistics Canada, Agriculture Division (2007) Biennial Program Report – FY 2005/06-2006/07.

Part V

DATA DISSEMINATION AND SURVEY DATA ANALYSIS

18

The data warehouse: a modern system for managing data

Frederic A. Vogel
The World Bank, Washington DC, USA

18.1 Introduction

There are fundamental forces sweeping across agriculture that are significantly affecting organizations that produce agricultural statistics. Statistical organizations like the National Agricultural Statistics Service (NASS) are increasingly being asked to supply information to aid in making public policy decisions on everything from the transition from a planned to a marketing economy to monitoring agriculture's effect on the environment and food safety.

The explosion of the Internet has opened the floodgates of people wanting access to information. The data users are quickly becoming increasingly sophisticated, wanting more data in greater detail, more quickly, and with improved accuracy and reliability. The globalization of trade and markets has only increased the need for data of higher quality than ever provided before.

The challenge to statistical organizations is that the additional demands for data are not always accompanied by appropriate resources to satisfy these needs using conventional methodology. Additionally, the changing structure of agriculture to fewer but larger farms is increasing their reporting burden and resulting in declining response rates.

A long-time dream of researchers and analysts has been to have the ability to link historical sample survey and census data for individual farms across long periods of time. The development of improved statistical estimators and analytical tools would be

Agricultural Survey Methods Edited by Roberto Benedetti, Marco Bee, Giuseppe Espa and Federica Piersimoni

enhanced by easy access to historical data. Data quality could be better monitored if a respondent's currently reported data could be readily linked to what was previously reported. Respondent burden could be reduced by eliminating redundant questions across survey periods and using previously reported data instead. Respondent burden could also be reduced if their data reported for administrative purposes could be used in lieu of survey responses.

The good news is that technology is coming to the rescue. There have been advancements in software and hardware technology that make easy and efficient access to current and historical data a reality.

The following sections will describe the data situation in the NASS that led to the need to develop a data warehouse. This will be followed by a brief technical discussion about the hardware and software systems that support the data warehouse. Then the chapter will include a review of how the data warehouse is now being used. The summary will include a discussion about the guiding principles that the statistical organization needs to adopt along with the high level of management commitment required to successfully implement a data warehouse system.

18.2 The data situation in the NASS

There are about 2.1 million farms in the United States. The size distribution is extremely skewed, with the 180 000 largest farms accounting for three-quarters of the annual sales of agricultural products. Only about 5000 farms account for one-quarter of the sales.

Over 400 reports are published annually by the NASS. These statistics are inclusive in their coverage of crop and livestock production on a weekly, monthly, quarterly, semi-annual, and annual basis. Some reports contain data on economic and environmental indicators. The official estimates in most of these reports result from a system of sample surveys of farm operators. Several hundred surveys a year are conducted that cover over 120 crops, 45 livestock species and associated economic and environmental items. In addition to the ongoing survey programme, the results of the census of agriculture are published at five-year intervals.

The skewed size distribution of farms results in many of them being included in multiple surveys during a year. The situation is compounded by the very nature of the NASS statistics programme. For example, one purpose of the crops estimating programme is to provide data for forecasting purposes. Before the crop planting period, a sample of farmers is asked to report the area by crop they intend to plant in the coming season. This is the basis for the March Acreage Intentions report published in late March each year. This is followed by another survey in June after crop planting to obtain crop areas planted and expected areas to be harvested for grain. Monthly surveys are then conducted during the growing season where sampled producers are asked to report the yield they expect to produce. After harvest, another sample of producers provides data on areas actually harvested and quantities produced. Four times a year, sampled producers are asked to report the quantities of grain stored on the farm. Table 18.1 provides examples of questions asked by survey period.

The sample surveys are designed to provide both direct estimates and ratio estimates of change. To reduce respondent burden, the samples are also designed so that producers rotate into and out of surveys based on replicated-rotating sample designs. Some farms, because of their sheer size, are included with certainty in every sample.

Table 18.1 Typical questions asked of producers.

Typical questions asked on a sample survey	March	June	August	September	September to December	December
How many acres of ... expect to plant?	X					
How many acres of ... have been planted?		X				
How many acres of ... expect to harvest?		X	X			
What do you expect the yield of ... to be?			X		X	
How many acres of ... did you harvest?						X
How many bushels of ... did you harvest?						X
How many bushels of ... are in storage?	X	X		X		X
How many pigs born the last 3 months?	X	X		X		X
How many sows are expected to farrow the next 3 months?	X	X		X		X

A brief summary of the issues and problems that provided the impetus to develop a data warehouse follows.

- Processing systems for various surveys have generally been determined by the size of the application. As a result, processing platforms include a mainframe, Unix platform, and personal computers connected into local area networks. Software includes SAS, Lotus, X-Base, and other systems. Therefore, it has been very difficult or nearly impossible to link data across survey systems, because of the variety of software and hardware platforms being used. Even when the data were in one program, such as SAS, separate files were maintained for each state and survey, resulting in thousands of separate files being created each year.
- Many farms, especially large farms, are selected in recurring sample surveys. As a result, they can often be asked questions about items they previously reported. The NASS surveys are voluntary, thus producers do not have to report every time, and some do not. Imputation for missing data would be greatly enhanced if previously reported information could be readily used.

- Data analysts find it difficult to access the myriad of systems for ad hoc analysis without the aid of computer programmers who are often too busy or reluctant to provide assistance. As a result, many questions raised during the analysis of survey data go unanswered. It is also difficult to link historical and current survey and census data to improve estimation and imputation procedures.

For these reasons and others, the NASS made a strategic decision in late 1994 to implement a database that would contain all individual farm level reported data across time.

18.3 What is a data warehouse?

The data warehouse, using the NASS situation, is a database system that contains all 'historical' individual farm-level data from the 1997 Census of Agriculture and major surveys from 1997 to the present. Also, the data warehouse now contains data from 'current' surveys so that one-stop shopping of current and previous data is provided. When the 2002 Census is being conducted, farm level reported data will be added to the data warehouse during the day, every day, to facilitate data review and analysis.

A guiding principle in the choice of the database and in its development was that the data be separated from the application systems. A driving force was to make the data easily accessible by all statisticians using simple access and analysis tools without requiring intervention of programmers or systems analysts. Another system requirement was that it be easy to track data across surveys/censuses and across time – a necessity to optimize use of historical data.

The database design needed to support fast data loading and speedy data access in contrast to transactional processing where the emphasis is on adding, deleting, or updating records rather than accessing them. The data warehouse must provide rapid read-only access to large chunks of data records across surveys and across time.

18.4 How does it work?

The intent of the data warehouse was to provide statisticians with direct access to current and historical data and with the capability to build their own queries or applications. Traditional transactional databases are designed using complex data models that can be difficult for anyone but power users to understand, thus requiring programmer assistance and discouraging ad hoc analysis. The database design results in many database tables (often over 100 tables), which result in many table forms when querying, so query speed is often slow. Again, technological developments in the industry came to the rescue. The goal to provide end users simple access and analysis capabilities quickly led to the development of the dimensional modelling concept. Dimensional modelling is the primary technique for databases designed for ad hoc query and analysis processing.

In effect, the data view involves two groups of tables including data and the metadata. The metadata tables contain the information that identifies, explains, and links to the data in the data table. Figure 18.1 shows the scheme of the overall dimensional model designed for the NASS data warehouse. Notice that there are only seven tables in the data warehouse – one data table and six metadata tables.

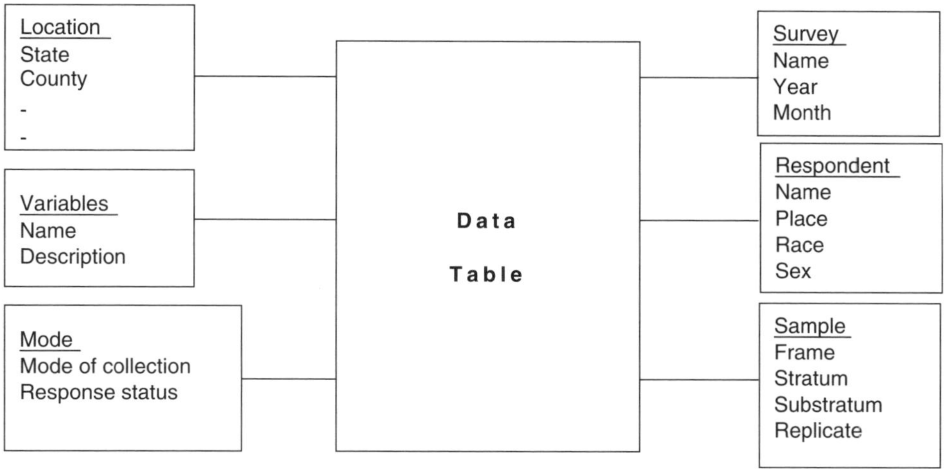

Figure 18.1 The data warehouse dimensional model.

The central data table contains the individual data responses for each question for every survey or census from 1997 onward, when available; a weight is also provided that can be a non-response adjustment or sampling weight so that direct estimates can be readily derived. The six surrounding tables contain the metadata describing the survey or census response. These six tables were designed in a way that reflects how the end user or commodity analyst views the data, which is by a given variable (corn acres planted), by location (state), by survey (June acreage), by sample frame (stratum), by respondent, and by mode (method of collection or response). Users simply browse the six metadata tables to define what data they want to access from the data table for their analysis.

Every statistician in the NASS can access the warehouse in a read-only format. The NASS has implemented an easy-to-use query tool (BrioQuery) purchased from Brio Technology. Using this query and analysis tool and the dimension table logic, statisticians build their own queries to generate tables, charts, listings, etc. Canned applications are also prepared for analysis situations that occur frequently.

Some simple examples of ad-hoc queries developed by users are the following:

- Provide all historical data as reported over time by one hog producer. The time period included every quarterly survey for the last three years.
- Provide all data over time for the 100 largest hog producers in the USA arranged from the largest to the smallest.
- Provide historical data for selected variables for all non-respondents to the current survey.
- Obtain a preliminary sample weighted total of hog inventory for the current survey.
- Obtain number of farms, acres, and production for all farms in the USA reporting corn yields of 10 bushels per acre or less.

These applications can be processed and returned to the end user in seconds. The ability to slice and dice the data is limited only by the statisticians' desires and ingenuity.

Figure 18.2 shows the overall data flow. Remember that one of the original guiding principles was that the data be separated from the applications. Figure 18.2 illustrates how the survey systems load data into the warehouse and how the warehouse feeds data into the analysis and production systems.

18.5 What we learned

Senior management support is crucial, because the development of a data warehouse containing all census and survey responses with metadata in one place and easily accessible to everyone strikes a blow to a traditional organizational structure and ways of doing business. There is great resistance to change by many, so senior management must realize its strategic value and help sell its merits to others in the organization.

Information technology (IT) personnel may actively oppose such ideas and view them as a threat or not understand the analysts need for data. Some reactions the NASS heard from its IT managers are as follows:

- Users do not know what they want.
- Users will not be able to do their own queries.
- It will not be compatible to 'our' system.
- We cannot pull that data together now.
- Why do you need to have previous data readily accessible?
- No one will use it because they are too busy.

It needs to be said that the primary reason why the NASS has a data warehouse is that it was developed by statisticians who understood the value of data. The NASS's data warehouse was developed and implemented without IT support. Even now, responsibility for the data warehouse resides in a programme division, not the IT Division. However, to promote more support in the future from the IT Division, one of the two designers of our data warehouse has been transferred to lead our IT Division.

The end users in the past (unless they transformed themselves into power users) were dependent on IT professionals to get at their data. The IT professionals often determined what data platforms and software would be used for an application, with little thought being given to the end user. The philosophy was 'Tell me what you want, and I will get it for you'.

End users need to be involved at the design stage so that the dimensional models reflect their way of viewing the data. Too rarely, end users are brought in at the end of the development chain, then held hostage by the system provided to them. For the data warehouse effort to succeed, senior managers need to ensure that end users are fully engaged in the developmental process.

The desire for simplicity should prevail. There are simple solutions to complex problems, and the end user view often points the way to the solution.

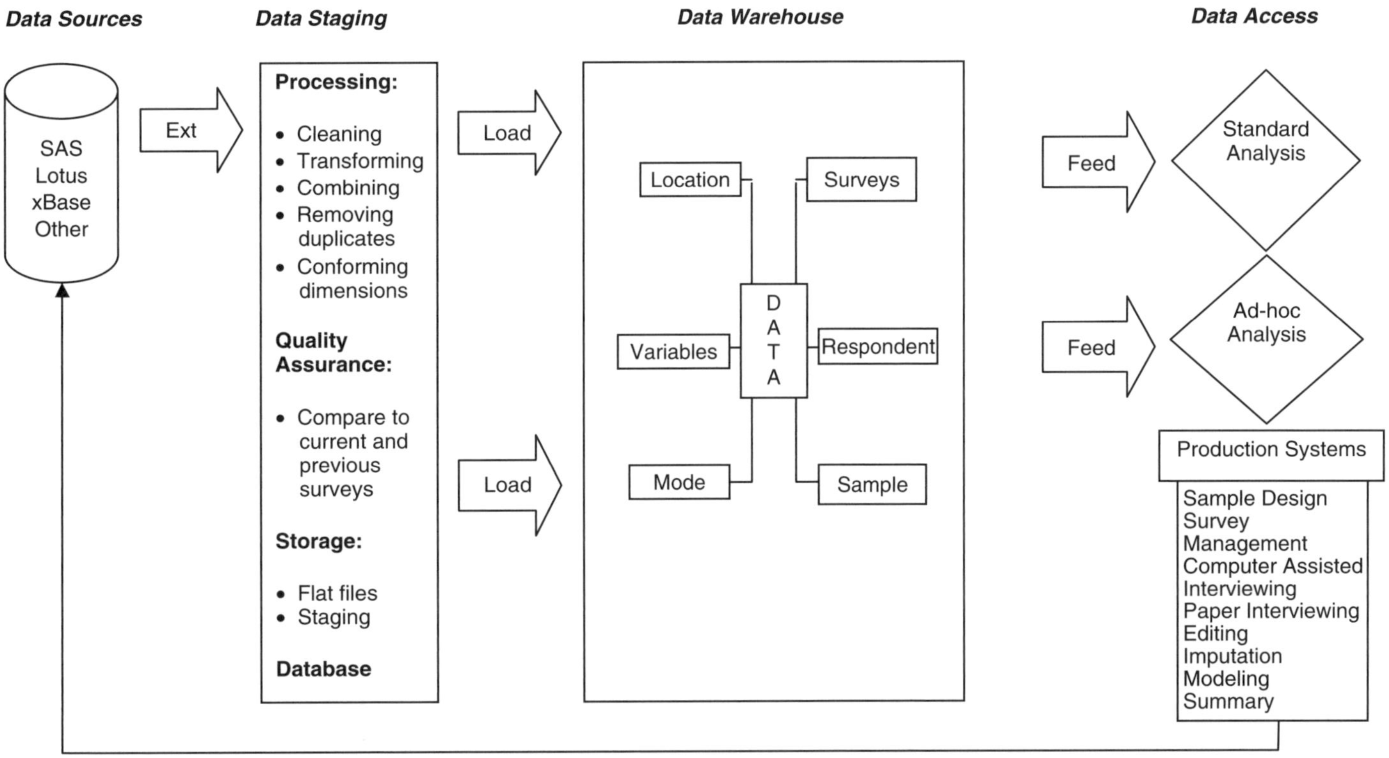

Figure 18.2 NASS architecture.

18.6 What is in store for the future?

The 400+ reports that the NASS publishes each year are not close to capturing all of the information embedded in the census and survey data files. As statisticians become more capable in their use of the data warehouse, they will be able to 'mine' the data for new relationships or conclusions that will enhance their reports.

Another major accomplishment would be to improve the storage of all of the summary level information in the 400+ reports across time in the data warehouse. Then data users could create their own data tables rather than being limited to the basic formats now being provided in the official releases.

A huge achievement will be the capability to use historical information to 'tailor' questionnaires for each respondent. Respondents will not be asked to provide information reported earlier, but only to update the information as needed. A recent development was to design a process to load current survey data into the data warehouse as soon as they pass through a preliminary edit. Thus, current data and estimation can benefit from the connection to historical data at the record level.

18.7 Conclusions

The development of the data warehouse concept ranks as one of the most significant technological and methodological that the NASS has ever achieved. The entire sequence of events required to produce official statistics will be affected by the warehouse concept and technology. This new technology also is rocking the very foundations of the organization's way of doing business. The hardware and software systems can be developed independently of the platform. This simplifies the developmental process. The most significant change is that the data management and data use activities revolve around the end user.

19

Data access and dissemination: some experiments during the First National Agricultural Census in China

Antonio Giusti
Department of Statistics, University of Florence, Italy

19.1 Introduction

More than twelve years ago, in the framework of the FAO Program for the World Census of Agriculture 2000, the First National Agricultural Census (FNAC) of the People's Republic of China (PRC) was implemented. In 2006 a Second National Agricultural Census was carried out. Considering the technologies available and the solutions adopted, the size of the country (in terms of population and area) and the fact that agricultural censuses are, in general, more complex than most other kinds of survey, the FNAC represented the most demanding statistical computing task ever accomplished: more than 7 million enumerators were used, about 220 million forms were filled, 220 gigabytes of data entered (about 80 gigabytes of compressed files), and so on. No census in the world (including the other censuses in China, i.e. both population and economic censuses) comes near these figures.

The FNAC was characterized by the heavy implementation of new techniques and technologies, thanks to international assistance (above all the support of four FAO projects: GCP/CPR/006-010-020-025/ITA), and the considerable ongoing development

Agricultural Survey Methods Edited by Roberto Benedetti, Marco Bee, Giuseppe Espa and Federica Piersimoni

of the PRC. Technical and organizational aspects were of particular importance: there was considerable use of computer systems, with more than ten years spent on training national statistical officers (both on hardware and software[1]), and the construction of a well-organized geographical network for enumeration and data processing activities.[2] During the pilot censuses (in 1991, 1995, 1996) different software products were tried for data entry, checking, and imputation: IMPS (developed by the Bureau of Census, USA), Blaise (developed by CBS Statistics Netherlands – a special version partially translated into Chinese). Other packages were evaluated but not implemented. In total, between the beginning of the training programme and the data analysis stage, about 5000 computers were used. A relevant characteristic of the FNAC was the introduction of optical character recognition (OCR) to conduct data entry for the 'Rural Household Holding' forms (about 210 million in number). A total of 573 scanners were used for census data acquisition.[3]

Checking and imputation activities were performed at different levels, with considerable use of automated checking. A nation-wide quality check on OCR data entry was also conducted, using a sampling scheme.

Data processing was organized at three levels: national, provincial and prefectural. At each level raw data were used.[4]

Thus census data were available for analysis at different levels: prefecture, province and National Bureau of Statistics (NBS). The NBS Computer Centre developed a database using Oracle. The database included three kinds of data: micro-data, metadata, and macro-data. Micro-data consisted of all data collected during the census; metadata comprised classifications and dictionaries; macro-data included summary statistics and tables. The client–server architecture was chosen. This database was reserved for internal use only. For a more detailed description of this subject, see Giusti and Li (2000). A database with a 1% data sample was also completed.[5]

In Section 19.2 we describe data access and dissemination of the FNAC in China, and then in Section 19.3 we present some general characteristics of SDA. In Section 19.4 we show a sample session using SDA. A final section is devoted to the conclusions.

19.2 Data access and dissemination

In the framework of the FNAC, data access and dissemination represented a very important opportunity to experiment with new techniques and technologies. The Second

[1] Training courses on hardware were concerned with personal computers, servers, Unix machines, networks, maintenance, while training courses on software dealt with operative systems (MS/DOS, Unix, Windows, Windows NT), programming languages, spreadsheets, word processors, databases, and statistical software (especially SAS).

[2] Many data processing centres were instituted in different provinces, municipalities and autonomous regions. Each centre was provided with servers, personal computers, printers, network facilities, etc. These centres played a very important role during the preparation and execution of the FNAC.

[3] Due to the introduction of OCR, available data entry software was not used. The PRC National Bureau of Statistics Computer Centre developed a software package for data entry, checking and imputation, including: address code management, optical data transfer, data entry, editing (verification), and tabulation. This software was implemented for both DOS and Unix operating systems. It was used both for OCR and keyboard data entry.

[4] To give an idea of the administrative entities involved, this was the specification of the administrative organizations in the PRC in 1996: 22 provinces, 4 municipalities, 5 autonomous regions, 1 special administrative region, 335 prefectures, 2849 counties, 45 484 towns or townships, 740 128 administrative villages.

[5] A systematic sample was selected from the list frame of households, sorted by province, prefecture, county, town, village and household number.

National Agricultural Census has certainly benefited from the experience accumulated during the FNAC, but by the end of February 2008 only six communiqués had been released by the NBS, with no information on methodological issues.

Data dissemination of the FNAC was carried out using both manual tabulation and computer data processing. Manual tabulation was used to quickly release information on general results: the NBS published five communiqués at national level. These communiqués were also available on the NSB website. At the same time, all the provinces released their own communiqués.

Computer data dissemination included macro-data (tables and graphics), micro-data (files) and interactive queries (to produce tables and files). Data dissemination was by traditional approaches (basically, paper publications). Advanced methods – website, CD-ROM and online databases – were also experimentally implemented.

Using data processing results, several books containing census data were published in 1998 and 1999, such as *Abstract of the First National Agricultural Census in China* (National Agricultural Census Office, 1999) and *Facts from the First National Census of Agriculture in China* (National Agricultural Census Office, 1998) , in English and Chinese. At the same time, all the provinces published census results in Chinese.

The evolution of data processing and the increasing diffusion of the Internet in the late 1990s had a very important impact on data dissemination activities carried out by national statistical institutes (NSIs). Many NSIs disseminated data in electronic form, using both optical devices and network facilities.

Using new technologies, data release could be done in several ways, depending on the type of data to be disseminated (micro- or macro-data).[6] Micro-data files could be released on CD-ROM or by file download via FTP (File Transfer Protocol) or other protocols, usually from the NSI website. Macro-data (tables, graphical representations, reports, etc.) could be disseminated in the same ways, but paper publication was still the preferred method.

A more advanced method of releasing both micro- and macro-data was through an interactive system that allowed the user to ask for specific analyses. Some NSIs allowed the user to produce tables and to download micro-data files. This new tool was defined as 'interactive Web data dissemination'.

With the cooperation of the NBS Computer Centre, the census results were disseminated by using these emerging methods of data dissemination. For example, the *Abstract* was released on CD-ROM, while some other publications, tables and micro-data files, were published on the NBS website.[7]

In order to permit the use of census data directly by the users and to supply the researchers with an instrument for a complete, safe and adequate data access to perform statistical analysis, the NBS decided to release the 1% sample data by using the net. For more details on the sample see Pratesi (2000).

In carrying out this task, we took into account the experiences of some NSIs (Bureau of Census, Statistics Canada, Istat, etc.) and the new technologies available for data management (data warehouse, multidimensional data management, etc.).

After detailed evaluation, it was decided to use SDA (Survey Documentation and Analysis), a set of programs developed and maintained by the Computer-Assisted Survey Methods Program (CSM) of the University of California, Berkeley (see the next section).

[6] In electronic data release very often micro- and macro-data files include metadata.

[7] See `http://www.stats.gov.cn/english`

For this purpose two national technical officers had two weeks of training in Florence. During this time they prepared a preliminary experimental data set using some records from the FNAC. The work was accomplished using the Unix version of SDA.

In July 1999 all 1% sample files from A601[8] were implemented at the Food and Agricultural Statistical Centre (FASC) of the NBS using SDA in the Windows NT environment (Giusti, 2000).

Disclosure issues were considered in the installation of SDA. Some experiments were carried out on the 1% sample database, using general principles accepted at international level.[9] The 1% sample data was also released on the web for download, but due to the huge data size this option was not used.

19.3 General characteristics of SDA

SDA is a set of computer programs for the documentation and web-based analysis of survey data.[10] There are also procedures for creating and downloading customized subsets of data sets. The software is maintained by the CSM.[11] Current version of SDA is release 3.2; at the time of our experiment version 1.2 was available. All the following information are related to version 1.2.

Data analysis programs were designed to be run from a web browser. SDA provides the results of the analysis very quickly – within seconds – even on large data sets; this is due to the method of storing the data and the design of the programs.

The features of SDA are as follows.

1. Browse the documentation for a data set or a questionnaire:
   - introduction files, appendices, indexes to variables;
   - full description of each variable.
2. Data analysis capabilities:
   - frequencies and cross-tabulations;
   - comparisons of means (with complex standard errors);
   - correlation matrix;
   - comparisons of correlations;
   - regression (ordinary least squares);
   - list values of individual cases.

---

[8] The Rural Household Holding form, divided into six sections.

[9] The first experiment was carried out at provincial level obtaining as a result the possibility of one identification for every 59 million inhabitants. The second experiment, at county level, showed a possibility of one identification for every 3 million inhabitants. These values are very low with respect to many NSI rules. So we concluded that, by disabling the download option of SDA, disclosure problems could be avoided. For more details see Franconi (1999).

[10] More information on this topic can be found in the SDA Manual (version 3.2), available from `http://sda.berkeley.edu`.

[11] The CSM also maintains the CASES software package for telephone and self-interviewing.

3. Create new variables:
   - recode one or more variables;
   - treatment of missing data;
   - compute new variables;
   - list newly created variables.
4. File handling:
   - make a user-specified subset of variables and/or cases, and download data files and documentation;
   - ASCII data file for the subset;
   - SAS, SPSS or Stata data definitions;
   - full documentation for the subset.
5. Other features under development:
   - logit and probit regression;
   - interface in various languages.

Online help for SDA analysis programs is available on the Web. For each program, an explanation of each option can be obtained by selecting the corresponding word highlighted on the form.

In particular, the SDA Frequencies and Crosstabulation Program generates univariate distributions or crosstabulations of two variables. If a control variable is specified, a separate table will be produced for each category of the control variable. After specifying the names of variables, the desired display options can be selected: these affect percentages (column, row or total), text to display, and statistics. Using filter variable(s), cases can be included or excluded from the analysis. The cases can be given different relative weights, using weight variables. Specifying both a row and a column variable, a set of bivariate statistics is generated (Pearson's chi-square and the likelihood-ratio chi-square, each with its p-value, Pearson correlation coefficient and eta, gamma and tau). Specifying a row variable only, a set of univariate statistics is generated (mean, median, mode, standard deviation, variance, coefficient of variation, standard error of the mean and its coefficient of variation).

In many programs (including the Frequencies and Crosstabulation Program), the table cells can be color coded, in order to aid in detecting patterns. Cells with more cases than expected (based on the marginal percentages) become redder, the more they exceed the expected value. Cells with fewer cases than expected become bluer, the smaller they are, compared to the expected value. The transition from a lighter shade of red or blue to a darker shade depends on the magnitude of the t-statistic.

The sample design can be specified for each study when the SDA data set is defined; this definition allows SDA to compute the standard errors of means for complex samples.

With SDA it is also very easy to create new variables using basic or composite expressions (with if, else if, else). In expressions, arithmetic and logical operators, arithmetic, random distribution, statistical and trigonometric functions can be used.

19.4 A sample session using SDA

The use of SDA is very simple ,and data processing time is not related to the computational power of the computer used to browse the website, but only with the computational power of the server used. In this section we show some aspects of an online session, during the Conference on Agricultural and Environmental Statistical Application (CAESAR) held in Rome in 2001, using the FNAC 1% sampling data, stored in Beijing, in the FASC computer division (Giusti, 2003).

We started the session, using a web browser, with the page `www.stats.gov.cn/en-glish/index.html` (now off-line). The page referred to the interactive electronic data release used for the FNAC. From this page, after another dialogue that provided access to the project documentation, it was possible to go to the study and action selection page (see Figure 19.1).

The study selection was necessary to access one of the six sections of the form (to access the file that we would like to analyse). The following actions could be selected: browse codebook; frequencies or cross-tabulation; comparison of means; calculation of correlation matrix; use a multiple regression program; list values of individual cases.

We assumed we knew the file structure (contents and characteristics of each variable recorded in the six sections) information that was necessary to go on with the requests. Otherwise, using the action 'Browse codebook', we had to give the name, the label, the categories and the 'question text' of each variable.

Choosing 'Frequencies and crosstabulations' led to the dialogue in Figure 19.2. In this example the data section used is the 'General Information of Households'. In Figure 19.2 a cross-tabulation is requested, using as row variable the 'province' (or equivalent administrative level) and as column variable the 'number of persons in the household'. We did not give any weight, recode or class definition. The results would appear in 3–4 seconds (see Figure 19.3), despite the large number of processed records

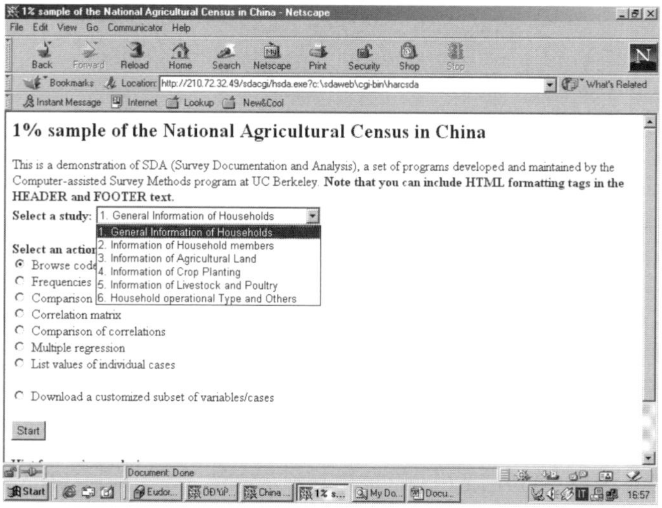

Figure 19.1 Study and action choice.

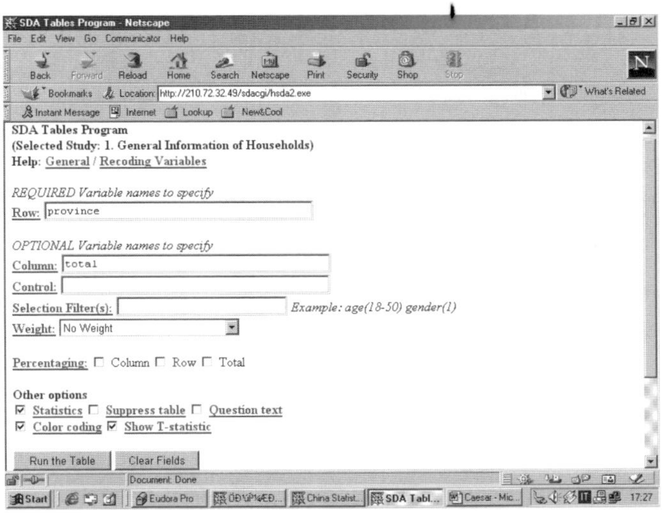

Figure 19.2 Tables program.

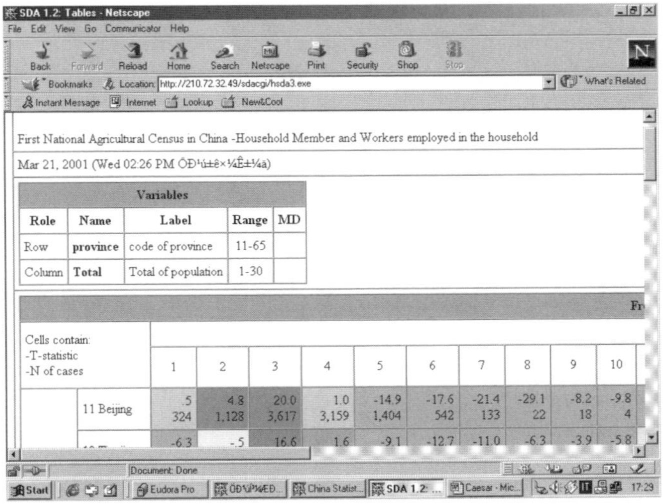

Figure 19.3 Tables Program results (I).

(2 138 275). After a description of the variables used, the table was presented. The use of colour coding allows a quick identification of the most common patterns.

Using the options in Figure 19.1, we were also able to analyse the 'Household members' data. Using the dialogue in Figure 19.2, we requested a crosstabulation using 'education' as row variable and 'gender' as column variable. In this situation 7 826 390 records were used to obtain the results. Data processing time remained under 10 seconds (see Figure 19.4).

The last example is devoted to the calculation of the 'total population' by 'province'. In Figure 19.5 we present the request, using the Means Program, while in Figures 19.6 and 19.7 we show the results obtained.

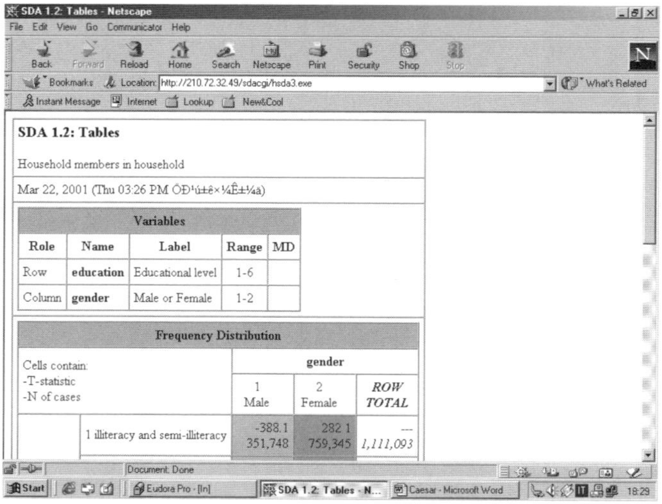

Figure 19.4 Tables Program results (II).

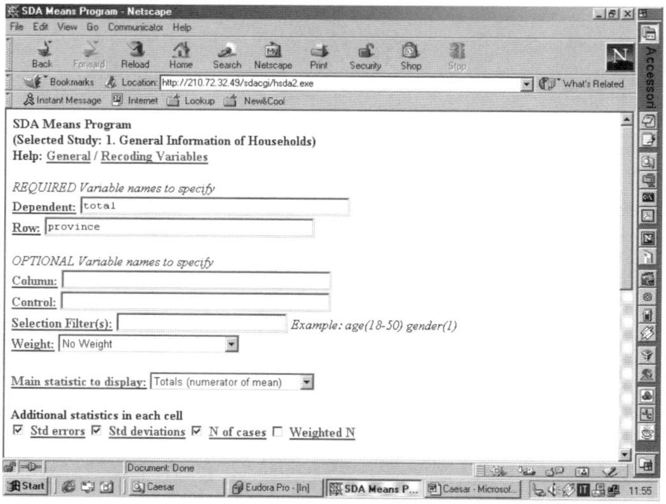

Figure 19.5 Means Program.

19.5 Conclusions

Data processing and dissemination constituted an important part of the FNAC activities. In order to permit the use of census data directly and to supply researchers with an instrument for complete, safe and adequate access to raw data in order to perform statistical analysis, the NBS decided to release a 1% sample data by also using SDA on the World Wide Web. With SDA it was possible to produce codebooks and to analyse the census data from any remote personal computer. The SDA experiment demonstrates the interest of China NBS, FASC and FAO in new technologies for data dissemination through the Internet.

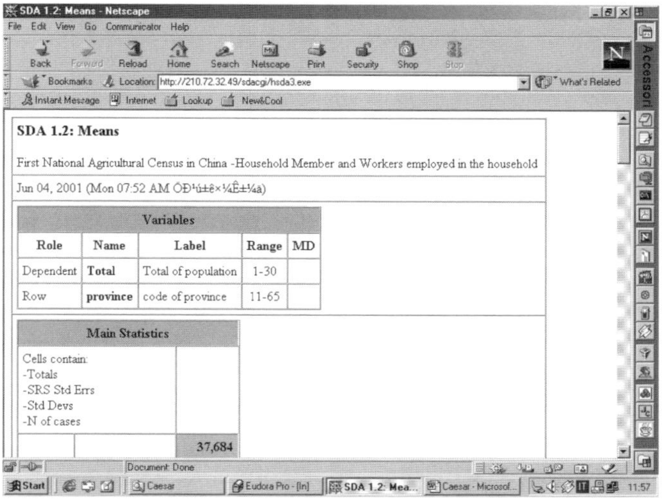

Figure 19.6 Means Program results (I).

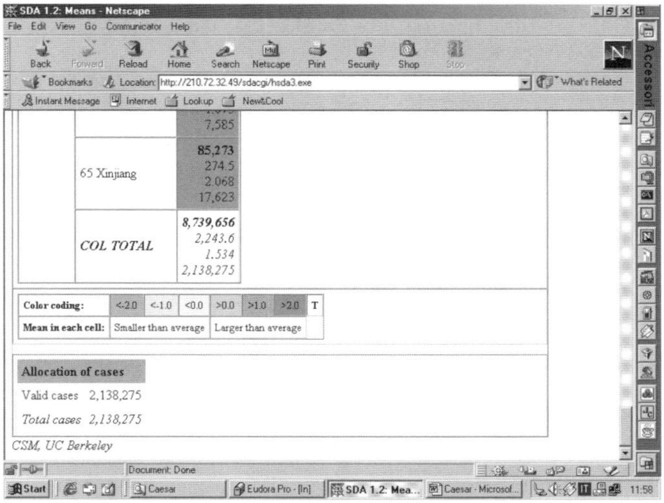

Figure 19.7 Means Program results (II).

The main idea was to integrate in the FNAC dissemination as many new technology tools as possible, to increase the accessibility and the utilization of a useful source of information.

At that time, we could envisage the widespread use of more advanced techniques of data analysis, including some form of interactive geographical information system. We also tried some experiments involving releasing maps on the Web. We expected a high degree of data integration with other sources of information. In our opinion, data integration and the continuous updating of agricultural statistics on the NBS website should have represented a very important and advanced tool for the knowledge of PRC agricultural information.

In the following years, unfortunately, we did not observe any new advances in these directions. At the time of writing we have no information on NBS dissemination plans for the PRC's Second National Agricultural Census. We hope that the NBS will learn from the important experiences acquired during the FNAC.

References

Franconi L. (1999) Report of the visit of the FASC by Luisa Franconi, (July 1999), FAO internal report.

Giusti A. (2000) Report of the visit of the FASC by Antonio Giusti, (July–August 2000), FAO internal report.

Giusti A. (2003) First agricultural census in China: data access and dissemination, *CAESAR – Conference on Agricultural and Environmental Statistical Application in Rome*, ISI-Fao-Eurostat-Istat, Rome.

Giusti A., Li W. (2000) Data processing and Dissemination, *International Seminar on China Agricultural Census Results*, FAO, Beijing.

National Agricultural Census Office (1998) *Facts from the First National Census of Agriculture in China*. Beijing: China Statistics Press.

National Agricultural Census Office (1999) *Abstract of the First National Agricultural Census in China*. Beijing: China Statistics Press.

Pratesi M. (2000) Notes on the dissemination of the 1% Sample Data from the Census of the Agriculture in China, FAO internal report.

20

Analysis of economic data collected in farm surveys

Philip N. Kokic[1], Nhu Che[2] and Raymond L. Chambers[3]

[1]*Commonwealth Scientific and Industrial Research Organisation (CSIRO), Canberra, Australia*
[2]*Australian Bureau of Agriculture and Research Economics (ABARE), Canberra, Australia*
[3]*School of Mathematics and Applied Statistics, University of Wollongong, Australia*

20.1 Introduction

The analysis of economic data collected in a farm survey extends well beyond the production of cross-classified tables. Econometric models are often required to address a variety resource and policy implications. For example, production levels in agriculture depend on the availability, allocation and quality of land, capital, labour and other resources, and the efficiency of allocating those for agricultural production relative to alternative activities. In turn, the profitability of agricultural activities depends on prices for farm outputs relative to farm inputs and the productivity that ensues when these inputs are transformed into agricultural products. Economic modelling of these relationships is crucial for determining the potential impact of agricultural policy changes.

One of the most important issues in agriculture over the last few decades has been a continuing decline in terms of trade (Knopke *et al.*, 2000) which has meant that farmers have had to improve the productivity of their farms to maintain profitability and international competitiveness. Consequently, governments, industry organizations and farmers have continued to develop policies and to invest capital in order to improve productivity. Analysis and statistical modelling of farm productivity data collected in farm surveys in

Agricultural Survey Methods Edited by Roberto Benedetti, Marco Bee, Giuseppe Espa and Federica Piersimoni

order to explore the drivers of productivity growth can play an important role in this policy development (Knopke *et al.*, 2000; Mullen, 2007; Mullen and Crean, 2007).

Farm surveys that focus on collection of economic data are intrinsically different from other business surveys in two distinct ways. First, a farm is typically a merging of a household and business unit.[1] Thus there is often the possibility of selecting sample units from a list frame or as a multistage area sample, which has consequences for the analysis of the sample data. Typically multistage sample data are more complex to analyse because the clustering of sample units means that the assumption of independent errors when modelling is less likely to hold. Second, farms are highly dependent on the natural resources available at or close to their physical location. For example, Gollop and Swinand (2001) point out that 'proper measures of productivity growth are barometers of how well society is allocating its scare resources', including natural resources. The dependence of farms on natural resources implies that econometric models need to take account of this dependency in analysis and modelling. This dependence has been utilized in farm system models using spatial and survey data (see Lesschen *et al.*, 2005, and the references therein), and in econometric models of land values and productivity (Davidson *et al.*, 2005; Gollop and Swinand, 1998, 2001).

The dependence on natural resources, such as weather and water availability, can have a major impact on the risk faced by farmers (Cline, 2007; Hill *et al.*, 2001). Furthermore, there are often significant economic and social implications faced by the rural communities in which these farm households reside (Anderson, 2003; Ellis, 2000; Hardaker *et al.*, 1997). Thus, integrated analysis of farm survey data that takes this broader social, natural resource and economic context into account has gained some momentum in recent years (Gollop and Swinand, 1998; Meinke *et al.*, 2006).

Furthermore, interest in local governance and regional issues (Nelson *et al.*, 2008) and the varying impact of government policy on regional economies (Nelson *et al.*, 2007) has increased the demand for small-area statistics and distribution estimates from farm economic surveys. There has been active research going on in these areas for some time; see Dorfman (2009) for an overview of research into estimation of distributions from sample survey data. Rao (2003) discusses methods for construction of small-area estimates based on mixed effects models, while Chambers and Tzavidis (2006) describe more recent research that uses M-quantile models for this purpose. The use of spatial information for constructing such estimates is described in Pratesi and Salvati (2008).

An essential requirement for economic analysis of farm survey data is the provision of a suitable database, within time and budget, that contains up-to-date and reliable information, and includes enough data and detail to enable relevant economic modelling and estimation of sufficient accuracy to address policy needs. The type of information stored on the database should enable the construction of key economic performance indicators such as profit and total factor productivity[2] (TFP: Coelli *et al.*, 1998). It should also include a range of basic financial, production and management data and any additional information that may be required for policy-relevant economic analysis. The units should be linked over time and include suitable geographic and industry classification information.

---

[1] In developing countries farm production is frequently used for farm household sustenance and bartering, and they often do not engage in substantial business activity.

[2] TFP is the conventional economic measure of on-farm productivity growth. It is a relative measure between two units or two points in time and is defined as the ratio of a volume change index of all marketable outputs relative to the same volume change index of all marketable inputs.

Various aspects of these issues are discussed in the following sections. First we outline the requirements of the survey to meet the needs of any subsequent economic analysis. The typical content in terms of information requirements is then addressed. Multipurpose weighting of survey data and the use of survey weights in modelling are then briefly discussed. Common errors are discussed such as sample size requirements and data issues such as outliers. Finally, to illustrate the complex statistical issues that need to be considered when undertaking econometric modelling of farm survey data, two case studies are described: a time series analysis of the growth in feed in the Australian cattle industry, and a cross-sectional analysis of farm land value in central New South Wales.

20.2 Requirements of sample surveys for economic analysis

One of the most important requirements of sample surveys in general, not only farm surveys, is that sample sizes are large enough to enable sufficiently accurate estimates to be produced for policy analysis. Working against this are two important objectives: the provision of timely results and the need to collect detailed economic data (Section 20.3), which often necessitates expensive face-to-face data collection methods. At the same time the sample often needs to be spread spatially (for regional-specific analysis) and across industries, which further increases running costs. Given a budget constraint, the design of agricultural surveys is primarily about a trade-off between precision of estimates and the degree of detailed information required (ABARE, 2003).

To address policy issues that have an impact on farm units over time it is usually necessary to build a database for econometric analysis by repeating the survey at fixed time intervals (e.g. once per year), or by carrying out a longitudinal survey. Non-response in farm economic surveys, typically due to some combination of high response burden, sample rotation policy, sensitivity of the questions and changing target population, can undermine the possibility of obtaining a 'balanced' data set for subsequent analysis. Furthermore, farm surveys are often stratified, with unequal sampling proportions in the different strata, in order to improve the precision of cross-sectional estimates of population aggregates. As a consequence, what is obtained is typically a sample that may be severely unbalanced with respect to known population distributions and so can be difficult to analyse. For statistical and econometric analysis, the sample need not be 'perfectly' balanced. However, this analysis is usually more straightforward if the distribution of the sample data 'reflects' the corresponding population distribution. Typically this requires inclusion of covariates in the model to represent the sample design. For example, inclusion of strata indicators or a size variable can help adjust for biases caused by imbalance due to non-proportional representation of farms in different strata or of farms of different sizes in the sample and enable one to correctly interpret significance levels (Chambers and Skinner, 2003).

Most econometric modelling makes the assumption of homogeneity of model structure across the population. However, farm survey data are typically observational rather than controlled experimental data, and so the contribution of causal variables in econometric models is likely to vary quite significantly between farms and over time, and often not all causal variables can be easily obtained or derived from the collected survey data. As a consequence, there has been a growing interest in modelling non-homogeneous parameters with multilevel and mixed effect models (Goldstein, 1995; Hsiao, 2003). Furthermore, the use of appropriate mixed effect statistical models (Laird and Ware, 1982) can partly overcome missing information, but it is very important to test the appropriateness of modelling assumptions.

Data requirements for econometric analysis are generally very detailed, with information typically required over a long period of time and with a suitable number of observations at each point in time. This leads to a significant issue with excessive response burden that needs to be addressed to avoid problems of non-response bias when modelling data and measurement error. To partly address the response burden issue it is often necessary to collect information in face-to-face interviews using experienced interview staff, questionnaires must be designed and tailored to minimize response burden (e.g. by eliminating irrelevant questions for specialist producers) and sample farms often need to be rotated out of sample after a few periods. To make the most of information collected, efficient estimation techniques, which calibrate the sample distributions of key variables to their corresponding population distributions, should be used (Deville and Särndal, 1992; Chambers, 1997).

List-based frames with extensive benchmark information are generally preferable to area-based frames for farm economic surveys as they enable the construction of efficient aggregate estimates and they often contain useful auxiliary information for modelling purposes. However, list frames are often poorly maintained, and then area based multistage surveys can be considered. An advantage of area frames is that one has greater control over the geographic spread of sample, and the ease of linking with scientific spatial data which can be very useful for economic modelling (Kokic *et al.*, 2007a). However, area frames are generally expensive to construct (there is usually a large initial cost in doing this) and it is often not possible to fully exploit significant efficiency gains that can be obtained through size stratification, except where this is related to the geography. Furthermore, if care is not taken, area frames can also be of poor quality and cause biases in estimation.

20.3 Typical contents of a farm economic survey

To obtain a basic economic profile of farms for construction of key economic variables such as major output and input variables, profit and TFP, and to undertake policy-relevant analysis, a large amount of information needs to be collected. In addition, construction of the survey database requires that certain basic information about the sample farms making up the database be collected and maintained. At a minimum, such 'paradata' should include a unique identifier so that data from the same farm can be linked over time, contact information for follow-ups and industry and size information for classification. Ideally, it will also include georeferencing data (Hill, 2006) so that surveyed farms can be linked to other spatial data if required or for use in regional stratification and in econometric modelling.

The essential economic content required from farm surveys include: the key output, and input variables both in terms of quantity and values (receipts and costs); changes in technology, farm size, production condition, management practices, on and off-farm assets including liquid assets, and debt, off-farm income, business structure etc. (ABARE, 2003). In particular, the output and input variables usually required in a farm survey are the following:

- *Outputs* – output quantity and value for each commodity produced as well as the information about change in quality. This will be specific to the type of agricultural industry covered, but for Australian broadacre agriculture it includes crops, livestock, wool and other production and, for dairy, milk production.
- *Land use* – area of land used area for production of each commodity, land care costs in details including fertilizer, insecticide and other crop chemicals used, water purchased, etc., and change in land use.
- *Capital use* – details of each capital item used for production in detail, its current value, average lifetime and depreciation rate. Also, information about livestock capital (value and number) and the change in this from the beginning to the end of the year, and similar information for grains stored on the farm. Also similar value and depreciation information for buildings and plant capital.
- *Labour use* – including quantity of labour undertaken on the farm (weeks) by the farm operator's family, partners, and hired labour and total costs of hired labour. Also quantity and values of stores and rations are often required.
- *Feeding* – quantity and value of fodder used.
- *Materials and services* – major material and service items used for production in quantity and value terms. These include, under materials, fuel, livestock materials, seed, fodder and other materials, and under services, rates and taxes, administrative costs, repairs and maintenance, veterinarian expenses, motor vehicle expenses, insurance, contracts and other services.

In some cases, specific variables necessary for carrying out the particular economic analysis of interest also need to be collected. For example, in order to measure production efficiency, technology variables are required, while for measuring the impact of climate, key proxy variables representing climatic conditions should be collected or linked to the data.

20.4 Issues in statistical analysis of farm survey data

20.4.1 Multipurpose sample weighting

Since the sample of farms that contribute to a farm survey is typically a very small proportion of the farms that make up the agricultural sector of an economy, it is necessary to 'scale up' the sample data in order to properly represent the total activity of the sector. This scaling up is usually carried out by attaching a weight to each sample farm so that the weighted sum of the sample values of a survey variable is a 'good' estimate of the sector-based sum for the same variable. Methods for constructing these weights vary depending on what is known about the sector and about the way the sample of farms has been selected. If the sample has been selected using some form of probability sampling, then one commonly used approach is to set a sample farm's weight equal to the inverse of the probability of its selection. The resulting weighted sum is often referred to as a Horvitz–Thompson estimator (Lohr, 1999). However, depending on the method of sampling, such weights can be quite variable and lead to inefficient estimates. They can also lead to estimates that are inconsistent with known sector characteristics – for example, an estimate of the total production of a particular agricultural commodity that is substantially different from the known production level. In order to get around this problem it is now common to use calibrated weights. These recover estimates of sector characteristics that agree with their known sector values. Calibrated weights are not unique and can be constructed in a variety of ways, but the most common is to choose them so that they are closest (typically in terms of a Euclidean metric) to the probability-based weights referred to above (Deville and Särndal, 1992).

A key advantage of calibrated weights is their multipurpose nature. An estimator for a particular farm economic characteristic will be most efficient if the weights that define it are calibrated to a farm variable that is highly correlated with this characteristic. Consequently, weights that are calibrated to a range of known sector quantities, including industry counts and outputs, should in principle define efficient estimators for a wide range of economic performance characteristics for the sector. Unfortunately, it turns out that one can over-calibrate, especially if the variables used for this purpose are highly correlated, ending up with weights that are considerably more variable than the original probability-based weights. In extreme cases the weights can even be negative. The reason for this is not difficult to ascertain: essentially, calibration to a set of known sector quantities is equivalent to assuming a linear specification for the regression of the economic performance variables of interest on the farm characteristics corresponding to these sector quantities. A large set of calibration 'constraints' is therefore equivalent to an overspecified linear model for any particular economic characteristic, implying a consequent loss in estimation efficiency for that characteristic. This loss of efficiency is reflected in more variable weights.

There are a number of ways of limiting weight variability while at the same time imposing sufficient calibration constraints to ensure that the implicit linear model assumption is at least approximately valid. For example, Deville and Särndal (1992) suggest searching for the set of calibrated weights that also lie between pre-specified limits. Unfortunately, there is no guarantee that such weights exist. In contrast, Chambers (1996) describes an approach that 'relaxes' some of the calibration constraints (particularly those felt to be least correlated with economic performance) in order to minimize mean squared error, rather than variance, and which then allows weight variability to be limited by choosing the degree to which calibration constraints are enforced. Another, more ad hoc, method involves setting weights that are either too high or too low to unity and repeating the calibration exercise with the corresponding sample farms excluded. We do not pursue this issue further here, being content to note that, before computing sector-level estimates from farm survey data, an analyst should always check that an adequate compromise exists between the calibration properties of the weights used in these estimates and their variability.

20.4.2 Use of sample weights in modelling

The weights referred to in the previous subsection are necessary for estimation of sector-wide characteristics, such as would be used in a cross-classified table of estimates. However, this is not always the case when fitting econometric models to the farm survey data. In fact, in many situations a model can be fitted ignoring the survey weights – provided the method of sampling is non-informative for that model. By non-informative we mean here that, conditional on the model covariates, the random variable corresponding to whether or not a farm is selected for the survey is independently distributed of the variable being modelled. Clearly, if sample selection depends on certain farm characteristics, and these characteristics are included in the model covariates, then the sampling method is non-informative for that model. Consequently, if the sample farms are randomly drawn from industry or farm size strata, then this method of sampling is non-informative for any model that includes industry classifiers and farm size in its set of covariates, and sample weights can then be ignored when fitting it. Note, however, that this does not mean that other aspects of sample design (e.g. clustering) can necessarily be ignored, since they impact on assessing the fit of the model. Thus, if some form of spatial clustering is used in the sample selection process then it may well be that even after accounting for variation in the model covariates within a cluster, different sample farms in the same cluster are spatially correlated. In such cases the econometric model should allow for this spatial correlation, either explicitly in the model fitting process, or implicitly by using a method of standard error estimation that is consistent even when model residuals for farms in the same cluster are correlated – for example, by using an 'ultimate cluster' method of variance estimation (Wolter, 2007).

However, there are situations where, for good economic reasons, it is not appropriate to include so-called 'sample design' covariates in the econometric model, or where the method by which the sample farms are selected is not well known to the analyst. In such cases inclusion of the sample weights in the model fitting process (e.g. via weighted least squares) can provide some protection against a sample bias in the model fit, that is, a bias due to the fact that the fitted model characterizes behaviour of the sample farms rather than in the sector as a whole. Unfortunately, the weights that provide this protection are generally not the efficient calibrated weights discussed in the previous subsection, but the uncalibrated probability-based weights (which are often not available), and furthermore, the cost of this protection can be large, in terms of increased variability in the estimated model parameters. On the positive side, many software packages for statistical and econometric analysis now include options that allow sample weights to be used in model fitting.

Efficient methods of model fitting under informative sampling, where the sample inclusion variable and the model variable are correlated even after conditioning on model covariates, requires specification of a joint model for sample inclusion and the variable of interest. For example, the Heckman model for sample selection (Heckman, 1979) assumes the existence of a latent variable that is correlated with the variable of interest and is such that sample selection occurs when the latent variable takes a value above an unknown threshold. Here efficient inference requires specification of both the joint distribution of the latent model and the variable of interest as well as a model for the threshold. Maximum likelihood inference under informative sampling is discussed in Chambers and Skinner (2003).

Finally, we note that although the discussion above has focused on the impact of sampling on model fitting, it could just as easily have been about the impact of non-response (and missing data more generally) in this regard. This is because non-response can be viewed as a stage in sample selection that is not controlled by the sampler but by the person or entity being sampled, in the sense of deciding whether or not to cooperate with the survey. Consequently, sample weights are usually calibrated to not only adjust for designed differences between sampled farms and the agricultural sector as a whole (typically driven by efficiency considerations in the sample design), but also to adjust for 'non-designed' differences between sampled farms and responding sampled farms

(Särndal and Lundström, 2005). Ideally the latter differences disappear once responding farms are placed in response strata, which are then included in the set of calibration constraints, or, if appropriate, in the set of model covariates. However, here we are much less sure about whether the non-response is informative or non-informative, and so joint modelling of the response status of a farm and its value of an economic characteristic of interest needs to be undertaken.

20.5 Issues in economic modelling using farm survey data

20.5.1 Data and modelling issues

Farm economic surveys may not capture or miss out on key output or input variables, which are essential for economic analysis purposes. If a database does not include the required economic information then it is of little use for economic analysis.

Most econometric modelling makes the assumption of homogeneity of parameters. A number of statistical models have been developed to deal with this situation. One of the mostly widely used is the mixed regression model (Goldstein, 1995; Laird and Ware, 1982). A mixed regression model incorporates heterogeneity by assuming that some or all of the regression parameters are random (typically normally distributed). Mixed regression models have often been used with farm survey data (Battese *et al.* 1988; Pan *et al.* 2004), although they can be challenging to validate. A related regression model that does not require the restrictive assumption of normality inherent in a mixed model and is also robust against outliers is a M-quantile regression model, which includes quantile regression as a special case (Breckling and Chambers 1988; Koenker 2005; Koenker and Bassett 1978). In recent years M-quantile regression models have been used for a variety of purposes with farm economic survey data: to calibrate micro-simulation models (Kokic *et al.* 2000), to make seasonal forecasts of farm incomes (Kokic *et al.*, 2007b), and as a small-area estimation tool (Chambers and Tzavidis 2006).

Economic analysis also requires homogeneity of characteristics, for example homogeneity in production size, natural resource condition, technology, output and inputs. This often dictates the way farm surveys are designed. For example, the use of stratification and collection of classification data is required so that homogeneity can be achieved in subsequent analysis. The usual way that economists deal with heterogeneity is to divide the sample into subgroups or strata in which modelling assumptions are valid. Sometimes, however, this approach is inadequate, particularly when relationships between the dependent and independent variables become quite complicated. Appropriate statistical tools that may be used in this case are tree-based models and non-linear statistical models. Regression and classification tree analysis (Breiman *et al.*, 1998) partition the data into subcategories that are as homogeneous as possible with respect to the dependent variables. Non-parametric non-linear models such as general additive models (Hastie and Tibshirani, 1990) are used to fit smooth non-linear functions to the data. Both of these methods have been utilized to model farm economic survey data, For example, Severance-Lossin and Sperlich (1999) has used general additive models as a non-parametric tool for production function modelling, and Nelson *et al.* (2004) have used tree analysis to explore the effectiveness of government programmes for natural resource management.

20.5.2 Economic and econometric specification

It is not uncommon for economists to depart from the proper economic and econometric specifications of the model because the observed variables are not precisely as required. For example, in an economic analysis of technology choice and efficiency on Australian dairy farms (Kompas and Che, 2006) a variable representing feed concentration was not available. Average grain feed (in kilograms per cow) was therefore used as a proxy for this variable in the inefficiency model. As a consequence, the economic interpretation of the resulting analysis was compromised. Policy implications drawn from this analysis should also be interpreted carefully because of this divergence between the fitted econometric model and the underlying economic theory.

In general, missing variables, or inclusion of incorrect variables in the econometric model, can cause results to be insignificant when they should be significant, and vice versa. For example, fodder used may be more easily measured in terms of value rather than quantity. However, value also depends on price, which varies from year to year, and from region to region. Therefore, economic analysis that uses value may not give the correct policy implication for the supply of feed to farms.

There are a number of widely available diagnostic tests that can be used to assess the statistical significance of econometric results. The most common of these tests are for the correct functional form, autocorrelation, heteroscedasticity, multicollinearity and normally distributed residuals. Correct functional form means both the inclusion of the correct variables (including interactions where necessary) as well as the appropriate scale to be used in modelling the relationship between the dependent and independent variables (log, linear, etc.). The econometric model specification is often limited by data availability, but the choice of functional form for the model may be quite flexible within the data range. If an econometric specification fails the diagnostic test for functional form, then it may not be appropriate for the specified economic analysis. However, even if the functional specification is appropriate for the observed data, there is always the danger of drawing false conclusions if other diagnostic checks are not performed.

Autocorrelation occurs when the model error term for an observation is correlated with similar error terms associated with other observations that are 'close' to it in space or in time. Thus, in a time series model one can assume that the error term at time t is related to the error term at time $t - 1$ (lag 1 autocorrelation), in the sense that it can be written in the form $\varepsilon_t = \rho\varepsilon_{t-1} + u_t$, where $|\rho| < 1$ is the autocorrelation coefficient between the two error terms, and u_t is a disturbance term whose distribution is the same at any time point and is uncorrelated across time. Under this type of autocorrelation structure, ordinary least square (OLS) estimators are unbiased and consistent, but inefficient. In particular, the true variance of estimators is inflated compared to the no-autocorrelation case ($\rho = 0$); second, estimated variances of the coefficient estimates are smaller (biased downward); third, the presence of autocorrelation causes an increase in the coefficient t statistics (biased upwards), making the estimate appear more significant than it actually is; and fourth, the presence of autocorrelation also causes the estimated fit of the model to the data to appear better than it actually is. Dealing with serial autocorrelation in linear regression is relatively straightforward (Greene, 2008; Maddala, 2001).

Many commonly used statistical techniques for model fitting are efficient provided a number of assumptions about the model and the underlying data hold true, and so it is important to be aware of these assumptions and of the consequences if they are incorrect. Thus, the OLS method of fitting a linear regression model is efficient when the error term has constant variance. This will be true if these terms are drawn from the same distribution. The data are said to be heteroscedastic when the variance of the error term varies from observation to observation. This is often the case with cross-sectional or time series data, and can arise when important variables are omitted from the economic model. Heteroscedasticity does not cause OLS coefficient estimates to be biased. However, the variance (and, thus, standard errors) of the coefficients tends to be underestimated, t statistics become inflated and sometimes insignificant variables appear to be statistically significant (Greene, 2008; Maddala, 2001). One approach to dealing with heteroscedasticity is to model it explicitly, as in Carroll and Ruppert (1988).

Multicollinearity occurs when the model used in a multiple regression analysis includes explanatory variables that are highly correlated, or where linear combinations of the explanatory variables are highly correlated. In such situations, parameter estimates and their variance estimates become highly unstable. Slight changes to the data used to fit the model, or removal of a statistically insignificant covariate, can result in radical changes to the estimates of the remaining coefficients. Belsey *et al.* (1980) contains an extensive discussion of this issue and methods for dealing with it. For example, one approach is to replace collinear variables by their principal components. Another approach when exclusion, or replacement, of covariates is unacceptable is ridge regression (Hoerl and Kennard, 1970).

It is quite common to observe highly non-normal residual distributions when modelling economic data. One of the main causes is the presence of outlying observations, which can be highly influential on the outcome of the modelling process. One approach, at least for OLS regression, is to identify and remove such observations from the analysis using Cook's D statistic (Cook, 1977). Because outliers are discarded, this approach can sometimes result in too optimistic an assessment of the fit of the model. An alternative approach is to use a robust fitting method that 'accommodates' outliers, such as M-regression (Huber, 1981), which automatically downweights outlying observations when fitting the regression model.

20.6 Case studies

20.6.1 ABARE broadacre survey data

The data used in these case studies was obtained from the annual broadacre survey run by the Australian Bureau of Agriculture and Resource Economics (ABARE) from 1977–78 to 2006–07 (ABARE, 2003). The survey covers broadacre agriculture across all of Australia (see Figure 20.1), but excludes small and hobby farms. Broadacre agriculture involves large-scale dryland cereal cropping and grazing activities relying on extensive areas of land. Most of the information outlined in Section 20.3 was collected from the surveyed farms in face-to-face interviews. The sample size over the 30 years of the collection varies from about 1000 to 1500, and due to sample rotation farms remain in sample for an average of between 3 and 4 years.

The population of farms in the ABARE's broadacre survey is stratified by region (Figure 20.1) and size. The sample is selected from a list supplied by the Australian Bureau of Statistics (ABS), and in varying proportions between strata to improve the precision of cross-sectional survey estimates. Survey estimates are calculated by appropriately weighting the data collected from each sample farm and then using these weighted

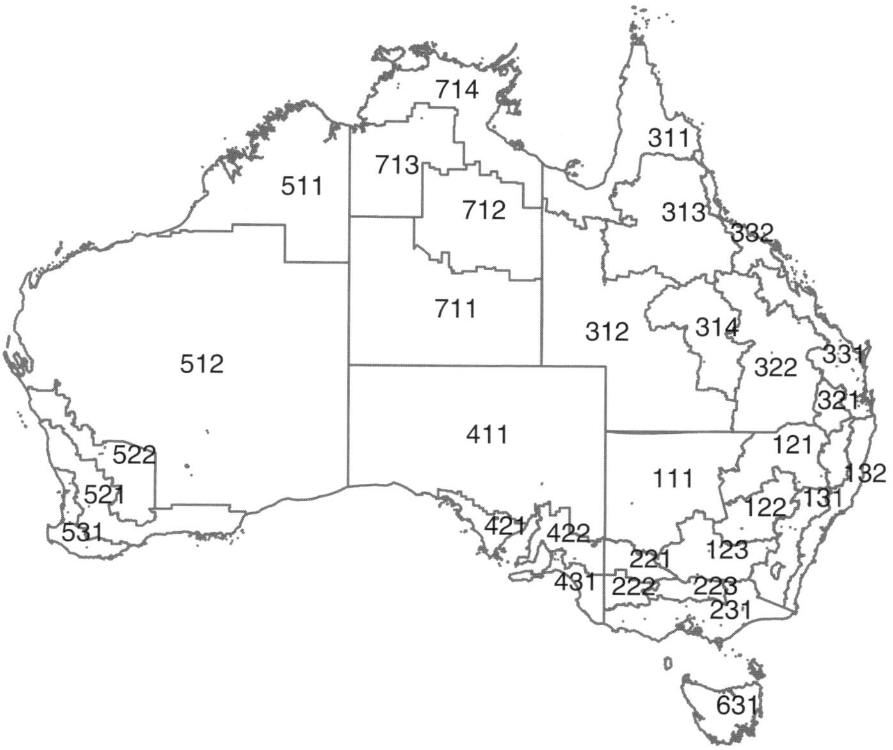

Figure 20.1 Survey regions. First digit represents state, second digit zone and third digit region.

data to calculate population estimates. Sample weights are calculated using a model-based weighting procedure (Chambers 1996) so that population estimates of the number of farms, areas of crops and numbers of livestock in various geographic regions correspond as closely as possible to known totals supplied by the ABS.

20.6.2 Time series model of the growth in fodder use in the Australian cattle industry

The Australian cattle industry makes an important contribution to the Australian economy and farm sector. Cattle production accounted for about $7.9 billion, or about 23% of the value of farm production, in 2006-07 (ABS, 2008). Cattle production in Australia is spread across New South Wales, Victoria, southern Western Australia and northern Australia.

The role of fodder is essential for the cattle industry as it not only is used as a supplement to dryland pasture in times of drought (Kompas and Che, 2006), but also has become increasingly important due to the growth in the livestock finishing industry in Australia. It is also an important input that contributes to farm productivity. Measuring the contribution fodder has had on the growth in productivity is therefore an important issue for the cattle industry.

Survey weighted estimates of mean fodder quantity per farm were produced for the northern and southern regions, separately, for the 30 years from 1977–78 to 2006–07.

The northern region includes all of Queensland and Northern Territory (states 3 and 7 in Figure 20.1) and region 511, and the south region is the remainder of Australia not covered by the north. A simple econometric specification for the growth rate of feed by the north and the south is given by:

$$\log(f_t) = \alpha + \beta t + \varepsilon_t,$$

where $\log(f_t)$ is the logarithm of the mean fodder quantity per farm at time t, β is the growth rate in fodder quantity and ε_t is a residual error term. The model above was fitted to each time series using OLS regression. The results of the regressions are presented in Table 20.1.

All estimates are highly statistically significant. The results indicate that on average the growth rate of mean fodder quantity per farm in the northern region is slightly higher than in the southern region (5% per year compared to 4% per year). This result may help explain the increasing growth rate of TFP for both regions since 1990. The F test is very significant, indicating that the overall fit of the independent variables in the regression model is good.

The R^2 goodness-of-fit statistic for the northern region is better than for the south. This fact is also illustrated in the plot of the actual and fitted values (see Figures 20.2(a) and 20.2(b)). This indicates that there are likely to be other time-independent factors influencing the use of fodder in the south more so than in the north (e.g. drought versus good years). To check whether the residuals are approximately normally distributed, quantile quantile (Q-Q) plots were produced (Figures 20.2(c) and 20.2(d)). These plots show that the residuals are closer to the normal distribution for the northern region than for the south, but in both cases they were reasonably acceptable. With economic time series data it is also important to check whether there is a change in the assumed functional form over time. Empirical fluctuation process (EFP) plots (Zeileis, 2005) are an appropriate graphical tool for examining this. Since in both EFP plots (Figures 20.2(e) and 20.2(f)) the cumulative sum of recursive residuals lies within the two critical bounds there is no statistical evidence that the functional form for either regression is invalid. Both time series were also tested for stationarity using the Dickey–Fuller test (Maddala, 2001) and for serial correlation using the Durban–Watson test (Maddala, 2001). In all cases these tests were found to be statistically insignificant, indicating no evidence for non-stationarity or serial correlation of the residuals in the time series model.

Table 20.1 Growth rate of fodder quantity by region.

	Northern region		Southern region	
Regression variable	Coefficient	t ratio	Coefficient	t ratio
Intercept	0.77*** (0.09)	7.90	0.63*** (0.15)	4.16
Growth rate of feed quantity	0.05*** (0.01)	9.78	0.04*** (0.01)	4.60
Number of observations	30		30	
R^2	0.77		0.43	
Adjusted R^2	0.76		0.41	
F-statistic	95.71***		21.18***	

***Statistically significant at the 0.01 level. Numbers in parentheses are asymptotic standard errors.

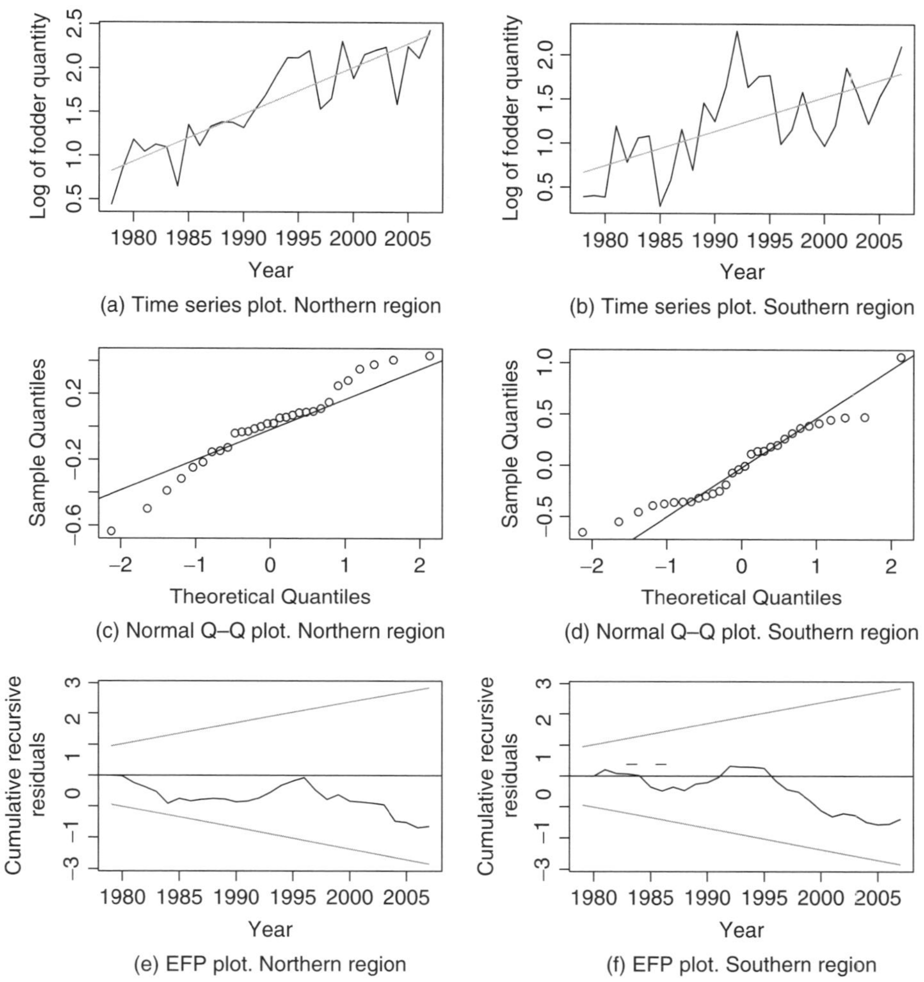

Figure 20.2 Model diagnostics for time series analysis of mean fodder quantity per farm. For (a) and (b) actual values are in black and fitted values are grey. For (e) and (f) the grey lines represent critical bounds at 5% significance level.

20.6.3 Cross-sectional model of land values in central New South Wales

From an agricultural perspective, the productivity of land can be highly dependent on access to natural resource factors such as water and vegetation. Where such access is regulated beyond what farmers consider to be optimal, land productivity may decrease. In the long run, changes in access to natural resources are ultimately expressed in changes to agricultural land values. In broadacre areas where the value of farm land is largely dependent upon its agricultural income earning potential, land value models are useful for estimating the long-run impacts of policy changes which affect access to natural resources.

In order to investigate how farm values are related to natural resources and other factors, in 2005 the ABARE surveyed about 230 broadacre farmers situated in central

New South Wales. This region is characterized by cattle, wool, prime lamb and extensive dryland cropping operations on farms with diverse levels of vegetation. Hedonic regression has long been used to explain broadacre land values in terms of land characteristics that vary between farms (Palmquist and Danielson 1989), as shown in the equation

$$\log(v_i) = \alpha + \sum_{j=1}^{J} \beta_j x_{ij} + \varepsilon_i,$$

where v_i is the value of broadacre agricultural land (in dollars per hectare) less depreciated values of the operator's house and other capital improvements on farm i, $\{x_{ij}\}$ is a set of explanatory covariates and ε_i is a random error term with mean zero and constant variance σ^2. The β_j coefficient represents what proportional impact on land values an increase of one unit of the land characteristic x_{ij} will have.

A brief description of the covariates that were considered for predicting v_i is given in Table 20.2. For a more detailed description refer to Kokic *et al.* (2007a). In addition, several variables were also considered to account for the sample design. Indicator (or dummy) variables for each of four broad geographic strata covered by the study were included. Size strata were determined according to the size variable: expected value of agricultural operations (EVAO), which was also tested in the regression. In the ABARE's farm survey, sample imbalance can also occur for a variety of reasons and this is adjusted

Table 20.2 Covariates considered for the prediction of land values per hectare.

Climate	Climate is measured by the forest productivity index, a relative climate index of long-term plant productivity.
Vegetation density	Cleared land tends to have greater value than uncleared land. This is measured as the average percentage cover across the farm.
Land capability	Land capability captures erosivity as well as acting as a proxy for other soil productivity characteristics, which can significantly affect land values. A measure was constructed so that larger values correspond to higher soil quality.
Transport costs	Access to services and businesses that support agricultural industries as well as other lifestyle choices can often be an important determinant of land value. Sometimes referred to as 'distance to town', access is best captured by a proxy measure of transport cost from each farm to the closest town of population (say) 5000 or greater.
Stream frontage	This metric captures productivity benefits associated with access to water. This is measured as a percentage of the operating area as described in Kokic *et al.* (2007a).
Land use intensity	Land use intensity is measured as the hectares of arable land relative to the operating area of the farm. Total arable land is based on sheep equivalent, where cropping land is defined as possessing a carrying capacity of 12 sheep equivalent per hectare and each cow is defined as 8 sheep equivalent.

Table 20.3 Regression results for the log of land value per hectare.

Regression variable	Coefficient	t ratio
Intercept	3.40*** (1.19)	2.86
log(climate)	1.23*** (0.17)	7.27
log(travel costs)	−0.41*** (0.05)	−8.15
log(land use intensity)	0.53*** (0.04)	13.60
log(land quality)	0.16*** (0.04)	3.69
stream frontage	0.02*** (0.01)	2.43
vegetation density	−0.02*** (0.00)	−6.03
SE region	0.21*** (0.06)	3.33
log(DSE)	−0.14*** (0.03)	−4.37
Number of observations	230	
R^2	0.89	
Adjusted R^2	0.88	
F statistic	221.81***	

***Statistically significant at the 0.01 level. Numbers in parentheses are asymptotic standard errors.

for by including production covariates when computing survey weights. An overall summary measure of production is sheep equivalent (DSE) described briefly in Table 20.2, and so this was also considered. Inclusion of such covariates in the regression model will help ensure that the regression is ignorable with respect to the design and sample selection process (Chambers and Skinner, 2003), and so in this case it is possible to fit and interpret regression parameter estimates and their associated significance statistics as if the sample was drawn at random from the entire population.

Estimation of the parameters in the land value model was by OLS regression. One highly influential observation was identified using Cook's D statistic (Cook, 1977), so this was removed from the subsequent analysis. The model estimates, standard errors and diagnostics summary are reported in Table 20.3. The model explained around 89% of the variation in the data between farms. All covariates from Table 20.3 included in the model had the sign expected and were highly significant. EVAO was found to be statistically non-significant so it was excluded from the regressions results. Interactions between the various covariates in Table 20.2 were tested and found to be either statistically insignificant, or highly correlated with one of the main effects, and so these were also excluded. Furthermore, only one region indicator was found to be significant (the south-east region in the study area), so the remainder were removed from the regression.

The Q-Q plot (Figure 20.3(a)) indicates that the residuals are very close to normally distributed. For cross-sectional regression the presence of heteroscedasticity can be a serious problem for the reasons mentioned in Section 20.5.2. The Breusch–Pagan test (Maddala, 2001) was statistically insignificant, indicating no evidence of heteroscedasticity. Visually, the absence of significant heteroscedasticity is further supported by the plot of observed against predicted values (Figure 20.3(b)). The strong linearity in this figure and the high value of the R^2 fit statistic strongly indicate that the functional form for the model is appropriate.

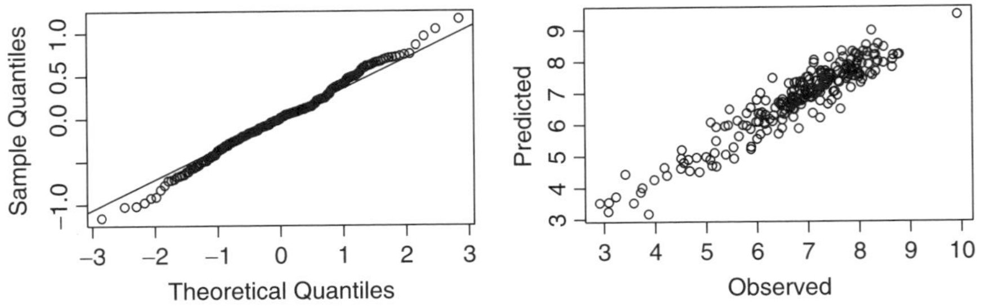

(a) Normal Q–Q plot (b) Predicted versus observed values

Figure 20.3 Model diagnostics for cross-sectional regression of land values in central New South Wales.

References

ABARE (2003) *Australian Farm Surveys Report 2003*. Camberra: ABARE. http://www.abare.gov.au/publications_html/economy/economy_03/fsr03.pdf.

Anderson, J.R. (2003) Risk in rural development: challenges for managers and policy makers. *Agricultural Systems*, 75, 161–197.

Australian Bureau of Statistics (2008) Value of principal agricultural commodities produced. February.

Battese, G.E., Harter, R.M. and Fuller, W.A. (1988) An error-components model for prediction of county crop areas using survey and satellite data. *Journal of the American Statistical Association*, 83, 28–36.

Belsey, D.A., Kuh, E. and Welsch, R.E. (1980) *Regression Diagnostics: Identifying Influential Data and Sources of Collinearity*. New York: John Wiley & Sons, Inc.

Breiman, L., Freidman, J., Olshen, R.A. and Stone, C.J. (1998) *Classification and Regression Trees*. Boca Raton, FL: CRC Press.

Breckling, J. and Chambers, R. (1988) M-quantiles. *Biometrika*, 75, 761–771.

Carroll, J.R. Ruppert, D. (1988) *Transformation and Weighting in Regression*. New York: Chapman & Hall.

Chambers, R.L. (1996) Robust case-weighting for multipurpose establishment surveys. *Journal of Official Statistics*, 12, 3–32.

Chambers, R.L. (1997) Weighting and calibration in sample survey estimation. In C. Malaguerra, S. Morgenthaler, and E. Ronchetti (eds) *Conference on Statistical Science Honouring the Bicentennial of Stefano Franscini's Birth*. Basle: Birkhuser.

Chambers, R. and Skinner C.J. (2003) *Analysis of Survey Data*. Chichester: John Wiley & Sons, Ltd.

Chambers, R. and Tzavidis N. (2006) M-quantile models for small area estimation. *Biometrika*, 93, 255–268.

Cline, W.R. (2007) *Global Warming and Agriculture*. Washington, DC: Centre for Global Development Peterson Institute for International Economics.

Coelli, T., Rao, D. and Battese, G.E. (1998) *An Introduction to Efficiency and Productivity Analysis*. Boston: Kluwer Academic Publishers.

Cook, R.D. (1977) Detection of influential observations in linear regression. *Technometrics*, 19, 15–18.

Davidson, A., Elliston, L., Kokic, P., Lawson K. (2005) Native vegetation: cost of preservation in Australia. *Australian Commodities*, 12, 543–548.

Deville, J.C. and Särndal, C.E. (1992) Calibration estimators in survey sampling. *Journal of the American Statistical Association*, 87, 376–382.

Dorfman, A.H. (2009) Inference on distribution functions and quantiles. In C.R. Rao and D. Pfeffermann (eds), *Handbook of Statistics*, Volume 29B: *Sample Surveys, Inference and Analysis*, Chapter 36. Amsterdam: Elsevier.

Ellis, F. (2000) *Rural Livelihoods and Diversity in Developing Countries*. Oxford: Oxford University Press.

Goldstein, H. (1995) *Multilevel Statistical Modelling*, 2nd edition. London: Edwin Arnold.

Gollop, F.M. and Swinand, G.P. (1998) From total factor to total resource productivity: an application to agriculture. *American Journal of Agricultural Economics*, 80, 577–583.

Gollop, F.M. and Swinand, G.P. (2001) Total resource productivity: accounting for changing environmental quality. In C. Hulten, M. Harper and E. Dean (eds) *New Developments in Productivity Analysis*. Chicago: University of Chicago Press.

Greene, W.H. (2008) *Econometric Analysis* 6th edn. Upper Saddle River, NJ: Prentice Hall International.

Hardaker J.B., Huirne R.B.M., Anderson J.R. (1997) *Coping with Risk in Agriculture*. Wallingford: CAB International.

Hastie, T. and Tibshirani, R. (1990) *Generalized Additive Models*. London: Chapman & Hall.

Heckman, J.J. (1979) Sample selection bias as a specification error. *Econometrica*, 47, 153–161.

Hill, H.S.J., Butler, D., Fuller, S.W., Hammer, G.L., Holzwort, D., Love, H.A., Meinke, H., Mjelde, J.W., Park, J. and Rosenthal, W. (2001) Effects of seasonal climate variability and the use of climate forecasts on wheat supply in the United States, Australia, and Canada. In C. Rosenzweig (ed.), *Impact of El Niño and Climate Variability on Agriculture*, ASA Special Publication 63, pp. 101–123. Madison, WI: American Society of Agronomy.

Hill, L.L. (2006) *Georeferencing*. Cambridge, MA: MIT Press.

Hoerl, A.E. and Kennard, R.W. (1970) Ridge regression: biased estimation for nonorthogonal problems. *Technometrics*, 12, 55–67.

Hsiao, C. (2003) *Analysis of Panel Data*, 2nd edition. Cambridge: Cambridge University Press.

Huber, P.J. (1981) *Robust Statistics*. New York: John Wiley & Sons, Inc.

Knopke, P., O'Donnell, V. and Shepherd, A. (2000) Productivity growth in the Australian grains industry. ABARE research report 2000.1, Canberra. `http://www.abare.gov.au-/publications_html/crops/crops_00/productivity_grains-industry.pdf`.

Koenker R. (2005) *Quantile Regression*. Cambridge: Cambridge University Press.

Koenker, R.W. and Bassett, G.W. (1978) Regression quantiles. *Econometrica*, 46, 33–50.

Kompas, T. and Che, T.N. (2006) Technology choice and efficiency on Australian dairy farms. *Australian Journal of Agricultural and Resource Economics*, 50, 65–83.

Kokic, P., Chambers, R., Beare, S. (2000) Microsimulation of business performance. *International Statistical Review*, 68, 259–276.

Kokic, P., Lawson, K., Davidson, A. and Elliston, L. (2007a) Collecting geo-referenced data in farm surveys. In *Proceedings of the Third International Conference on Establishment Surveys*, Montréal, Canada, 18–21 June.

Kokic, P., Nelson, R., Meinke, H., Potgieter, A. and Carter, J. (2007b) From rainfall to farm incomes – transforming advice for Australian drought policy. I. Development and testing of a bioeconomic modelling system. *Australian Journal of Agricultural Research*, 58, 993–1003.

Laird, N.M. and Ware, J.H. (1982) Random effects models for longitudinal data. *Biometrics*, 38, 963–974.

Lesschen, J.P., Verburg, P.H. and Staal, S.J. (2005) Statistical methods for analysing the spatial dimension of changes in land use and farming systems. LUCC Report Series 7, Wageningen University, The Netherlands.

Lohr, S.L. (1999) *Sampling: Design and Analysis*, Pacific Grove, CA: Duxbury Press.

Maddala, G.S. (2001) *Introduction to Econometrics*. Chichester: John Wiley & Sons, Ltd.

Meinke, H., Nelson, R., Kokic, P., Stone, R., Selvaraju, R. and Baethgen, W. (2006) Actionable climate knowledge: from analysis to synthesis. *Climate Research*, 33, 101–110.

Mullen, J. (2007) Productivity growth and the returns from public investment in R&D in Australian broadacre agriculture. *Australian Journal of Agricultural Economics*, 51, 359–384.

Mullen, J.D. and Crean, J. (2007) *Productivity Growth in Australian Agriculture: Trends, Sources and Performance*. Surry Hills, NSW: Australian Farm Institute.

Nelson, R., Alexander, F., Elliston, L. and Blias, F. (2004) Natural resource management on Australian farms. ABARE report 04.7, Canberra.

Nelson, R., Kokic, P. and Meinke, H. (2007) From rainfall to farm incomes-transforming advice for Australian drought policy. II. Forecasting farm incomes. *Australian Journal of Agricultural Research*, 58, 1004–1012.

Nelson, R., Howden, S.M. and Stafford Smith, M. (2008) Using adaptive governance to rethink the way science supports Australian drought policy. *Environmental Science and Policy*, 11, 588–601.

Pan, W.K.Y., Walsh, S.J., Bilsborrow, R.E., Frizzelle, B.G., Erlien, C.M. and Baquero, F. (2004) Farm-level models of spatial patterns of land use and land cover dynamics in the Ecuadorian Amazon. *Agriculture, Ecosystems and Environment*, 101, 117–134.

Palmquist, R., Danielson, L. (1989) A hedonic study of the effects of erosion control and drainage on farmland values. *American Journal of Agricultural Economics*, 1, 55–62.

Pratesi, M. and Salvati, N. (2008) Small area estimation: the EBLUP estimator based on spatially correlated random area effects. *Statistical Methods and Applications*, 17, 113–141.

Rao, J.N.K. (2003) *Small Area Estimation*. Hoboken, NJ: John Wiley & Sons, Inc.

Särndal, C.E. and Lundström, S. (2005) *Estimation in Surveys with Nonresponse*. Chichester: John Wiley & Sons, Ltd.

Severance-Lossin, E. and Sperlich, S. (1999) Estimation of derivatives for additively separable models. *Statistics*, 33, 241–265.

Wolter, K.M. (2007) *Introduction to Variance Estimation*, 2nd edition. New York: Springer.

Zeileis, A. (2005) A unified approach to structural change tests based on ML scores, *F* statistics, and OLS residuals. *Econometric Reviews*, 24, 445–466.

21

Measuring household resilience to food insecurity: application to Palestinian households

Luca Alinovi[1], Erdgin Mane[1] and Donato Romano[2]

[1]*Agricultural Development Economics Division, Food and Agriculture Organization of the United Nations, Rome, Italy*

[2]*Department of Agricultural and Resource Economics, University of Florence, Italy*

21.1 Introduction

Most research in the field of food security has focused on developing and refining methods of analysis in order to more accurately predict the likelihood of a food crisis. The emphasis of such work has been on the development of early warning systems, using 'behavioural patterns' in an economy to judge whether a crisis is about to happen, from the value change of selected indicators (Buchanan-Smith and Davies, 1995).

In the last decade, the collaboration between environmental and social scientists concerned with the sustainability of jointly determined ecological-economic systems has brought about a potentially fruitful concept, well known in the ecological literature, but quite new as applied to socio-economic systems: the concept of resilience, that is, the ability of a system to withstand stresses and shocks and to persist in an uncertain world (see Adger, 2000). This is a concept related to but different from vulnerability which, as applied to food insecurity, refers to the likelihood of experiencing future loss of adequate food.

Agricultural Survey Methods Edited by Roberto Benedetti, Marco Bee, Giuseppe Espa and Federica Piersimoni

In the field of social-ecological systems, vulnerability lies not in exposure to risk or hazards alone. Building on the social ecology literature, Adger (2006) argues that it also resides in the sensitivity and adaptive capacity of the system experiencing such hazards, that is, its resilience. There is not a common agreed-upon definition of the concepts of vulnerability and resilience. However, most authors agree on a common set of parameters that are used in both of them: the shocks and stresses to which a social-ecological system is exposed, and the response and adaptive capacity of the system. Nevertheless, vulnerability analysis often tends to measure only the susceptibility to harm of an individual or household and the immediate coping mechanisms adopted. Such an approach risks oversimplifying the understanding of the systemic component of the interaction between vulnerability and resilience.

This is why some scholars and practitioners (Folke *et al.*, 1998, 2002), as well as some international organizations such as the FAO (Hemrich and Alinovi, 2004), have proposed the use of the resilience concept in food security analysis. The implicit idea is that this concept could enrich the approach adopted in the early warning systems (EWS) framework. Indeed, the EWS approach tries to predict crises, coping mechanisms, and immediate responses, while the resilience framework tries to assess the current state of health of a food system and hence its ability to withstand shocks should they occur.

In pursuing the overall objective of sustainability of a social-ecological system (SES), a resilience approach may offer potential additional advantages compared to the standard vulnerability approach, analysing the different responses adopted by a system to respond both in negative and positive terms, and capturing the dynamic components and the different strategies adopted. A resilience approach investigates not only how disturbances and change might influence the structure of a system but also how its functionality in meeting these needs might change. This focus offers the chance to identify other ways of retaining the functionality of a SES while its components and structure may change. This ability to identify new ways to ensure system functionality cannot be achieved by utilizing only the standard vulnerability approach, which focuses more on how a system is disturbed in its structure and functionality.

Though such statements look sound and promising, many steps must be taken before the implications of adopting the resilience approach are fully clarified and operationalized. There is indeed a need to clarify the meaning and scope of resilience as applied to the analysis and management of food systems. There are many relevant questions to be addressed in pursuing this objective. What is the meaning of resilience as applied to food systems? What is the relationship between resilience and other concepts such as vulnerability? How should resilience in a food system be measured? This study addresses these questions in the specific context of household food security. The choice of this level of analysis is justified on the grounds that it is at this level that most risk management and risk coping strategies are implemented, and especially so in the case of informal strategies which often are the only ones readily available to the poor (World Bank, 2001). Therefore, we will not analyse other important issues that pertain to different levels of analysis such as the relationships between households and the broader system they belong to (and their implications for household resilience to food crises), how to measure food system resilience at the regional, national or global level, etc.

Therefore, the objective of this chapter is twofold: (a) to clarify the meaning of the resilience concept as applied to food insecurity, critically analysing its relationships with other competing and/or complementary concepts and the potential benefits of its use; and

(b) to propose a methodology for measuring household resilience to food insecurity, as well as to discuss and highlight data needs so that future surveys can eventually make the required data available. Although fully acknowledging the limited scope of an application at this stage, we will make an attempt to test the proposed methodology using the data from the 11th Palestinian Public Perception Survey.

The definition of resilience directly affects the methodology adopted for its measurement. In our model, resilience is considered as a latent variable defined according to four building blocks: income and food access, assets, access to public services, and social safety nets. Furthermore, stability and adaptive capacity are two other dimensions that cut across these building blocks and account for households' capacity to respond and adapt to shocks. These dimensions of resilience are also latent variables and, in order to estimate the relationships between resilience and its underlying dimensions, two approaches are presented. The first measures each dimension separately using different multivariate techniques (factor analysis, principal components analysis and optimal scaling) and then estimates a resilience index. The classification and regression tree (CART) methodology has also been used for the understanding of the process. The second approach measures all the dimensions simultaneously through structural equation models, and is based on normality assumptions on observed variables. As most of the variables in resilience measurement are ordinal or categorical, the first approach has been adopted in this study.

Section 21.2 describes the concept of resilience and its relationship to food security related concepts (e.g. vulnerability and risks). Section 21.3 sets out the framework for measuring the different dimensions of resilience to food security, and presents methodological approaches for measuring it at household level. Section 21.4 reports on how the proposed methodology was implemented using data from the 11th Palestinian Public Perception Survey. Section 21.5 validates the model through the CART methodology, discusses the implications of resilience measurement for understanding food security outcomes and provides a simplified model for forecasting resilience. Finally, Section 21.6 summarizes the main findings and policy implications and provides some hints for future research.

21.2 The concept of resilience and its relation to household food security

21.2.1 Resilience

The concept of resilience was originally described in the ecological literature (Holling, 1973) and has recently been proposed as a way of exploring the relative persistence of different states in complex dynamic systems, including socio-economic ones (Levin *et al.*, 1998). The concept has two main variants (Holling, 1996). *Engineering* resilience (Gunderson *et al.*, 1997) is a system's ability to return to the steady state after a perturbation (O'Neill *et al.*, 1986; Pimm, 1984; Tilman and Downing, 1994). It focuses on efficiency, constancy and predictability, and is the concept that engineers look at to optimize their designs ('fail-safe' designs). *Ecological* resilience is the magnitude of disturbance that a system can absorb before it redefines its structure by changing the variables and processes that control behaviour (Holling, 1973; Walker *et al.*, 1969). This concept focuses on disruptions to the stable steady state, where instabilities can flip a system into another behaviour regime (i.e. into another stability domain).

Both variants deal with aspects of the stability of system equilibria, offering alternative measures of a system's capacity to retain its functions following disturbance. The two definitions emphasize different aspects of stability, however, and so 'can become alternative paradigms whose devotees reflect traditions of a discipline or of an attitude more than of a reality of nature' (Gunderson *et al.*, 1997).[1] The two definitions reflect two different world-views: engineers aim to make things work, while ecologists acknowledge that things can break down and change their behaviour. The question now is how economists will define resilience. Traditionally, economists have tended to consider conditions close to a single stable state,[2] but the issue of ecological resilience is also beginning to emerge in economics, following the identification of multi-stable states caused by path dependency (Arthur, 1988), 'chreodic' development (Clark and Juma, 1987) and production non-convexities such as increasing return to scale (David, 1985).

Levin *et al.* (1998) argue that resilience offers a helpful way of viewing the evolution of social systems, partly because it provides a means of analysing, measuring and implementing the sustainability of such systems. This is largely because resilience shifts attention away from long-run equilibria towards a system's capacity to respond to short-run shocks and stresses constructively and creatively.

Within a system, key sources of resilience lie in the variety that exists within functional groups, such as biodiversity in critical ecosystem functions, flexible options in management, norms and rules in human organizations, and cultural and political diversity in social groups.[3] Resilience also comes from accumulated capital, which provides sources for renewal. Increasingly, resilience is seen as the capacity not only to absorb disturbances, but also to reorganize while undergoing changes, so as to retain the same functions, structures and feedbacks (Folke, 2006; Walker *et al.*, 2004). In ecological systems, this capacity includes mechanisms for regeneration, such as seeds and spatial recolonization, or soil properties. In social systems, it is the social capital of trust, networks, memory and relationships, or the cultural capital of ethics, values and systems of knowledge.

In addition, the kindred discipline of system ecology acknowledges that critical ecosystem organizational or 'keystone' (Paine, 1974) processes create feedbacks that interact, reinforcing the persistence of a system's temporal and spatial patterns over specific scales. In social-ecological systems, many factors contribute to this – including institutions, property rights and the completeness and effectiveness of markets – making the functions of critical organizational processes robust.

---

[1] For instance, engineering resilience focuses on maintaining the efficiency of functions, while ecological resilience focuses on their existence. This means that the former explores system behaviour in the neighbourhood of the steady state, while the latter explores the properties of other stable states, focusing on the boundaries between states. These attitudes reflect different traditions; engineering resilience was developed in accordance with deductive mathematics theory; ecological resilience stems from inductive disciplines such as applied mathematics and applied resource ecology.

[2] For instance, in partial equilibrium analysis, multiple equilibria are excluded by constructing convex production and utility sets; or when multiple equilibria theoretically exist, their number is reduced by means of individuals' strategic expectations and predetermined normative and social institutions.

[3] Diversity does not support stability but it does support resilience and system functioning (Holling, 1973; Holling, 1986), while rigid control mechanisms that seek stability tend to erode resilience and facilitate system breakdown.

21.2.2 Households as (sub) systems of a broader food system, and household resilience

Households are components of food systems and can themselves be conceived as (sub)systems. The household definition is consistent with Spedding (1988)'s definition of a system as 'a group of interacting components, operating together for a common purpose, capable of reacting as a whole to external stimuli: it is affected directly by its own outputs and has a specified boundary based on the inclusion of all significant feedback'. Moreover, as the decision-making unit, the household is where the most important decisions are made regarding how to manage uncertain events, both ex ante and ex post, including those affecting food security such as what income-generating activities to engage in, how to allocate food and non-food consumption among household members, and what strategies to implement to manage and cope with risks.

Households can therefore be viewed as the most suitable entry point for the analysis of food security. This does not mean disregarding the important relationships between the household and the broader food system it belongs to (e.g. the community, the market chain), which contribute to determining the household's food security performance, including its resilience to food insecurity. Systems comprise hierarchies, each level of which involves a different temporal and spatial scale, and a system's behaviour appears to be dominated by key structuring processes (see the previous subsection) that are often beyond the reach of its single components (e.g. households) and so are assumed as given by those components at a specific scale and in a specific time frame (e.g. the short run). In other words, household strategies for managing and coping with risks prove to be more effective in a given neighbourhood (the household livelihood space) and over a finite time span.

The multidimensionality of the food security and poverty concepts and the complexity of the conduit mechanisms for food insecurity make the household a system that faces largely unpredictable exogenous shocks. This implies that a household should be considered as a complex adaptive system. The survival of a household as a system depends less on the stability of its individual components than on its ability to maintain self-organization in the face of stresses and shocks – in other words, on its resilience. In a resilient household, change can create opportunities for development, novelty and innovation. As resilience declines, a progressively smaller external event can cause catastrophe. A household with low resilience may maintain its functions and generate resources and services – thereby seeming to be in good shape – but when subject to disturbances and stochastic events, it will exceed a critical threshold and change to a less desirable state.

Application of the concept of resilience to household food security therefore seems promising: it aims to measure households' ability to absorb the negative effects of unpredictable shocks, rather than predicting the occurrence of a crisis (as most of the vulnerability literature does).

21.2.3 Vulnerability versus resilience

According to Dercon (2001):

> Households and individuals have assets, such as labour, human capital, physical capital, social capital, commons and public goods at their disposal to

> make a living. Assets are used to generate income in various forms, including earnings and returns to assets, sale of assets, transfers and remittances. Households actively build up assets, not just physical capital but also social or human capital, as an alternative to spending.
>
> Incomes provide access to dimensions of well-being: consumption, nutrition, health, etc., mediated by information, markets, public services and non-market institutions. Generating incomes from assets is also constrained by information, the functioning of markets and access to them, the functioning of non-market institutions, public service provision and public policy....
>
> Risks are faced at various steps in this framework.

Assets, their transformation into income and the transformation of income into dimensions of well-being are all subject to risk.

According to this framework, well-being and its dimensions such as food security or poverty are *ex post* measures of a household's decision-making about its assets and incomes when faced with a variety of risks. Vulnerability to food insecurity describes the outcome of this process *ex ante*, that is, considering the potential outcomes rather then the actual outcome. Food insecurity is measured at a specific point in time, as a 'snapshot'; vulnerability is essentially forward-looking, based on information about a particular point in time. Vulnerability is the propensity to fall below the (consumption) threshold, and its assessment therefore deals not only with those who are currently poor but also those who are likely to become poor in the future (Chaudhuri *et al.*, 2002). Vulnerability to food insecurity is determined by:

- the risks faced by households and individuals in making a living;
- the options available to households (individuals, communities) for making a living (including assets, activities, market and non-market institutions, and public service provision);
- the ability to handle this risk.

This framework is not very different from that proposed in the global change literature (Ahmad *et al.*, 2001), which defines vulnerability as a function of:

- exposure to risks, that is, the magnitude and frequency of stress experienced by the entity;
- sensitivity, which describes the impacts of stress that may result in reduced well-being owing to the crossing of a threshold (below which the entity experiences lower well-being);
- adaptive capacity, which represents the extent to which an entity can modify the impact of a stress, to reduce its vulnerability.[4]

In the food security literature (FIVIMS/FAO, 2002), vulnerability to food insecurity is seen as a function of the nature of risks and the individual's or household's responses

[4] The first and third items in these two lists are the same. The second item in the second list (sensitivity) can be regarded as the outcome of the second item in the first list (options available to the household).

to such risks. In this chapter, however, vulnerability is a function of a household's risk exposure and its resilience to such risks (Løvendal *et al.*, 2004); an output-based analysis framework is adopted, which is in the same vein as the asset–income–outcome causal chain suggested by Dercon (2001). Therefore, household resilience to food insecurity, defined as a household's ability to maintain a certain level of well-being (food security) in the face of risks, depends on the options available to that household to make a living and on its ability to handle risks. It refers therefore to *ex ante* actions aimed at reducing or mitigating risks, and *ex post* actions to cope with those risks. It also covers both short-term actions (e.g. coping) and actions that have a longer-term impact (e.g. adaptation to structural changes to ensure household functioning). Empirical application focuses on how to measure resilience to food insecurity as a contribution to vulnerability assessment.[5]

21.3 From concept to measurement

21.3.1 The resilience framework

Figure 21.1 summarizes the rationale for attempting to measure resilience to food insecurity. Consistent with Dercon's (2001) framework, it is assumed that the resilience of a given household at a given point in time, T_0, depends primarily on the options available to that household to make a living, such as its access to assets, income-generating activities, public services and social safety nets. These options represent a precondition for the household response mechanisms in the face of a given risk (its ability to handle risk).

Assume that between T_0 and T_1 some shocks occur, which may be endogenous, if related to the household's control of its options, or exogenous, if beyond the household's control. Whether the shocks are endogenous or exogenous, the household reacts to them by using available response mechanisms and its absorption and adaptive capacities. The reactions to some shocks (e.g. systemic shocks) occur through policy support by decision-makers other than the household (e.g. government or international institutions), and such reactions might themselves be the causes of external shocks.

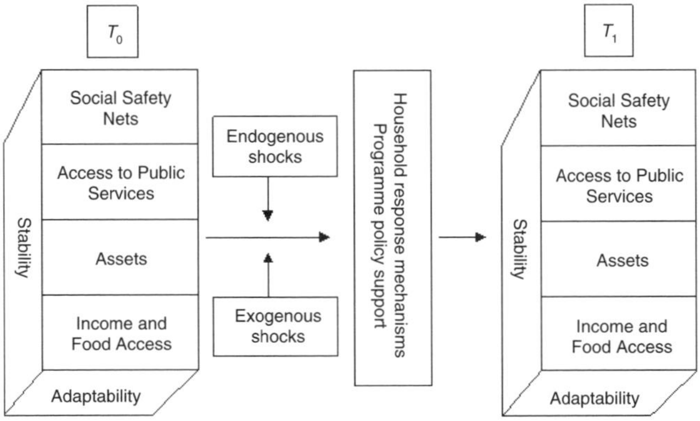

Figure 21.1 Resilience conceptual framework.

[5] This chapter therefore does not focus on the broader issue of how to measure vulnerability (Chaudhuri, 2004; Dercon, 2001; Scaramozzino, 2006; Ligon and Schechter, 2004), but on a subset of such a framework.

A third dimension of resilience is stability, which is the degree to which a household's options vary over time. Households showing high adaptability and high stability will likely have high resilience to food insecurity, while those showing low adaptability and low stability will have low resilience. Intermediate cases will likely determine medium-level household resilience.[6]

Every component at time T_0 is estimated separately, to generate a composite index of household resilience. The different components of the resilience observed at time T_1 then reflect how all these factors produce change in a household's resilience. In algebraic terms, the resilience index for household i can be expressed as

$$R_i = f(IFA_i, S_i, SSN_i, APS_i, A_i, AC_i), \tag{21.1}$$

where R = resilience, S = stability, SSN = social safety nets, APS = access to public services, A = assets, IFA = income and food access and AC = adaptive capacity (or adaptability). Resilience is not observable *per se*, and is considered a latent variable depending on the terms on the right-hand side of the equation. To estimate R, it is therefore necessary to separately estimate IFA, S, SSN, APS, A, AC, which are themselves latent variables because they cannot be directly observed in a given survey, although it is possible to estimate them through multivariate techniques.[7]

21.3.2 Methodological approaches

The framework described above can be estimated through multivariate analysis models. Equation (21.1) is a hierarchical model in which some variables are dependent on the one side and independent of the other. Unobservable (i.e. latent) variables also have to be dealt with. Figure 21.2 shows the path diagram of the model concerned.

In the causal models literature (Spirtes *et al.* 2000), circles represent latent variables and boxes represent observed variables. Most of the hierarchical or multi-level models

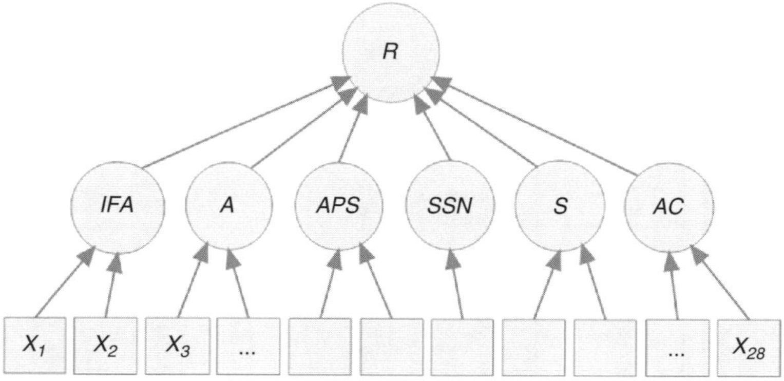

Figure 21.2 Path diagram of the household resilience model.

[6] Recalling the discussion of the different meanings of resilience in the previous section, it can be argued that high stability and low adaptability create a status closer to the definition of engineering resilience, while ecological resilience is closer to a status characterized by high adaptability and low stability.

[7] For example, *IFA* includes not only household income (which is observable), but also a series of estimated variables related to food consumption and expenditure and to the household's perception of food access and dietary diversity, which are context- and data-specific.

studied in the literature deal with measured variables, so the regression properties are extended. One of the innovative contributions of this research is the estimation of latent variable models in complex survey data.

Considering the complexity of the model concerned, two alternative estimation strategies could be adopted for the estimation of household resilience: structural equation modelling and multi-stage modelling.

Structural equation models (SEMs) are the most appropriate tool for dealing with the kind of model illustrated in Figure 21.2. Structural equation modelling combines factor analysis with regression. It is assumed that the set of measured variables is an imperfect measure of the underlying latent variable of interest. Structural equation modelling uses a factor analysis-type model to measure the latent variables via observed variables, while simultaneously using a regression-type model to identify relationships among the latent variables (Bollen, 1989). Generally, the estimation methods developed for SEMs are limited to normally distributed observed variables, but in most cases (including this one), many variables are nominal or ordinal. The literature has proposed ways to broaden the SEMs to include nominal/ordinal variables, but there are difficulties regarding computational aspects (Muthén, 1984). It is also possible to use generalized latent variable models (Bartholomew and Knott, 1999; Skrondal and Rabe-Hesketh, 2004) to model different response types. A major concern in using SEMs for measuring resilience is that the algorithms of SEM procedures are usually totally data-driven, but this chapter seeks to include some prior knowledge on deterministic relations among measured variables. Bayesian procedures could be used for their flexibility, but the number of variables to be used in this case make parameter identification problems likely to emerge, requiring careful consideration to incorporate proper prior information. In the last decade, the use of Markov chain Monte Carlo (MCMC) simulation methods (Arminger and Muthén, 1998; Lopes and West, 2003; Mezzetti and Billari, 2005; Rowe, 2003) has alleviated the computational concerns.

The other approach explored is a multi-stage strategy for estimating the latent variables separately, based on the relevant observed variables. This involves the use of various sets of observed variables (represented as squares in Figure 21.2) to estimate the underlying latent variables (circles in Figure 21.2). In other words, the circles represent the common pattern in the measured variables. The methods used for generating these latent variables depend on the scales of the observed variables. Traditional multivariate methods are based on continuous variables, but most of the variables in household-level surveys are qualitative (nominal, ordinal or interval), so it is necessary to use different techniques for non-continuous types of variables. The main multivariate techniques relevant for this analysis are SEMs or LISREL methods,[8] factorial analysis, principal components analysis, correspondence analysis, multidimensional scaling, and optimal scaling, all of which are usually combined with deterministic decision matrices that are based on prior knowledge of variables.[9]

In the household surveys considered, most of the variables are categorical or ordinal, and measured at different scales, which makes the situation even more complex and

[8] This should not be confused with the holistic SEMs proposed in the first approach. A single latent variable may be estimated through simpler cases of the general structural equations for the latent variable model: $\eta = B\eta + \Gamma\xi + \zeta$, where η is a vector of latent endogenous variables and ξ is a vector of exogenous variables.

[9] This chapter does not explain these methods, which are described in all multivariate analysis manuals. The following sections provide the information necessary for an understanding of the procedures.

requires a mechanism to identify the optimal scaling. The variables require the use of optimal scaling methods, and scaling categorical variables is an important issue. Usually, these variables are coded with natural numbers (1, 2, ..., n), but they may be coded as a, b, c, etc. In many cases, the distance between 1 and 2 may not be equal to the distance between 2 and 3, so categorical variables cannot be treated as quantitative variables. In this chapter, the principal components analysis by alternating least squares (PRINCALS) algorithm (De Leeuw and Van Rijckevorsel, 1980) is used to estimate optimal scaling and principal components simultaneously. The estimation minimizes the following objective function (loss function):

$$\sigma(X, Y) = m^{-1} \sum_j SSQ(X - G_j Y_j),$$

where SSQ is the sum of squares, m is the number of variables, X is the matrix of object scores, G_j is the indicator matrix for variable j, and Y_j (scaling) is the matrix of category quantifications for j.

There are two main reasons for adopting the second (multi-stage) estimation strategy: the available variables are not all normally distributed, so may require the use of different multivariate techniques; and measuring the different components separately makes the model more flexible, allowing the inclusion of prior information and thus reducing the parameter identification problem.

After estimation of the different components and resilience, the CART methodology is adopted to test the validity of the adopted model and identify the contributions of the original variables to the resilience index.

21.4 Empirical strategy

21.4.1 The Palestinian data set

The Palestinian Public Perception Survey (PPPS) is an inter-agency effort aimed at building understanding of the socio-economic conditions in the West Bank and Gaza Strip. The University of Geneva implemented the 11th PPPS in 2007, with the collaboration of several agencies, including the FAO for the food security component. Responsibility for the data collection rests with the Palestinian Central Bureau of Statistics. The PPPS provides a very rich data set, including key indicators relevant for defining and analysing household food security status and its dynamics.

The data are repeated cross-sections, but it is not possible to use the surveys carried out before 2007 owing to some changes in the food security section of the questionnaire, which make the last survey incompatible with the previous ones. In 2007, the sample size was 2087 households and the sampling design was a two-stage stratified cluster. The sampling weights (the reciprocal of selection probability) were adjusted to compensate for non-response and to satisfy the population size estimates, particularly in the analysis disaggregated by region, sex and age group.

The questionnaire has the following sections for each household in the survey:

- demographic, occupational and educational status
- security/mobility

- labour market
- economic situation (including the food security module)
- assistance/assistance priorities
- infrastructure
- coping strategies
- health
- children/women
- politics and peace/managing security/religion.

Preliminary data screening took place during the preparatory phase of the analysis, to select the variables related to each component in (21.1).

21.4.2 The estimation procedure

The empirical strategy follows a three-step procedure: (1) identification and measurement[10] of the variables selected for each resilience block; (2) estimation of the latent variable representing each block, and of the resilience index, using multivariate methods (factor analysis, principal components analysis, optimal scaling, etc.); and (3) application of the CART methodology to estimate precise splitting rules based on a regression tree, to improve understanding of the whole process.[11]

The selection of relevant variables for estimating the latent indicator for each block is particularly complex. The multidimensionality of the concept gives rise to concerns about having variables that are relevant for more than one block, when a variable cannot be included more than once in the model. The conceptual framework described in the previous chapter simplifies this issue.

This subsection describes the separate estimation process for each component of the resilience framework.

Income and food access (*IFA*)

This indicator is directly related to a household's access to food. Economic access to food is considered the main food insecurity concern in Palestine (Abuelhaj, 2008; Mane *et al.*, 2007). The traditional indicator for measuring food access capacity is income, but this study includes two additional indicators: the dietary diversity and food frequency score (*DD*) as a nutritional indicator; and the household food insecurity access scale (*HFIAS*) as an indicator of the household's perception of food security. Estimation of the *IFA* indicator involves use of the following variables:

- *Average income per person per day* (in Israeli new shekels (NIS)). This is an aggregated value of the different sources of income measured by the PPPS.

[10] Not all the observed variables are taken directly from the raw data. Some are measured using complex procedures (e.g. the household food insecurity access scale and the dietary diversity and food frequency score).

[11] The use of CART also made it possible to validate the decision process and identify the original variables (indicators) that play major roles in the different blocks defining resilience.

- *DD*. This is computed using specific methodologies for food security assessments applied to the weekly consumption of 20 food items. It can also be used as a proxy indicator for food access (Hoddinott and Yohannes, 2002).
- *HFIAS*. This indicator of a household's perception of food insecurity is measured through a set of nine questions developed by the Food and Nutrition Technical Assistance (FANTA) project[12] (Coates *et al*., 2006).

Missing values in these variables are imputed using regression techniques.

All these indicators aim to measure food access, so the high correlation among them should produce a latent variable that fits the common pattern in the data. Because of this, to estimate the *IFA* latent indicator, a factor analysis is run using the principal factor method and the scoring method suggested by Bartlett (1937):

$$\hat{f}_B = \Gamma^{-1}\Lambda'\Psi^{-1}x,$$

where $\Gamma = \Lambda'\Psi^{-1}\Lambda$, Λ is the unrotated loading matrix, Ψ is the diagonal matrix of uniquenesses, and x is the vector of observed variables. The estimates produced by this method are unbiased, but may be less accurate than those produced by the regression method suggested by Thomson (1951).[13] The regression-scored factors have the smallest mean square error, but may be biased. The first factor produced is quite meaningful and can be considered the underlying latent variable for food access. Table 21.1 shows the eigenvalue for each factor, and Table 21.2 shows the factor loadings for the original variables. The three indicators play almost the same role in estimating the *IFA* indicator, because their correlation coefficients are similar. As expected, *HFIAS* has a negative correlation because its score increases as food security decreases.

Table 21.1 Eigenvalues.

Factor	Eigenvalue
Factor 1	1.54613
Factor 2	0.75363
Factor 3	0.70024

Table 21.2 Factor loadings and correlations.

Variable	Factor 1	IFA
Income	0.4466	0.6779
DD	0.4786	0.7308
HFIAS	−0.4860	−0.7431

[12] The FANTA project supports integrated food security and nutrition programming to improve the health and well-being of women and children. It is managed by the Academy for Educational Development and funded by the United States Agency for International Development (USAID).

[13] The formula for Thompson's regression method is $\hat{f}_T = \Lambda'\Sigma^{-1}x$, where Σ is the correlation matrix of x.

Access to public services (*APS*)

Public service provision is beyond households' control, but is a key factor for enhancing a household's resilience, such as by improving the effectiveness of that household's access to assets. As a result, better access to public services affects a household's capacity to manage risks and respond to crises. The following public services are considered in the analysis:

- *Health*. Measurement of health involves two indicators: physical access to health (need unmet, and need met within time limit, and need met after time limit); and the health care quality score (based on the quality of services provided in different health areas).
- *Quality of education system* (ordinal scale from 1 to 6).
- *Perception of security* (ordinal scale from 1 to 4). For this, a proxy index based on the general perception of security is constructed.
- *Mobility and transport limitations* (ordinal scale from 1 to 3). Different sources are used to generate an indicator from the mobility restriction questions in the questionnaire.
- *Water, electricity and telecommunications networks*. An indicator is developed for the number of services available.

A characteristic of the variables related to *APS* is that households in the same neighbourhood are expected to be highly correlated in terms of having the same availability of services. Missing values in these variables are therefore treated using the governorate-level mean.

In this case, use of the traditional multivariate methods (factor or principal components analysis) is impossible because the observed variables are not continuous. It was therefore decided to use the optimal scaling technique. The first factor (the *APS* indicator) obtained through PRINCALS is very satisfactory. Table 21.3 shows that all the original variables (transformed using optimal scaling) are, as expected, positively correlated with the estimated *APS*.

Social safety nets (*SSN*)

Social safety nets are a crucial aspect of the mitigation of crises in Palestine. More and more households are becoming dependent on assistance from international agencies,

Table 21.3 Correlation of *APS* with transformed variables.

Transformed variable	APS
Physical access to health	0.6040
Health care quality	0.5984
Education system quality	0.4329
Perception of security	0.5317
Mobility constraints	0.5451
Water, electricity and telecoms	0.2838

charities and non-governmental organizations. Help received from friends and relatives is also substantial. Safety nets can therefore be taken to represent the system's capacity to mitigate shocks, and a general indicator for them has to be included in the estimation of resilience. The variables used to generate the *SSN* indicator are:

- *Amount of cash and in-kind assistance* (NIS/person/day).
- *Quality of assistance* (ordinal scale from 1 to 4).
- *Job assistance* (binary yes/no response).
- *Monetary values of first and second types of assistance* (NIS/person/day).
- *Evaluation of the main type of assistance* (ordinal scale from 1 to 4).
- *Frequency of assistance* (times assistance received in the last six months).
- *Overall opinion of targeting* (assistance targeted to the needy; to some not needy; or without distinction).

Missing values are treated using the governorate-level means and the level of income. It was decided to show the measurement scale of the variables as this is a crucial aspect of choosing the method for estimating the latent variable. The list above shows that types of variables were mixed, so for *SSN*, use of the PRINCALS algorithm is required. The first factor obtained is acceptable for estimating the latent variable *SSN*. Table 21.4 illustrates the correlation between estimated *SSN* and the transformed variables. All the variables are positively correlated with *SSN* and play important roles in the estimation. Nevertheless, the degree of satisfaction with the assistance received is the core of the generated variable.

Moreover, the correlation matrix for the various resilience blocks shows that *SSN* is negatively related to the others. This is because social cohesion increases as poverty increases. This aspect will be explained in more detail in the discussion of the results at the end of this subsection.

Assets (*A*)

Assets are part of a household's capital, and their availability is an important coping mechanism during periods of hardship. They therefore have to be considered as a key

Table 21.4 Correlation of *SSN* with transformed variables.

Transformed variable	SSN
Amount of cash and in-kind assistance	0.1669
Quality of assistance	0.7347
Job assistance	0.3794
First and second types of assistance	0.7223
Evaluation of main assistance	0.7304
Frequency of assistance	0.6775
Opinion of targeting	0.4462

factor in estimating resilience. Information on assets was not available from the PPPS data set; it was decided not to use proxies, so as not to contaminate the estimates.

Adaptive capacity (AC)

The adaptive capacity indicates a household's capacity to cope with and adapt to a certain shock, enabling that household to carry on performing its key functions. In other words, AC represents households' capacity to absorb shocks. For example, having more coping strategies means having a greater probability of mitigating food insecurity after, say, losing a job. The characteristic of adaptability is the buffer effect on household key functions. AC is measured by the following indicators:

- *Diversity of income sources* (count from 0 to 6). This indicates the number of income sources from different sectors (public, private, etc.); during a crisis, the more sources of income, the less the risk of losing the essential basis of the household's livelihood (i.e. income).
- *Coping strategy index* (count from 0 to 18). This represents the number of available coping strategies that have not yet been used. It does not consider whether or not a specific coping strategy has been adopted by the household (*ex post*), but instead how many coping mechanisms are available to the household (*ex ante*).
- *Capacity to keep up in the future* (ordinal scale from 1 to 5). This is based on a household's perception of its own capacity to keep up in the future, considering the current socio-economic shocks in Palestine. It is a forward-looking variable, which allows households' expectations to be taken into account.

The variables are not continuous, even in this case, so the PRINCALS methodology is needed. Table 21.5 shows the correlation of the estimated AC with the transformed variables. The correlation coefficients correspond to the loadings of component 1; the coping strategy index and the capacity to keep up in the future are twice as important as the diversity of income sources.

Stability (S)

Stability is a widely used concept in the food security literature, although it is usually used to describe the stability of food supply. This chapter considers stability to be a cross-sectoral dimension of resilience. For example, an index of income stability may be its variability (increased, decreased or the same) over the last six months. The following variables are used to measure stability:

- *Professional skills* (count). The number of household members with at least a diploma[14] is used as a proxy.
- *Educational level* (continuous). This is measured by the average number of years of schooling of household members.
- *Employment ratio* (from 0 to 1). This is the ratio of the number of employed household members to the household size.

[14] A weighting system is used to give major relevance to household members with a master's or PhD degree.

Table 21.5 Correlation of *AC* with transformed variables.

Transformed variable	AC
Diversity of income sources	0.3659
Coping strategy index	0.7551
Capacity to keep up in the future	0.7800

- *Number of household members to have lost jobs* (count). This is the number to have lost their employment in the last six months.
- *Income stability* (increased; the same; decreased). This is measured by income variation over the last six months.
- *Assistance dependency* (ratio from 0 to 1). This is the ratio of the monetary value of assistance to total income.
- *Assistance stability* (increased; the same; decreased). This is the variation in the quality of assistance over the last six months.
- *Health stability* (count from 1 to 8). This is measured by the number of institutions providing medical care.
- *Education system stability* (increased; the same; decreased). This is the variation in the quality of education over the last six months.

The variables have mixed measurement scales, so the PRINCALS algorithm is used to generate the *S* index. The correlations in Table 21.6 show that professional skills, educational level and employment ratio play the major roles in estimating *S*. Obviously, the correlations with the number of household members to lose their jobs and assistance dependency are negative: stability decreases as dependency on assistance increases, and as more household members lose their jobs. More surprising is the negative correlation with stability of the education system. This is because households with higher educational levels perceive a greater worsening of the education system than those with lower educational levels.

Table 21.6 Correlation of *S* with transformed variables.

Transformed variable	S
Professional skills	0.7234
Educational level	0.7930
Employment ratio	0.6786
Household members to lose jobs	−0.0609
Income stability	0.2112
Assistance dependency	−0.3723
Assistance stability	0.3116
Health stability	0.1198
Education system stability	−0.1315

Estimation of resilience (R)

The variables estimated above become covariates in estimation of the resilience index. Considering that all the estimated components are normally distributed with mean 0 and variance 1, it is easy to apply traditional factor or principal components analysis.[15] A factor analysis is run using the iterated principal factor method, which re-estimates communalities iteratively.[16]

The results obtained are very satisfactory. Table 21.7 shows that factor 1 alone explains more then 75% of the variance, with factors 2 and 3 accounting for 16% and 8%, respectively. Table 21.8 presents some interesting results in terms of interpretation, especially of the first two factors. The first factor seems to represent fairly well the household's level of well-being. Of the resilience building blocks, only *SSN* is not positively related to the first factor, because it is negatively correlated with the other variables. This is obvious given that social cohesion usually increases as households become poorer. Social safety nets are a positive feature of resilience, however, so they are captured in the second factor, where *SSN* becomes positive. Even adaptive capacity (AC) assumes a positive value in the second factor. It can be imagined that when a household becomes poorer it acquires adaptive capacities that it did not have when it was richer (e.g., household members consider doing lower-level jobs than they would have done before). This second factor probably captures mechanisms – mainly non-economic and informal – that are triggered when a household is under stress (i.e. when it is poorer and in need, income

Table 21.7 Eigenvalues and σ^2 explained.

Factors	Eigenvalue	% of σ^2
Factor 1	1.51766	0.7529
Factor 2	0.32620	0.1618
Factor 3	0.15795	0.0784
Factor 4	0.01399	0.0069
Factor 5	−0.00018	−0.0001

Table 21.8 Factor loadings.

Var.	Factor 1	Factor 2	Factor 3
IFA	0.8077	−0.0089	0.1396
AC	0.6039	0.3130	−0.1380
S	0.5753	−0.1407	−0.2049
APS	0.2808	0.1146	0.2759
SSN	−0.3011	0.4419	−0.0363

[15] It is assumed that all variability in an item should be used in the principal components analysis, while only the variability in an item that is common to other items is used in factor analysis. In most cases, these two methods yield very similar results, but principal components analysis is often preferred as a method for data reduction, and principal factors analysis for cases when the goal of the analysis is to detect the structure in the data.

[16] Communality is the proportion of the variance of a particular item that is due to common factors (i.e. shared among several items).

and stability become negatively correlated). The third factor is rather more difficult to interpret. It seems to have something in common with the concept of utilization in food security because it is positively related to public services and level of income.

The factor analysis shows that resilience cannot be a one-dimensional concept. As explained, the positiveness of *SSN* is not captured by the first factor but by the second. Even if the first factor explains more then 75% of the variance, the other two factors have to be included in the measurement of resilience. It is possible to have a weighted sum of three scored factors because they are orthogonal to each other, so the risk of multicollinearity is avoided. To estimate resilience, the three factors must therefore first be generated using Thompson's regression method and then each must be multiplied by its own proportion of variance explained:

$$Resilience = 0.753 \times Factor1 + 0.162 \times Factor2 + 0.078 \times Factor3.$$

Factor analysis produces a resilience index that is normally distributed with mean 0 and variance depending on the variance and covariance of the scored factors. This is a common feature of factor and principal components analysis, and is a limitation when resilience indices are to be compared among different countries or, more generally, different samples. Nevertheless, the role of the resilience index becomes crucial for policy-makers seeking to analyse subsamples of the original population, such as regions or social groups.

Discussion of results

This section concludes by presenting some of the estimates of the resilience index and its components in the five subregions of Palestine. Figure 21.3 shows the Epanechnikov

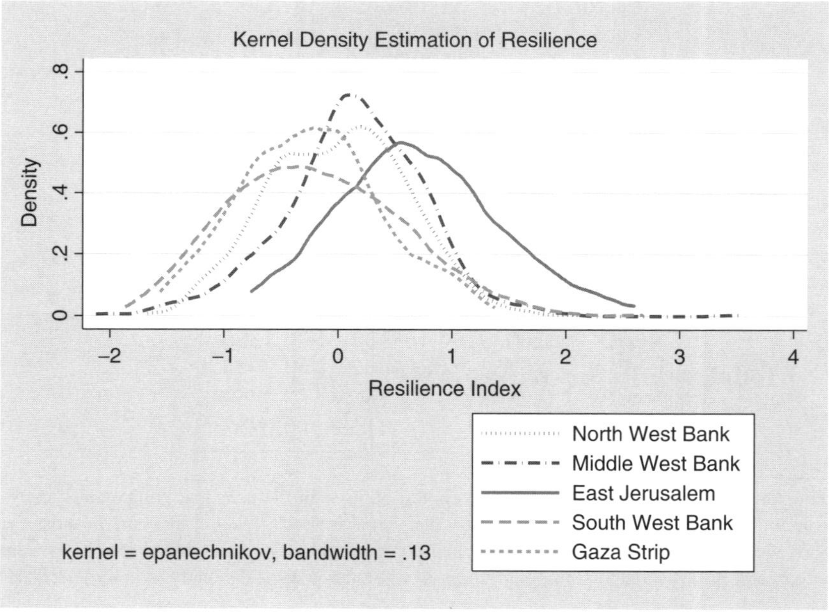

Figure 21.3 Distribution of resilience in the five Palestinian subregions.

kernel density estimates of the resilience distribution. The initial presentation of the results is based on a non-parametric method, because such methods are normally more informative.

Figure 21.3 shows a clear difference in resilience distribution between East Jerusalem[17] and the other subregions. However, South West Bank (WB) and Gaza Strip seem to have more or less the same resilience level, so the parametric approach is used to improve understanding of the differences among subregions and to obtain the relevant significance level. Table 21.9 shows the means and standard deviations (SDs) for resilience and its standardized components. The matrix in Table 21.10 shows the t statistics for pairwise comparison among the means of the different subregions.

The differences among subregional resilience levels are all significant except for that between Gaza Strip and South West Bank. The level of resilience in Gaza Strip would have been much lower if social safety nets had been excluded from the analysis. Figure 21.4 shows the differences among subregions, not only for resilience but also for its different components.

Jerusalem has the highest level of resilience, which depends mainly on income capacity, access to services and stability. Adaptive capacity seems to remain relatively similar to that of other subregions, however. At the opposite extreme, Gaza Strip shows very limited access to food and income, and a high level of dependency on safety nets (from both family ties and external assistance) and on services provided by the Palestinian Authority. South West Bank, with a low level of resilience that is equally distributed among the five pillars, shows a substantially lower level of adaptive capacities. North and Mid-West Bank seem to have relatively balanced distributions across the five pillars and more stable structures of their resilience levels.

The analysis also increases understanding of the current situation in different areas of West Bank and Gaza Strip, particularly Gaza Strip's dependence on external aid. The availability of and access to food depend substantially on the capacity of social safety net mechanisms, both traditional household-based ones and those provided by the international community. Physical access to markets also has a major influence on the overall capacity of Gaza Strip households to bounce back after acute crises.

The study of resilience according to its different components is crucial, especially for policy decisions. A more detailed analysis will therefore be conducted in the following section to explore the policy implications.

21.5 Testing resilience measurement

21.5.1 Model validation with CART

A cross-validation process is used to assess whether the procedures adopted for estimating the resilience index are meaningful. The cross-validation process tests the original hypothesis that sets of different variables and indicators belonging to different dimensions of food insecurity, the social sector and public services are correlated with (i.e. contribute to) the overall resilience index. The CART methodology (see Steinberg and Colla, 1995; Breiman *et al.*, 1984) is used to estimate the resilience decision tree and related splitting

[17] Villages outside the wall of Jerusalem are included in Mid-West Bank.

Table 21.9 Means and standard deviations for resilience and its components.

	Freq.	Resilience		IFA		APS		SSN		AC		S	
Region	**N**	**Mean**	**SD**	**Mean**	**SD**	**Mean**	**SD**	**Mean**	**SD**	**Mean**	**SD**	**Mean**	**SD**
North WB	648	0.0045	0.5987	−0.021	1.264	−0.240	0.966	0.017	1.017	0.106	1.020	0.054	0.961
Mid-WB	614	0.1436	0.6165	0.434	1.389	0.058	0.936	−0.19	0.830	0.077	0.980	−0.036	0.981
Jerusalem	93	0.7055	0.6931	1.880	1.509	0.951	1.021	−1.014	0.395	0.119	1.070	0.452	1.226
South WB	408	−0.1922	0.7540	−0.370	1.625	0.080	1.089	−0.143	0.992	−0.372	0.930	−0.053	1.079
Gaza Strip	324	−0.2415	0.609	−0.854	1.377	0	0.862	0.797	0.874	0.075	0.940	−0.104	0.898
Total	**2087**	**0**	**0.676**	**0**	**1.525**	**0**	**1**	**0**	**1**	**0**	**1**	**0**	**1**

Table 21.10 Mean differences across subregions (t-statistic tests: * not significant).

	North WB	Mid-WB	Jerusalem	South WB	Gaza Strip
North WB	0				
Mid-WB	−4.066	0			
Jerusalem	−10.343	−8.0543	0		
South WB	4.693	7.7906	10.5135	0	
Gaza Strip	6.004	9.1352	12.8082	0.9563*	0

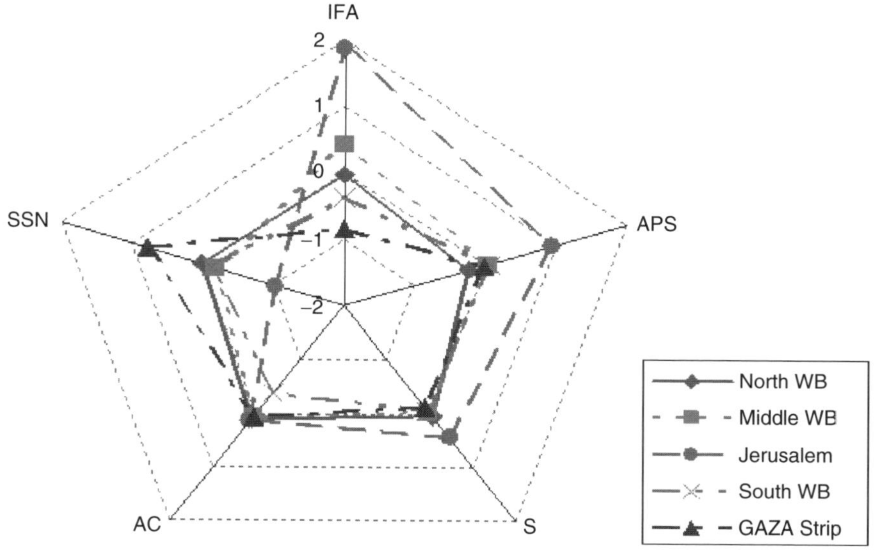

Figure 21.4 Radar graph for the components of resilience.

rules. The greatest advantage of CART is its cross-validation procedure, which allows the measurement of any errors in the model estimation.[18]

The target variable for the model implemented with CART is the resilience index. Because this is a continuous variable, CART produces a regression tree. (When the target variable is categorical, CART produces a classification tree.) As predictors, the model includes all the original variables used in the empirical approach. The weights deriving from the sample design are also considered. The optimal tree has 281 terminal nodes, and a relative cost (error) equal to 0.123. The approximate R^2 can be calculated using the formula: $1 - \textit{resubstitution error}$ (i.e. $1 - 0.019 = 0.981$). This is very important in confirming that the variables used in the multi-stage approach explain most of the resilience variability when a different estimation method is used (i.e. a non-parametric instead of a parametric one). The CART procedure includes use of the Gini splitting criterion and tenfold cross-validation for testing.[19]

[18] Other advantages of using CART are: it is a robust non-parametric tool; it has the capacity to handle complex data structures; it does not require probability density function assumptions; it works regardless of heteroscedasticity and multicollinearity; its testing procedures are more accurate; it has the capacity to deal with missing values; and decision rules can be transferred to new observations.

[19] The cross-validation test takes place over ten subsamples from the learning sample.

Table 21.11 Variables and their importance.

Code	Description	Importance	Code	Description	Importance
IFA1	Income	100.000	*SSN4*	Monetary values of 1st and 2nd type	2.069
IFA3	HFIAS	91.543	*APS3*	Education system	1.854
IFA2	DD	86.325	*SSN2*	Quality of assistance	1.458
AC2	Coping strategies	72.959	*S6*	Assistance dependency	1.450
AC3	Capacity to keep up in the future	65.945	*S9*	Education system stability	1.241
S3E	Employment ratio	53.730	*APS4*	Perception of security	1.210
S2	Educational level	15.162	*S7*	Assistance stability	1.168
SSN5	Evaluation of main assistance	5.315	*APS1*	Physical access to health	1.161
S1W	Professional skills	4.502	*SSN6*	Frequency of assistance	0.917
APS2	Health service quality	3.770	*APS5*	Mobility constraints	0.689
AC1	Diversity of income sources	3.590	*S5*	Income stability	0.623
SSN1	Cash and in-kind assistance	3.304	*SSN7*	Opinion on targeting	0.609
S8	Health stability	2.670	*S4*	No. of household members to lose jobs	0.194
APS7	Water, electricity and telecoms	2.078	*SSN3*	Employment assistance	0.070

The ranking of the variables according to their importance is shown in Table 21.11, and explains the role of each variable in defining resilience. This ranking takes into consideration main splitters, competitors and surrogates.[20]

In summary, the main advantage of CART is its capacity to capture variables that are relevant for specific subgroups of the population, which an ordinary least squares (OLS) estimator does not consider relevant for the whole population.

Moreover, as the approximated R^2 is very high, the decision rules established by the regression tree can be transferred to future applications on the same population to make predictions. The ranking by importance of variables makes it possible to return to the earlier steps of the analysis to build up more parsimonious models and select the most important variables for making predictions. In the next subsection, a regression model for forecasting resilience is built up.

21.5.2 The role of resilience in measuring vulnerability

In most studies of poverty, vulnerability indicators consider the probability distribution of household consumption as their objective (Dercon, 2001). They consider consumption as a stochastic variable, try to estimate the deterministic part of it through regression models, and then calculate the probability of falling below a certain threshold, usually the poverty line or a proxy for food security. Other theoretical studies consider vulnerability as a function of people's exposure to risks and their resilience to these (Løvendal *et al.*, 2004). This subsection assesses the role of the estimated resilience index in measuring vulnerability. As vulnerability to food insecurity is the issue, the study focuses on food consumption. The following regression model can be built, with food consumption on the logarithmic scale as a dependent variable, and resilience and other household characteristics as independent:

$$LogFoodCons = \alpha + \beta_1 R + \beta_2 Hsize + \beta_3 gender + \beta_r \overline{regions} + \epsilon$$

The aim of this is to construct a parsimonious model, from which all the variables previously used to estimate resilience are excluded. Subregions are included in the model through dummies. To avoid collinearity, one of the subregions, North West Bank, is left out because it has a medium level of consumption. The household characteristics included in the model are household size (number of household members) and a dummy variable for the gender of the head of household. The OLS estimates are shown in Table 21.12.

Resilience is the most important regressor and the variance explained by the model ($R^2 = 0.528$) is quite satisfactory. The OLS estimator produces consistent estimates and the disturbances are homoscedastic. The Breusch–Pagan test is used to detect heteroscedasticity, and produces a χ^2 equal to 0.50 (p-value 0.48), so the hypothesis of homoscedasticity is accepted.

The regression model in Table 21.12 shows the relevance of resilience as a key indicator in food security and vulnerability analysis. A unit increase of the level of resilience implies an increase of 0.381 in the logarithmic scale of total food consumption. The opposite is observed for household size, where a unit increase implies a decrease of 0.084. The coefficient of gender of household head confirms the argument that female-headed households are more food secure. In fact, female-headed households have a food

[20] When a variable is considered a competitor, CART finds the best split that reduces node heterogeneity; when the variable is considered a surrogate, it is constrained in mimicking the primary split.

Table 21.12 Regression of total food consumption.

OLS estimates: log total food consumption	Coefficients
Resilience	0.381 (0.000)**
Household size	−0.084 (0.000)**
Gender of head of household (dummy; female = 1)	0.112 (0.037)*
Mid-West Bank (dummy)	0.128 (0.001)**
East Jerusalem (dummy)	0.465 (0.000)**
South West Bank (dummy)	0.053 (0.199)
Gaza Strip (dummy)	−0.327 (0.000)**
Constant	2.281 (0.000)**
Observations	2087
R^2	0.528

Robust p-values in parentheses: *significant at 5%; **significant at 1%.

consumption level that is 0.112 logarithmic scale units higher then that of male-headed households. Interpretation of subregional coefficients is more complex. Positive values mean the level of consumption is higher than in the omitted subregion (North West Bank), and negative values mean a lower level of consumption. In fact, East Jerusalem and Mid-West Bank, where conditions are better, have positive values, while Gaza Strip, where conditions are very poor, has a negative value. South West Bank also has a positive coefficient, but it is not statistically significant.

21.5.3 Forecasting resilience

The CART output shown in Table 21.11 is very important in identifying the most important variables for defining resilience. The multivariate models applied so far have a major limitation: they always produce normalized indicators with mean 0, so it is difficult to compare the level of resilience over time. To make such comparison feasible, this subsection presents an OLS regression model, constructed using resilience as a dependent variable and the variables with a CART importance index of more than 15:

$$R = \alpha + \beta_1 IFA1 + \beta_2 IFA2 + \beta_3 IFA3 + \beta_4 AC2 + \beta_5 AC3 + \beta_6 S2 + \beta_7 S3 + \epsilon. \quad (21.2)$$

The error term ϵ represents the negligible information of the variables used to estimate resilience but excluded from the regression model in (21.2). Changes to the variables on the left-hand side imply changes to the level of resilience. This allows comparison of the resilience level over time.

Table 21.13 Regression model of resilience.

OLS estimates: resilience	Coefficients
Income (IFA1)	0.011
	(0.000)*
DD (IFA2)	0.011
	(0.000)*
HFIAS (IFA3)	−0.031
	(0.000)*
Coping strategies (AC2)	0.029
	(0.000)*
Capacity to keep up in the future (AC3)	0.104
	(0.000)*
Employment ratio (S3)	0.382
	(0.000)*
Educational level (S2)	0.025
	(0.000)*
Constant	−1.466
	(0.000)*
Observations	2087
R-squared	0.967

* = p-value significant at 1%.

Only seven of the 28 original variables are used in the forecasting model, and these variables explain almost 97% of the variance ($R^2 = 0.967$); see Table 21.13. The Breusch–Pagan test produces a χ^2 of 3.07 (p-value 0.08), so the null hypothesis (constant variance) is accepted. Multicollinearity is a concern in this model. The variance inflation factor (VIF)[21] is used to detect collinearity. As a rule of thumb, multicollinearity exists when individual VIF is greater than 10 or mean VIF is greater than 6. In our case, individual VIFs and the mean were all very low,[22] so the hypothesis of multicollinearity was rejected.

This model also increases understanding of the impact that right-hand-side variables have on the household's resilience, and is an important instrument for decision-making. Elasticities or semi-elasticities can be studied to identify the most appropriate policy interventions.

21.6 Conclusions

Vulnerability to food insecurity depends on a household's risk exposure and resilience to such risks. Most risks are unpredictable, which makes it difficult to measure vulnerability. This study therefore focuses on household resilience to food insecurity, which is defined as a household's ability to maintain a certain level of well-being (food security) in the face of risks, depending on the options available to that household to make a

[21] $VIF = 1 - R_k^2$, where R_k^2 is the partial R^2 of the variable k on all other variables in the model.

[22] The VIF of *HFIAS* was the highest (1.4) and the VIF was *DD* the lowest (1.17). The average VIF was 1.26.

living and its ability to handle risks. The study tests a conceptual framework that was developed for this purpose. Analysis of the 11th PPPS seems to confirm the validity of the conceptual framework adopted. The results are meaningful and the resilience indices in the five subregions show significant differences, as do the five components of the resilience model. Specifically, the overall structure of resilience is very different in each of the five subregions. The analysis shows that while in the richest and more stable area (East Jerusalem), most resilience depends on income and food access capacity, in Gaza Strip most depends on social safety nets. However, this analysis has limitations owing to the static nature of the available database. Such analyses should be carried out with panel data, as soon as panel databases become available in the future. It would also be interesting to extend the analysis to other case studies to assess the robustness of the proposed analytical framework and any emerging patterns of resilience.

Although the methodology adopted gives significant results, it is necessary to test the alternative methodology proposed in this research – the structural equation models with Bayesian networks – before making a final decision on which is more appropriate. Further work is also necessary on how to use the resilience index to identify the key determinants needed for designing adequate food insecurity responses and policies, and for strengthening households' economic resilience in crisis situations.

References

Abuelhaj, T. (2008) Food security in the Occupied Palestinian Territory. FAO, Rome, and Geneva University.

Adger, N.W. (2000) Social and ecological resilience: are they related? *Progress in Human Geography*, 24, 347–364.

Adger, N.W. (2006) Vulnerability. *Global Environmental Change*, 16, 268–281.

Ahmad, Q.K., Warrick, R.A., Downing, T.E., Nisioka, S., Parikh, K.S., Parmesan, C., Schneider, S.H., Toth, F. and Yohe, G. (2001) Methods and tools. In J. McCarthy, F.O. Canziani, N.A. Leary, D.J. Dokken and K.S. White (eds), *Climate Change 2001: Impacts, Adaptation and Vulnerability*. Cambridge: Cambridge University Press.

Arminger, G. and Muthén, B.O. (1998) A Bayesian approach to nonlinear latent variable models using the Gibbs sampler and the Metropolis–Hastings algorithm. *Psychometrika*, 63, 271–300.

Arthur, W.B. (1988) Self-reinforcing mechanisms in economics. In P. Anderson, K. Arrow and D. Pines (eds), *The Economy as an Evolving Complex System*, pp. 9–32. Redwood City, CA: Addison-Wesley.

Bartholomew, D.J. and Knott, M. (1999) *Latent Variable Models and Factor Analysis*. London: Arnold.

Bartlett, M.S. (1937) The statistical conception of mental factors. *British Journal of Psychology*, 28, 97–104.

Bollen, K.A. (1989) *Structural Equations with Latent Variables*. New York: John Wiley & Sons, Inc.

Breiman, L., Friedman, J., Olshen, R. and Stone, C. (1984) *Classification and Regression Trees*. Belmont, CA: Wadsworth International Group.

Buchanan-Smith, M. and Davies, S. (1995) *Famine Early Warning and Response: The Missing Link*. London: Intermediate Technology.

Chaudhuri, S. (2004) Incidence of child labour, free education policy, and economic liberalisation in a developing economy. *Pakistan Development Review*, 43, 1–25.

Chaudhuri, S., Jalan, J. and Suryahadi, A. (2002) Assessing household vulnerability to poverty: a methodology and estimates for Indonesia. Discussion Paper No. 0102-52, Department of Economics, Columbia University.

Clark, N. and Juma, C. (1987) *Long-Run Economics: An Evolutionary Approach to Economic Growth*. London: Pinter Publishers.

Coates J., Swindale A. and Bilinsky P. (2006) *Household Food Insecurity Access Scale (HFIAS) for Measurement of Food Access: Indicator Guide*. Washington, DC: USAID, FANTA.

David, P.A. (1985) Clio and the economics of QWERTY. *American Economic Review Proceedings*, 75, 332–336.

De Leeuw, J. and Van Rijckevorsel, J. (1980) HOMALS and PRINCALS: Some generalizations of principal components analysis. In E. Diday *et al.* (eds), *Data Analysis and Informatics*, pp. 231–241. Amsterdam: North-Holland.

Dercon, S. (2001) Assessing vulnerability to poverty. Report prepared for the Department for International Development (DFID). www.economics.ox.ac.uk/members/stefan.dercon/Vulnerability%20-to%20Poverty note%201 %20framework.pdf.

FIVIMS/FAO (2002) *FIVIMS Tools and Tips: Understanding Food Insecurity and Vulnerability*. Rome: FAO.

Folke, C. (2006) Resilience: the emergence of a perspective for social-ecological systems analyses. *Global Environmental Change*, 16, 253–267.

Folke, C., Berkes, F. and Colding, J. (1998) Ecological practices and social mechanism for building resilience and sustainability. In F. Berkes and C. Folke (eds), *Linking Social and Ecological Systems*, pp. 414–436. Cambridge: Cambridge University Press.

Folke, C., Carpenter, S., Elmqvist, T., Gunderson, L., Holling, C.S. and Walker, B. (2002) Resilience and sustainable development: building adaptive capacity in a world of transformations. *Ambio*, 31, 437–440.

Gunderson, L.H., Holling C.S., Peterson, G. and Pritchard, L. (1997) Resilience in ecosystems, institutions and societies. Beijer Discussion Paper No. 92, Beijer International Institute for Ecological Economics, Stockholm.

Hemrich, G. and Alinovi, L. (2004) Factoring food systems' resilience in the response to protracted crisis. In FAO, *The State of Food Insecurity in the World 2004*. Rome: FAO.

Hoddinott, J. and Yohannes, Y. (2002) *Dietary Diversity as a Household Food Security Indicator*. Washington, DC: Food and Nutrition Technical Assistance Project, Academy for Educational Development.

Holling, C.S. (1973) Resilience and stability of ecological systems. *Annual Review of Ecology and Systematics*, 4, 1–23.

Holling, C.S. (1986) Resilience of ecosystems; local surprise and global change. In W.C. Clark and R.E. Munn (eds), *Sustainable Development of the Biosphere*, pp. 292–317. Cambridge: Cambridge University Press.

Holling, C.S. (1996) Engineering resilience versus ecological resilience. In P.C. Schulze (ed.), *Engineering within Ecological Constraints*, pp. 31–44. Washington, DC: National Academy Press.

Levin, S.A., Barrett, S.A., Anyiar, S., Baumol, W., Bliss, C., Bolin, B., Dasgupta, P., Ehrlich, P., Folke, C., Gren, I., Holling, C.S., Janson, A., Janson, B., Mäler, K., Martin, D., Perrings, C. and Sheshinski, E. (1998) Resilience in natural and socioeconomic systems. *Environment and Development Economics*, 3, 222–235.

Ligon, E. and Schechter, L. (2004) *Evaluating different approaches to estimating vulnerability*. Social Protection Discussion Paper Series No. 0410. Washington, DC: Social Protection Unit, Human Development Network, World Bank.

Lopes, H. and West, M. (2003) Bayesian model assessment in factor analysis. *Statistica Sinica*, 14, 41–67.

Løvendal, C.R., Knowles, M. and Horii, N. (2004) Understanding vulnerability to food insecurity lessons from vulnerable livelihood profiling. ESA Working Paper No. 04-18, Agricultural and Development Economics Division, FAO, Rome.

Mane, E., Alinovi, L. and Sacco, E. (2007) Profiling food security in Palestinian governorates: Some classification techniques based on household-level data. Paper presented at the International Conference on Statistics and Development, Ramallah (Palestine), 27–28 March.

Mezzetti, M. and Billari, F.C. (2005) Bayesian correlated factor analysis of socio-demographic indicators. *Statistical Methods and Applications*, 14, 223–241.

Muthén, B.O. (1984) A general structural equation model with dichotomous, ordered categorical and continuous latent indicators. *Psychometrika*, 49, 115–132.

O'Neill, R.V., De Angelis, D.L., Waide, J.B. and Allen, T.F.H. (1986) *A Hierarchical Concept of Ecosystems*. Princeton, NJ: Princeton University Press.

Paine, R.T. (1974) Intertidal community structure: experimental studies on the relationship between a dominant competitor and its principal predator. *Oecologia*, 15, 93–120.

Pimm, S.L. (1984) The complexity and stability of ecosystems. *Nature*, 307, 321–326.

Rowe, D.B. (2003) *Multivariate Bayesian Statistics: Models for Source Separation and Signal Unmixing*. Boca Raton, FL: CRC Press.

Scaramozzino, P. (2006) Measuring vulnerability to food insecurity. ESA Working Paper No. 06-12, Agricultural and Development Economics Division, FAO, Rome.

Skrondal, A. and Rabe-Hesketh, S. (2004) *Generalized Latent Variable Modeling: Multilevel, Longitudinal, and Structural Equation Models*. Boca Raton, FL: CRC Press.

Spedding, C.R.W. (1988) *An Introduction to Agricultural Systems*, 2nd edition. London: Elsevier Applied Science.

Spirtes, P., Glymour, C. and Scheines, R. (2000) *Causation, Prediction and Search*. Cambridge, MA: MIT Press.

Steinberg, D. and Colla, P. (1995) *CART: Tree-Structured Non-parametric Data Analysis*, San Diego, CA: Salford Systems.

Thomson, G.H. (1951) *The Factorial Analysis of Human Ability*. London: University of London Press.

Tilman, D. and Downing, J.A. (1994) Biodiversity and stability in grasslands. *Nature*, 367, 363–365.

Walker, B.H., Ludwig, D., Holling, C.S. and Peterman R.M. (1969) Stability of semi-arid savanna grazing systems. *Ecology* 69, 473–498.

Walker, B.H., Holling, C.S., Carpenter, S.R. and Kinzig, A.P. (2004) Resilience, adaptability and transformability in social-ecological systems. *Ecology and Society*, 9, 5. www.ecologyandsociety.org/vol9/iss2/art5/.

World Bank (2001) *World Development Report 2000/01. Attacking Poverty*. New York: Oxford University Press.

22

Spatial prediction of agricultural crop yield

Paolo Postiglione[1], Roberto Benedetti[1] and Federica Piersimoni[2]

[1]*Department of Business, Statistical, Technological and Environmental Sciences, University 'G. d'Annunzio' of Chieti-Pescara, Italy*

[2]*Istat, National Institute of Statistics, Rome, Italy*

22.1 Introduction

Spatially dependent observations arise in a wide variety of applications, including archaeological and agricultural studies, the spread of epidemics, and economic and environmental analysis (Cressie, 1993). The hypothesis of independence among observations assumed by traditional statistical techniques is a very convenient one and often allows statistical theory to be more tractable. This conjecture is obviously violated in the present context, given that in all geographical and territorial studies, as stated in Tobler's (1970) first law of geography, 'everything is related to everything else, but near things are more related than distant things'.

To take into account this multidirectional dependence among nearest sites, a general class of well-established models has been introduced in the statistical literature. For earlier developments see, for example, the seminal papers by Besag (1974) and Strauss (1977); for a review of this topic see Cressie (1993). As a consequence of the introduction of the Markov random field (MRF: Besag, 1974) in the spatial data analysis context, a great effort has been devoted in recent decades to the analysis of spatial dependence and to the development of a large body of methods to estimate models under the assumption of

Agricultural Survey Methods Edited by Roberto Benedetti, Marco Bee, Giuseppe Espa and Federica Piersimoni

an MRF data-generating process. Dependence has thus become the conceptual basis for spatial data analysis.

Surprisingly, in the spatial literature we do not find a similar research effort in the analysis of the non-stationarity concept. Non-stationarity is a relevant, but not always fully recognized, characteristic of spatial data. The consideration that the relationship between a response variable and explanatory variables is not always constant across an area has given rise to methods allowing for spatially varying coefficients (Wheeler and Tiefelsdorf, 2005; Wheeler and Calder, 2007): for example, geographically weighted regression (Fotheringham *et al.*, 2002), Bayesian regression models with spatially varying coefficient (Gelfand *et al.*, 2003), and local linear regression models (Loader, 1999).

There are many reasons why we might expect estimated parameters to change over space (Fotheringham *et al.*, 2002). First, a possible cause of the presence of spatial non-stationarity may be related to sampling variation. If we consider spatial subsamples of a data set, it is reasonable to expect the estimated parameters to vary over space. Second, we may observe spatial non-stationarity because of possible misspecification of the model that describes the reality, due to their being some omitted variables. It might not be possible to remove this misspecification by adding some new variables in the global model: this information might be available only locally. Finally, spatial variations might exist in attitudes and preferences of the people who generate different behaviour over space in presence of the same input.

The main aim of this research is to produce a spatial prediction of agricultural crop yields for unsampled sites. Accurate and timely monitoring of agricultural crop conditions and estimating potential crop yield are essential processes for operational programmes. Assessment of particularly decreased production caused, for example, by a natural disaster, can be critical for countries where the economy is dependent on the crop harvest. Early assessment of yield reductions could avert a disastrous situation and help in strategic planning to meet demand. The timely evaluation of potential yield is increasingly important because of the huge economic impact of agricultural products on world markets and strategic planning.

From a statistical point of view, a prediction or interpolation problem can be summarized in the following main steps. Suppose that the number of units N in the finite population is known and that each unit is associated with a realization y_i. The general goal is to select a sample of units, observe the y_i and use these realizations to estimate this value for all the units of population via some function.

Now consider the units as geographically distributed. So, the spatial interpolation is the procedure of estimating the value of properties at unsampled sites within an area where observations exist. In agricultural surveys, data are available at specific spatial locations and the goal is to predict the values of the variable of interest in unknown locations. The unknown locations are often mapped on a grid, and the predictions are used to produce surface plots or contour maps. Note that in the present research the term 'prediction' is used in an unusual way, in the sense of making a statistical estimate on the y_i that are not known; not in the literal sense of forecasting future values as, for example, in the field of time series analysis.

The methods of spatial interpolation can be classified into two different groups: areal based interpolation and point based interpolation. Areal interpolation can be defined as 'the transfer of data from one set (source units) to a second set (target units) of overlapping, non-hierarchical, areal units' (Langford *et al.*, 1991, p. 56). Although the importance of areal interpolation has been widely recognized (Fotheringham and Rogerson, 1993), no single technique can be identified as better than the others.

On the other hand, point-based interpolation techniques have been developed within a new independent discipline known as geostatistics (Cressie, 1993). Geostatistics is a collection of statistical methods which were traditionally used in geo-sciences that predict some attribute of variables under investigation on the entire spatial domain using the information available in some sampled points. Originally, the term geostatistics was coined by Georges Matheron and his colleagues of the Centre de Géostatistique et de Morphologie Mathématique at Fontainebleau (France) to describe their work addressing problems of spatial prediction arising in the mining industry (Matheron, 1963, 1971). Finally, we note that many of the spatial interpolation techniques are two-dimensional extensions of the one-dimensional methods originally developed for time series analysis.

One of the simplest and most popular spatial prediction methods is kriging (see Wackernagel, 1995, for an introduction). The aim of kriging is to estimate the value of an unknown real-valued function through a best linear unbiased estimator, based on a stochastic model of the spatial dependence measured by the variogram, the expectation, and the covariance function of the random field (Cressie, 1993). However, classical kriging computes the prediction by using only an estimation function for the entire spatial domain.

Also in standard regression theory, spatial units have been traditionally represented as identical members of the same population:

$$y_i = f(\mathbf{x}_i; \boldsymbol{\beta}) + \epsilon_i, \tag{22.1}$$

where y_i is the dependent response variable, $f(.; \boldsymbol{\beta})$ is the general regression function, $\mathbf{x}_i = (x_{i1}, x_{i2}, \ldots, x_{ih})$ is the explanatory variables vector of dimension h, $\boldsymbol{\beta}$ is the parameter vector, $\epsilon_i \sim N(0, \sigma_\epsilon^2)$ is the error term, and $i = 1, 2, \ldots, n$ are the geographical zones under investigation.

Statistical units within a geographical domain can be assumed as spatially dependent by introducing an autoregressive term for the response variable y_i and/or by assuming the error terms matrix $\boldsymbol{\epsilon} \sim N(\mathbf{0}, \boldsymbol{\Sigma}_\epsilon)$, where $\boldsymbol{\Sigma}_\epsilon$ is any positive definite covariance matrix. In order to take into account the non-stationarity of the spatial parameters, model (22.1) can be modified as

$$y_i = f(\mathbf{x}_i; \boldsymbol{\beta}_{k_i}) + \epsilon_i,$$

where $\boldsymbol{\beta}_{k_i}$ are the locally stationary parameters vectors, ϵ_i are the error terms and $\mathbf{k} = (k_1, k_2, \ldots, k_i, \ldots, k_n) \in \mathcal{K}^n$ is the label vector representing the local stationarity zone related to each single unit i, $k_i \in \{1, 2, \ldots, K\}$, with K non-stationarity zones.

From a theoretical point of view, the above specification of non-stationary regression parameters leads to the study of a non-trivial combinatorial optimization problem. In fact, provided that the number K of local stationarity zones is fixed, the aim is to minimize an objective function (i.e. the classical sum of squares of the regression residuals) defined over the entire set of possible partitions of the study area. Simulated annealing represents a suitable solution for the above-described problem by identifying the zone of non-stationarity. Finally, our proposed approach uses this zone in order to make spatial predictions that consider the local stationarity highlighted by spatial data.

The layout of the chapter is as follows. Section 22.2 describes the proposed approach based on the simulated annealing algorithm. Furthermore, we test the algorithm on simulated data. In Section 22.3 we describe our method by applying it to spatial prediction of the durum wheat yield in 2005 in a province of southern Italy: Foggia. Finally, in Section 22.4 we make some concluding remarks and outline a future research agenda.

22.2 The proposed approach

Our spatial prediction technique is based on the simulated annealing (SA) algorithm. SA is a method to approximate the solution of difficult combinational optimization problems.

The publication of an article in *Science* by Kirkpatrick *et al.* (1983), who reported results based on sketchy experiments, had a remarkable impact on the optimization community. Cerny (1985) developed the same method independently. Since its formulation, the SA method has been applied to various different scientific fields, among them the design of electronic circuits, image processing and the collection of household garbage. It has become one of the best-known and most often used optimization strategies to solve complex combinatorial problem.

SA is a generalization of a Monte Carlo method (Metropolis *et al.*, 1953) and is based on the analogy between the simulation of annealing of solids and an optimization problem. The central idea of SA can be illustrated as follows. A material is heated by giving high energy to it. Then the material is slowly cooled so that the system is approximately in thermodynamic equilibrium. If the decrease in temperature is too fast, the material can evidence some defects that can be eliminated by local reheating. This strategy leads to a crystalized solid and stable state corresponding to an absolute minimum of energy. If the initial temperature of the process is too low or cooling is done insufficiently slowly the system may become unstable or, in other words, trapped in a local minimum energy state. Thus, the temperature becomes the most important parameter of the strategy.

The generalization of this approach to combinatorial problems is straightforward (Kirkpatrick *et al.*, 1983). The current state of the thermodynamic system is analogous to the current solution of the combinatorial problem, the energy equation is analogous to the objective function, and the solid state is analogous to the global minimum. Obviously, in this simulated method a control parameter needs to be introduced that has the same role as the temperature in real annealing.

The most important difference with the other conventional algorithms is that SA explores the function's entire surface and tries to carry on the optimization process by moving both uphill and downhill. In this way, it can avoid local optima and go to the global optimum of the objective function. Also, it is largely independent of the choice of the starting point that represents a critical issue in the descending optimization algorithm. In addition, SA lets us work with complex objective functions, including non-linear and discontinue functions. Finally, it has been demonstrated that by choosing a large initial temperature and an appropriate temperature schedule, SA reaches a globally optimum solution. For further details, see Geman and Geman (1984) and van Laarhoven and Arts (1987).

More formally, in the approach of Geman *et al.* (1990), a spatial combinatorial optimization problem can be viewed as an MRF that is a stochastic process first introduced in theory of probability by Dobrushin (1968). This class of processes is neither homogeneous nor isotropic (Geman and Geman, 1984) and is described, adopting the Gibbs representation, through the probability measure as

$$\pi(\mathbf{y}, \mathbf{x}, \mathbf{k}) = \frac{\exp\{-U(\mathbf{y}, \mathbf{x}, \mathbf{k})/T\}}{\sum_{\mathbf{k}' \in \mathcal{K}^n} \exp\{-U(\mathbf{y}, \mathbf{x}, \mathbf{k})/T\}}, \tag{22.2}$$

where $\mathbf{y}$ and $\mathbf{x}$ are the observed data, $\mathbf{k}$ is the label vector of non-stationarity zones, $U(\mathbf{y}, \mathbf{x}, \mathbf{k})$ is the energy function that plays a fundamental role in the optimization procedure, and T is a control parameter called temperature. It can be established that $\lim_{T\to 0} \pi(\mathbf{x}, \mathbf{y}, \mathbf{k}) \equiv \pi_{\text{lim}}(\mathbf{x}, \mathbf{y}, \mathbf{k})$, where π_{lim} is uniform on the points which minimize the energy $U(\mathbf{y}, \mathbf{x}, \mathbf{k})$ and $\lim_{T\to 0} \pi(\mathbf{x}, \mathbf{y}, \mathbf{k}) = 0$ elsewhere. The denominator of equation (22.2) is called the normalization constant of the model.

The energy function $U(\mathbf{y}, \mathbf{x}, \mathbf{k})$ of an MRF is expressed as the sum of two terms: an interaction term $I(\mathbf{y}, \mathbf{x}, \mathbf{k})$, which depends on the observed data and the labels, and a penalty term $V(\mathbf{k})$, which depends only on $\mathbf{k}$. The interaction between the labels and the data is represented by a distance measure, while the penalty function is useful to discourage forbidden configurations.

So far, at the jth iteration, the energy function is defined as

$$U_j(\mathbf{y}, \mathbf{x}, \mathbf{k}) = I_j(\mathbf{y}, \mathbf{x}, \mathbf{k}) + V_j(\mathbf{k}),$$

while the interaction term is given by

$$I_j(\mathbf{y}, \mathbf{x}, \mathbf{k}) = \sum_{i=1}^{n} e_{i,j}^2,$$

where $e_{i,j} = y_i - f(\mathbf{x}_i; \widehat{\boldsymbol{\beta}}_{k_i,j})$ are the squares of the distance among the observations and the estimated regression function at the jth iteration.

The penalty term is useful to formalize the adjacency constraint for which, following Sebastiani (2003), we adopt a Potts model:

$$V_j(\mathbf{k}) = -\lambda \sum_{s=1}^{n} \sum_{v=1}^{n} \mathbf{c}_{sv} \mathbf{1}_{(k(j)_s = k(j)_v)},$$

where $\mathbf{c}_{sv}$ is the (s, v)th element of a binary contiguity matrix, $\mathbf{1}_{(k(j)_s = k(j)_v)}$ is the indicator function of the event $k(j)_s = k(j)_v$, and $\lambda > 0$ is a parameter to be estimated. Note that the sum is taken over all pairs of units s and v. Therefore, in probabilistic terms the Potts model favours configurations with high residuals and contiguous zone.

Let us define $U(S)$ as the objective function observed in any possible label configuration $S \in \mathcal{K}^n$, where $\mathcal{K}^n$ represent all the possible configurations. The classical algorithm uses a logic in which given a configuration S_j at the jth iteration, another configuration, say S_{j+1}, is chosen according to a visiting schedule, with an exchange of status if and only if $U(S_{j+1}) < U(S_j)$. In our algorithm each spatial unit i is selected in turn and then a label randomly chosen from the set $k_{i,j+1} \in \{1, 2, \ldots, K\}$ is assigned. For any suitable choice of the stopping criterion, the final configuration is therefore obtained. Generally, this result represents a local minimum and not necessarily a global one. In order to avoid entrapment in local minima, it is necessary to define positive probability for the change of configuration even when an increase in the objective function $U(S)$ is obtained.

Therefore, given a starting configuration S_0, the algorithm replaces the solution obtained at the jth iteration S_j with a new solution S_{j+1} chosen randomly with probability

$$p = \begin{cases} 1 & \text{if } U\left(S_{j+1}\right) < U\left(S_j\right), \\ \exp\left(-\frac{U(S_{j+1})-U(S_j)}{T_j}\right) & \text{otherwise.} \end{cases}$$

This means that a move from a configuration S_j to a worse configuration S_{j+1} is allowed with a probability that depends on the value of the parameter $T(j)$, where $T(j)$ indicates the temperature at the jth step of the procedure. It is clear that larger values of the parameter T are expected to move the solution away from a local optimum, while smaller values of T decreasing very slowly to zero, in analogy with the metallurgical process, are certainly appropriate to reach a global optimum. We used a classical approach to let $T(j) = T(0)\rho^{j-1}$ for some updating constant $\rho < 1$ (Givens and Hoeting, 2005).

This version of SA allows us to identify zones of non-stationarity or, in other words, to classify geographical sites in K groups that are similar in their regression parameters. These zones represent the basis for the subsequent spatial prediction process. In the next subsection, in order to validate our algorithm a simulated exercise is presented.

22.2.1 A simulated exercise

The proposed algorithm has been applied to simulated data to test its performances. The first stage of the simulation experiment consisted in generating 100 bivariate observations from a variable (X, Y) on a 10×10 regular lattice, with X uniformly distributed $U(0,1)$ and Y generated according to

$$Y_i = a + bX_i + \epsilon_i,$$

with $\epsilon_i \sim N(0, \sigma_\epsilon = 1/\sigma)$, considered henceforth as the true error distribution. In more detail, we obtain different observations that are classified in a certain number K of groups, by using different values of the parameters a and b. The objective of the proposed algorithm is to identify these K different groups of observations, by recognizing the zones of non-stationarity that have been generated in the first phase. Note that the approach followed in this chapter is similar to that used in the analysis of remote sensed images.

First, we have chosen to generate four groups ($K = 4$) of bivariate observations for different values of the regression parameters ($a = 0, b = 1; a = 0.75, b = -0.5; a = 0.25, b = 0.5; a = 1, b = -1$) of the variable Y and of the standard deviation of the error σ_ϵ (see Figure 22.1(a), (c), (e) and (g)). Looking at these figures, we can observe a pattern of points that globally shows an uncorrelated trend.

The results of our SA algorithm are reported in Figure 22.1(b), (d), (f) and (h). It is clear how a pair of variables apparently uncorrelated (see Figure 22.1(a), (c), (e) and (g)) could be efficiently modelled if the concept of non-stationarity is introduced. We obtain an efficiency that, if measured in terms of errors of identification of the true zone, is of course proportional to the standard deviation of the errors introduced in each regression. In fact, there are some units that are not classified in the correct way with $\sigma_\epsilon = 1/5$ (see Figure 22.1(b), misclassification rates equal to 0.26), while we do not have any errors when σ_ϵ decreases.

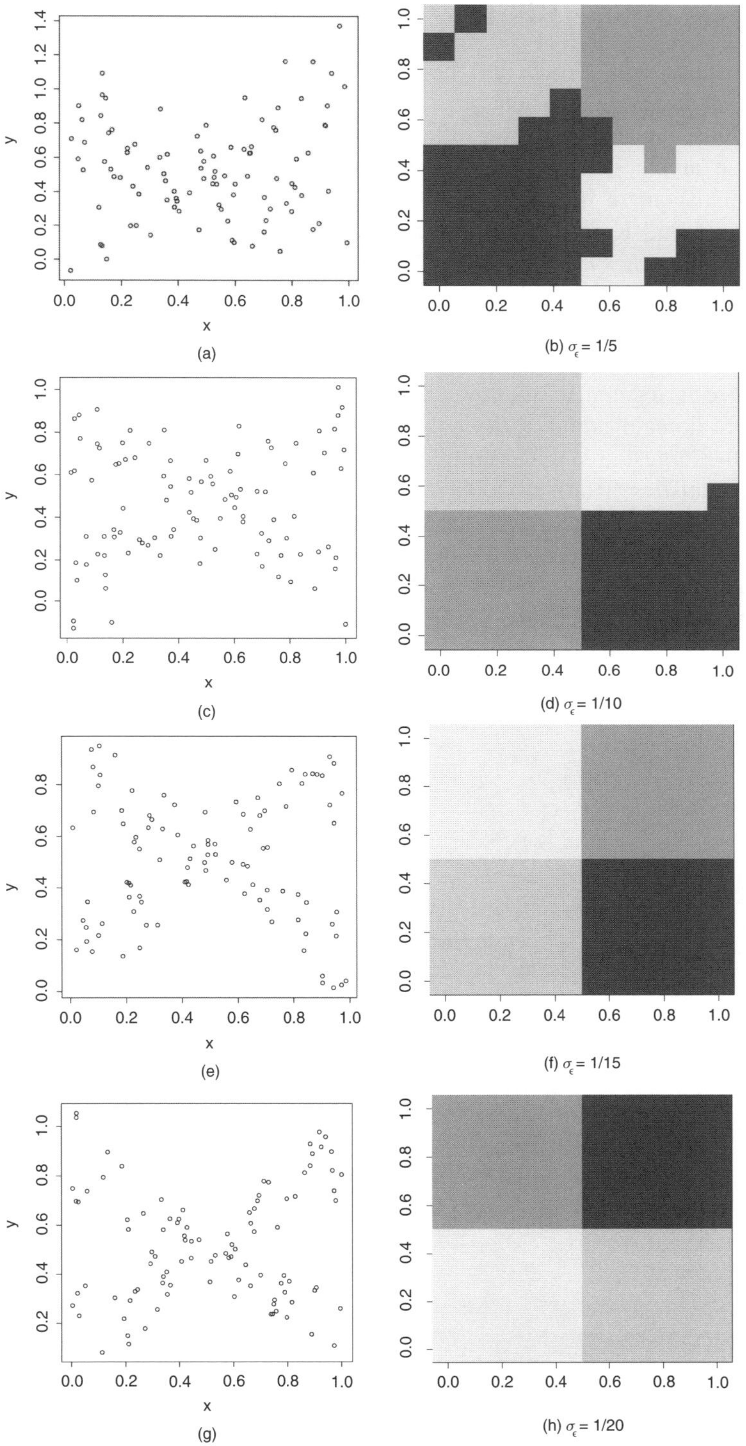

Figure 22.1 Simulated annealing partition of a regular 10×10 grid: (a), (c), (e), (g) X/Y scatter; (b), (d), (f), (h) local stationarity map, $K = 4$.

22.3 Case study: the province of Foggia

Forecasting agriculture production is of critical importance for policy-makers as for market stakeholders. In the context of globalization, it is of great value to obtain as quickly as possible accurate overall estimates of crop production on an international scale. For example, timely and accurate estimates of durum wheat yield are an important management instrument for the European Commission's Directorate-General for Agriculture and Rural Development (Baruth *et al.*, 2008). The main objective of a crop yield forecasting system is to provide precise, scientific, traceable, independent and timely forecasts for the main crops yields at many different geographical levels. These considerations constitute a valid motivation for our research.

Our prediction application is devoted to a particular province of Apulia: Foggia. This province is located in southern Italy (see Figure 22.2). The choice of Apulia as our reference domain is due to its position as the biggest producer of durum wheat in Italy, with the largest area under cultivation. According to data provided by the National Institute of Statistics (Istat), in 2008 Apulia accounted for 23.29% of the total area and 22.09% of the total production of durum wheat in Italy.

The data on durum wheat in Foggia province are drawn from the AGRIT survey of 2005. The next subsection describes the AGRIT survey in some detail. Then, our proposed approach is tested for the spatial prediction of the durum wheat yield in Foggia.

Figure 22.2 Map of domain under investigation: Foggia province.

22.3.1 The AGRIT survey

AGRIT is a sample survey the aim of which is to produce estimates, taking into account the economic situation, on areas and yields of the main crops and on the main land uses (see also Chapter 13 of this book). This survey is carried out using techniques of spatial sampling, particularly of point typology. Generally speaking, the method is based on the integration of data collected in ground samples, and data acquired through remote sensing. The list is constituted by a set of points (*point frame*), theoretically dimensionless, which exhaustively covers the territory under investigation.

In order to carry out the survey, each point is endowed with an *operational dimension area*: a circle of radius 3 metres centred on the reference point (thus with a total of about $30\,m^2$); in the presence of association between different land uses, the observation area is extended to about $700\,m^2$ by increasing the radius to 15 metres. If there are different land uses, each of these, up to a maximum of three, is assigned to the point in the correct proportion (*pro-rata criterion*). The adopted sampling design is stratified in three phases.

The first-stage sample consists of an aligned spatial systematic selection of geographical units. This reference list of units for the production of estimates is defined as the AGRIT sampling frame and is composed of a regular grid of points (about 1.2 million in number), with a spacing of 500 metres defined on a Gauss–Boaga coordinate system, for complete coverage of the area of interest. Secondary sampling units are based on a preliminary identification of areas of specific interest for the particular problem under investigation at the provincial, regional and national levels of administrative division Finally, the yield estimates, which the third sampling phase is devoted to, are significant with respect to the same geographical levels as already described, but for only 12 different crops.

Generally, an archive of land uses prepared for agricultural, forestry or environmental surveys may arise from three different sources: ground surveys; interpretation of digital ortho-photos; and the classification of remotely sensed images. Here, it was decided to use photo interpretation of remotely sensed images acquired from aircraft. This activity, which produced the hierarchical nomenclature of reference (on two levels for a total of 25 labels), was developed within the POPOLUS (Permanent Observed POints for Land Use Statistics) project. The 25 POPOLUS classes were aggregated into six groups, only four now sampled.

Concerning the preparation of the archive from which the second-phase sample was selected, the first step was the stratification of the primary sampling units. In this regard, the points were stratified by 103 provincial codes and six classes of land use, by obtaining 618 non-empty strata. The strata codes for land uses are reported in Table 22.1.

All points that show agricultural activity compose the reference population for the selection of the second-phase sample. The criterion for the definition of the strata is conservative: all points where a class of potential agricultural interest is present, even if reduced by the *pro-rata* criterion, are included in strata 1 and 3. For the subpopulations that are not considered interesting from an agricultural point of view (codes 4, 8), the second-phase sample overlaps with the first-phase sample. A similar strategy was adopted for the points in stratum 3* that were classified through the interpretation of aerial photos. Therefore, the stratum codes that are of interest for the AGRIT 2005 survey are 1, 2, 3 and 5 (see Table 22.1). From these strata and for each of the 103 Italian provinces a

Table 22.1 Strata codes obtained by aggregating POPOLUS classes.

Strata codes	Description
1	Arable land
2	Permanent crops
3	Permanent fodder land (altitude $\leq$ 1200 m)
3*	Permanent fodder land (altitude > 1200 m)
4	Wooded land
5	Isolated trees and agricultural buildings
8	Other (artificial surfaces, water, non-vegetated natural surfaces)

random second-phase subsample was extracted. Overall, approximately 150 000 points were to be collected on the ground.

The AGRIT 2005 was a multipurpose survey and was calibrated in such a way as to produce estimates with a predetermined sample error for a set of variables (i.e. areas of different crops). Hence, the allocation procedure is necessarily multivariate and its implementation (Bethel, 1989), in the case of stratified sampling, requires a generalization of Neyman's classic formulas for calculating the optimal sample size (Neyman, 1934; Cochran, 1977).

Define Ω as the finite population consisting of N points of the POPOLUS frame (i.e. about 1.2 million units), divided into strata according to the codes of one or more known variables and consider $\Omega_1, \Omega_2, \ldots, \Omega_h, \ldots, \Omega_H$ as the subsets of Ω that constitute the partition induced by H in the population strata. Denote by $N_1, N_2, \ldots, N_h, \ldots, N_H$ the corresponding size of these subsets such that $N = \sum_h N_h$. The technique of multivariate allocation determines the sample size n_h in each stratum h, so that the expected error of the estimates for each parameter of interest h is less than certain levels χ_h. Briefly, this allocation technique involves setting an arbitrary upper bound, positive and constant χ_h for each coefficient of variation for 12 different crops of the estimate of the total $\widehat{t_h}$. It can be shown that for the optimization problem considered here, there is always an optimal solution, which can be detected through certain convex programming methods. The final crop areas estimates were produced via a Horvitz–Thompson estimator (Särndal *et al.*, 1992).

Finally, the third-phase sample produces the yield estimates. The applied solution for the determination of sample size was to use Bethel's (1989) algorithm by using the variances of the yield of the previous AGRIT surveys. The sample size was about 60 000 points, located in the different regions of Italy.

22.3.2 Durum wheat yield forecast

In this subsection a regression model is presented and used for the spatial prediction of the yield of durum wheat in the province of Foggia. The idea is very simple: the estimated regression equation obtained with a sample of observations is used for predicting the value of y_i for given values of x_i – in other words, the geographical information (i.e. covariates) is available for each point of spatial domain under investigation, and by using this information and the estimated model, a spatial prediction $\hat{y}_i$ is computed.

However, it is worth noting that we are dealing with sample survey data (see the previous subsection for details). As well stated by Chambers and Skinner (2003), a great number of generalizations of regression analysis are frequently applied to survey microdata such as those analysed in this chapter. These methods are usually formulated in a statistical framework that is not specific to sample surveys. In other words, the methods should explicitly consider the sample nature shown by the data and so the corresponding statistical methods should be formulated and used appropriately for the analysis of these particular data: the method of sample selection is not completely irrelevant to the results obtained. In this last case, the selection method is called *non-ignorable* or *informative*, because how the units are selected should not be ignored when making inference. Otherwise, the selection methods are called *ignorable* or *non-informative* (Chambers and Skinner, 2003). Ignorable and non-ignorable sampling have received much attention in the context of superpopulation models (see Valliant *et al.*, 2000, for more details). In the context of the present research, the simplest general way to consider the effect of the selection method is to attach sampling weights to the regression estimates (see Pfeffermann, 1993; see also Chapter 20 of this book). However, our particular sampling survey can be considered *non-informative*, because the greater part of the sample points derive from stratum 1, and so all the points assume the same sampling weight. Therefore, following these considerations, our predictive procedure can be correctly applied.

The regression model used here is based only on three purely geographical covariates, the two coordinates and the elevation, and on a spectro-agro-meteorological (SAM) yield forecasting model:

$$y_i = \alpha + \boldsymbol{\beta}^t \mathbf{x}_i + \epsilon_i, \tag{22.3}$$

where y_i are the yields of durum wheat, t denotes vector transpose, α is the intercept, $\boldsymbol{\beta} = (\beta_1, \beta_2, \beta_3, \beta_4)$ is the vector of regression parameters, $\mathbf{x}_i = (x_{i1}, x_{i2}, x_{i3}, x_{i4})$ is the vector of four covariates representing respectively the elevation, the two geographical coordinates, and the SAM model, ϵ_i is the error term, and finally $i = 1, 2, \ldots, n$ are the zones under investigation. The SAM is an agro-meteorological model for the estimation of agricultural yield used in the AGRIT project in Italy and is determined on the basis of seasonal trends and of pedological and hydrological characteristics of the soil (see Carfagna *et al.*, 1992). In particular, for each crop, the production capacity is obtained by applying specific numerical models of phenological development, taking into account the seasonal weather to assess both the energy supply to the plant and the water stress and thermal intensity that, if they occur in certain phases of the growing season, cause significant reductions in agricultural production compared to the final production achievable in optimal weather conditions.

Model (22.3) was estimated by classical OLS using a sample of 993 points of the province of Foggia. Figure 22.3 is a map of the sample points. Note that the white coloured points represent our sample. Looking at Figure 22.3, it is evident that the greater part of sample points is in stratum 1, as previously stated. The main results obtained are reported in Table 22.2; p-values of the t test on the parameters of the model are shown in parentheses.

As already noted, by using the covariates available for each point, the spatial prediction $\hat{y}_i$ could be computed. Unfortunately, the goodness of fit in terms of R^2 of the model estimated is not very satisfactory and so the spatial prediction of the yield could be affected by some imprecision. For this reason we attempted to improve the modelling in this direction. Our idea was that by explicitly considering the non-stationarity highlighted by geographical data, the results obtained could be considerably improved. In order to

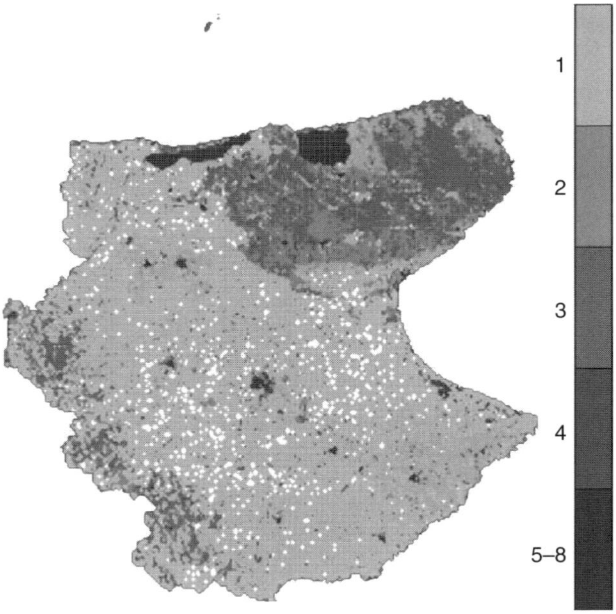

Figure 22.3 Sample points in the province of Foggia.

Table 22.2 Regression on durum wheat yield data in Foggia.

N	R^2	α	β_1	β_2	β_3	β_4
993	0.259	1881.948 (0.000)	−0.031 (0.000)	−42.320 (0.000)	−21.458 (0.000)	0.549 (0.000)

test this hypothesis, we applied our SA based algorithm to the Foggia data. The SA algorithm was computed for different values of the number of classes K, thus identifying K different zones of non-stationarity.

The first question in our analysis was the identification of the optimal number K of zones of non-stationarity. For each number K of classes, the ratio between the sum of squares of the residuals and the total sum of squares (i.e. the deviance of y) was calculated as

$$\Phi = \sum_{i=1}^{n} e_{ik}^2 / Dev(y).$$

The function Φ is a measure of the discrepancy between the data and an estimation model. A small value of Φ indicates a close fit of the model to the data. The values of Φ obtained for different values of K are graphed in Figure 22.4. This plot shows two different trends: the first is greatly decreasing, while the second is nearly constant. The point where these two paths meet is often called the elbow. After $K = 5$, the gain in terms of diminution of Φ is very small, or in other words the reduction in discrepancy between the model and the data is negligible; so, in our analysis we identify five zones of

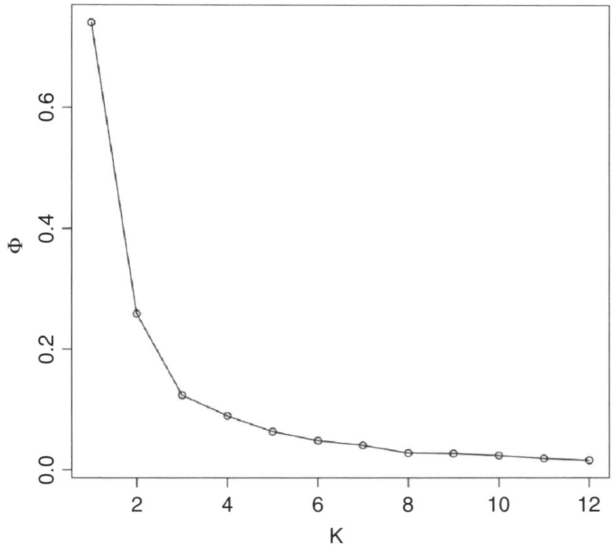

Figure 22.4 Plot of Φ against K.

non-stationarity for the sample under investigation. The map of local stationarity zones for durum wheat yield data in Foggia is shown in Figure 22.5. The zones are displayed in different grey levels.

Therefore, in order to take the spatial non-stationarity into account, the estimated model (22.3) becomes

$$y_i = \alpha_{k_i} + \boldsymbol{\beta}^t_{k_i}\mathbf{x}_i + \epsilon_i, \tag{22.4}$$

where y_i are again the yields of durum wheat, t denotes vector transpose, α_{k_i} are the intercepts, $\boldsymbol{\beta}_{k_i} = (\beta_{1k_i}, \beta_{2k_i}, \beta_{3k_i}, \beta_{4k_i})$ are the vectors of regression parameters, with $k_i = 1, 2, \ldots, 5$ identifying the different non-stationarity zones, $\mathbf{x}_i = (x_{i1}, x_{i2}, x_{i3}, x_{i4})$ are the vectors of four covariates, ϵ_i are the error terms, and finally $i = 1, 2, \ldots, n$ are the zones under investigation.

The results of the estimation process for different values of K are reported in Table 22.3; p-values of the t test on the parameters of the model are given in parentheses. If we compare the results of the global model (22.3) (see Table 22.2) with those obtained with the partitioned model (22.4) (see Table 22.3), we can observe that the goodness of fit in the second case is always improved. In fact, the R^2 computed on the entire sample is only equal to 0.259, while the R^2 related to the partitions identified by the different classes K are always greater. This evidence shows that the results can be greatly improved if we take into account the presence of non-stationarity of the spatial observations. For $K = 5$, for example, all the groups give R^2 values of 0.716 or greater (the smallest R^2 is achieved by group 2; see Table 22.3).

The parameters are all significant for all partitions. Furthermore, we note that the algorithm identifies zones where the relationships between variables are different; in fact, in some groups we have that the sign of the parameters is negative, while in other regions the corresponding parameter is positive (note, for example, in the model estimated for

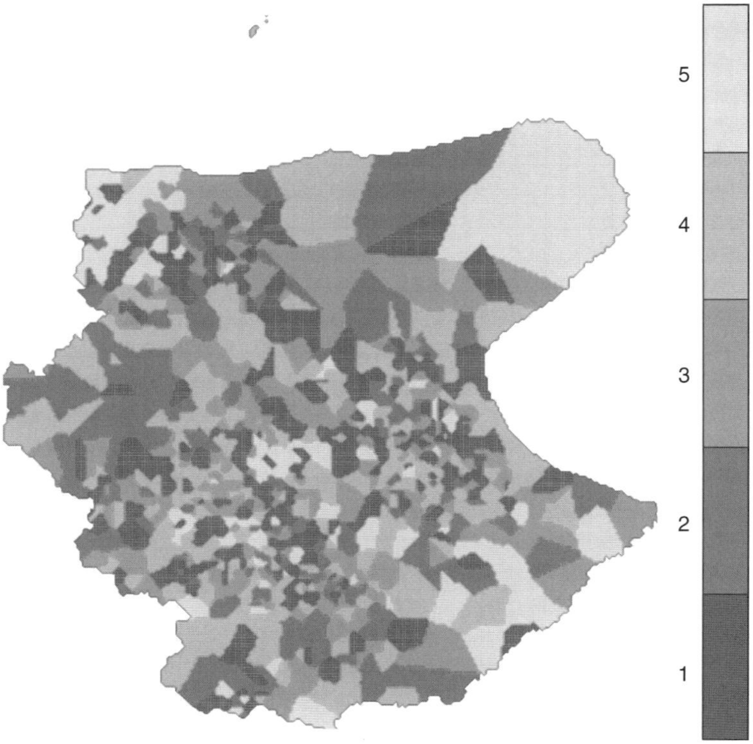

Figure 22.5 Zones of non-stationarity.

$K = 5$, that the coefficient β_4 is negative only in group 3 and positive otherwise). This observation shows once again that it is not appropriate to estimate the model in a global way, but it is more appropriate to consider geographical zones where the relationship among variables is defined in a different way.

Finally, by using the estimated regressions obtained for $K = 5$ and the values of covariates, the response variable $\hat{y}_i$ is predicted for all 28 000 points of the spatial domain under investigation, obtaining a more accurate spatial prediction of the yield of durum wheat. In this sense, we can argue that by considering the non-stationarity in the geographical data we can improve the agricultural crop forecast.

However, we are aware that the most appropriate tool for spatial prediction is kriging interpolation. Consider, for example, universal kriging. It is well known that this is similar to a standard regression model with an error term that is not independent for geographical site $i \neq j$. In other words, if the spatial dependence is negligible, the classical regression prediction can be used without loss of accuracy.

In order to test this hypothesis, some correlograms were calculated; in the analysis of spatial series, a correlogram, or autocorrelation plot, is a commonly used tool for checking randomness in a data set. If random, such autocorrelations should be near zero for any and all spatial-lag distances. The calculated correlograms are reported in Figure 22.6. The first correlogram in Figure 22.6 shows the dependence path of the yield of durum wheat in the observed sample, while the second correlogram highlights the trend of dependence of the global error term calculated in the partitioned model (22.4). As is evident from looking at

Table 22.3 Regression on durum wheat yield data in Foggia for different partitions identified by the simulated annealing algorithm.

	Partition	N	R^2	α_{k_i}	β_{1k_i}	β_{2k_i}	β_{3k_i}	β_{4k_i}
$K = 2$	1	333	0.633	2285.835	−0.023	−63.851	−18.755	−0.089
				(0.000)	(0.000)	(0.000)	(0.000)	(0.571)
	2	660	0.407	1609.871	−0.022	−37.671	−17.567	0.101
				(0.000)	(0.000)	(0.000)	(0.000)	(0.183)
$K = 3$	1	404	0.530	1549.604	−0.017	−31.918	−19.050	0.360
				(0.000)	(0.000)	(0.000)	(0.000)	(0.000)
	2	425	0.732	2239.811	−0.024	−39.831	−30.457	0.869
				(0.000)	(0.000)	(0.000)	(0.000)	(0.000)
	3	164	0.806	4084.137	−0.048	−70.820	−56.321	1.035
				(0.000)	(0.000)	(0.000)	(0.000)	(0.000)
$K = 4$	1	275	0.620	2049.486	−0.018	−21.533	−34.501	0.447
				(0.000)	(0.000)	(0.000)	(0.000)	(0.000)
	2	255	0.838	1436.434	−0.019	−57.306	−5.187	0.133
				(0.000)	(0.000)	(0.000)	(0.000)	(0.066)
	3	147	0.835	4482.508	−0.054	−71.450	−64.778	1.309
				(0.000)	(0.000)	(0.000)	(0.000)	(0.000)
	4	316	0.818	2435.173	−0.024	−44.641	−32.515	0.887
				(0.000)	(0.000)	(0.000)	(0.000)	(0.000)
$K = 5$	1	229	0.884	3257.037	−0.039	−48.870	−48.508	1.342
				(0.000)	(0.000)	(0.000)	(0.000)	(0.000)
	2	172	0.716	1135.717	0.002	−16.448	−17.455	1.116
				(0.000)	(0.480)	(0.000)	(0.000)	(0.000)
	3	209	0.883	−343.067	−0.029	−33.863	23.679	−0.752
				(0.000)	(0.000)	(0.000)	(0.000)	(0.000)
	4	265	0.876	3182.161	−0.030	−38.723	−51.420	0.590
				(0.000)	(0.000)	(0.000)	(0.000)	(0.000)
	5	118	0.771	3431.404	−0.043	−56.290	−48.601	1.105
				(0.000)	(0.000)	(0.000)	(0.000)	(0.000)

the first graph, the yield of durum wheat shows negligible dependence: the autocorrelation becomes near zero within a short distance. In the second graph the autocorrelation of the residuals of model (22.4) is half that of the durum wheat yield and is almost always close to zero. Generally speaking, we can argue that the modest dependence of the standard regression model is dramatically reduced by considering the joint effect of the covariates and of the spatial non-stationarity of the data, identified by the K zones. Following these observations, we can usefully predict the yield of durum wheat for unsampled points by using the partitioned regression model (22.4).

The results of the spatial prediction of the yield of durum wheat in Foggia are shown in Figure 22.7. Note that the points that are not included in the 993 sample points have been classified into one of the five classes identified with the proposed approach according to the minimum distance criterion.

Finally, the predictive capacity of the proposed method is compared with the official AGRIT 2005 estimate of the durum wheat yield. Recall that in AGRIT 2005 the yields are estimated via a Horvitz–Thompson estimator. In AGRIT 2005 the estimated mean

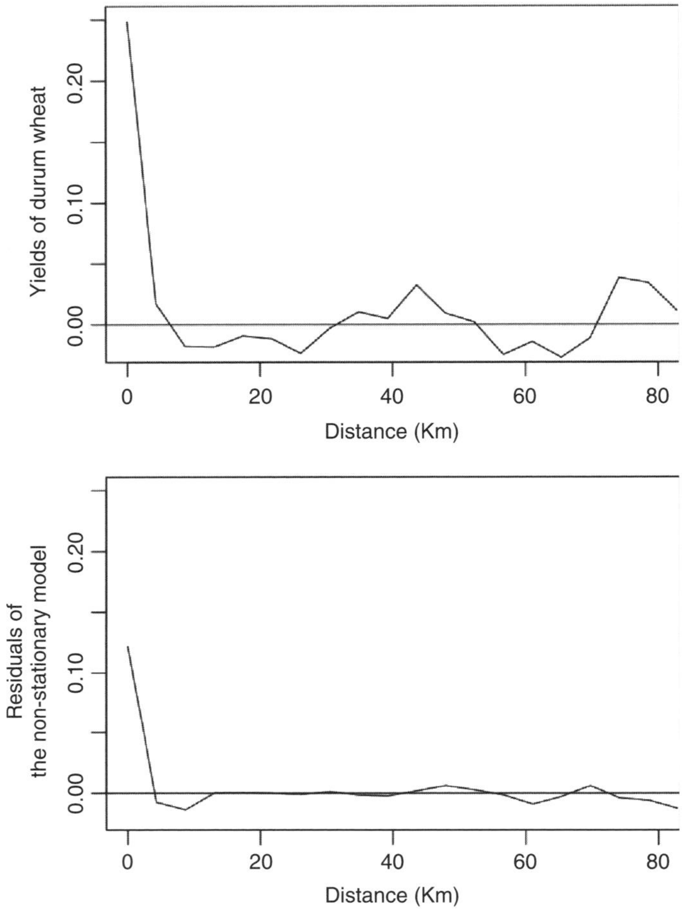

Figure 22.6 Correlograms.

yield of the durum wheat in Foggia is 35.92 quintals per hectare, with a coefficient of variation equal to 1.21%. The estimated mean of our predictive approach is calculated via the best linear unbiased predictor estimator (see Theorem 2.2.1 in Valliant *et al.*, 2000, p. 29 for further details). In this case, the estimated mean is 34.97 quintals per hectare (i.e. very similar to that obtained with Horvitz–Thompson estimator), but with a coefficient of variation of 0.63%. In other words, our prediction method not only gives an estimate similar to the official value but also permits a reduction in variability and a consequent gain in the precision of the estimates.

22.4 Conclusions

The problem of spatial non-stationarity is often neglected in the empirical analysis of geographical data. This neglect can have a noticeable effect on model estimates. In this chapter, we study this issue and show that by considering the presence of local stationarity in the data, we can greatly improve the results obtained.

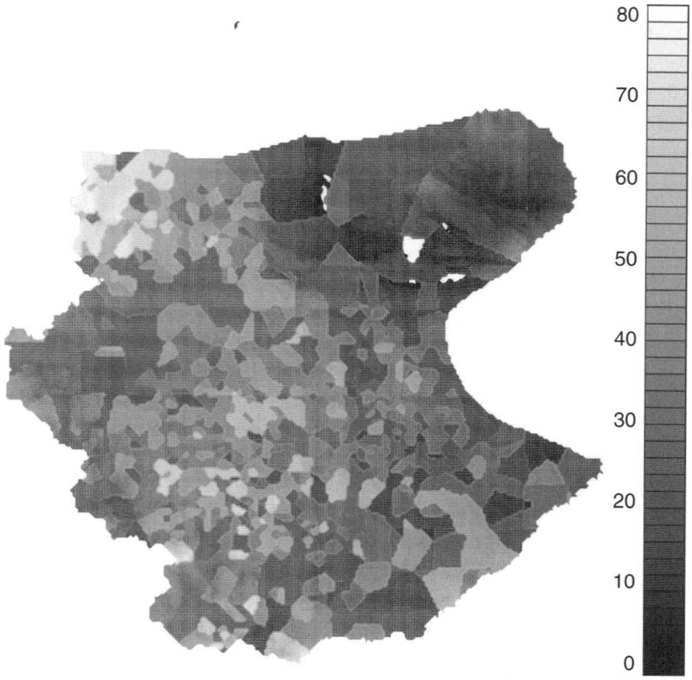

Figure 22.7 Spatial prediction of the yield of durum wheat in Foggia.

In particular, we analyse the problem of spatial prediction of crop yields by using a regression model, where the covariates are three purely geographical covariates (the two coordinates and the elevation), and an agro-meteorological model estimated yield (the SAM model). Furthermore, we propose a simulated annealing based algorithm for the identification of zones of non-stationarity of the data. The regression coefficients estimated on these subsamples of data are better in terms of goodness of fit, measured by R^2, than those estimated on the entire sample. By using this estimated model, we can predict the yield of the durum wheat for each point of the spatial domain, where the information is not available. The precision of the estimated mean yield with our predictive approach is greater than that obtained in the official AGRIT 2005 estimates.

The central concern of the procedure is the determination of the correct number K of zones of local stationarity: this choice influences the subsequent analysis. The researcher should evaluate in the best way the trade-off between gain in terms of goodness of fit and increasing complexity of the estimated model due to a larger number of classes K. Further research is needed on this topic.

Finally, we observe that the computational effort was not prohibitive even when dealing with large data sets (approximately 30 minutes for 1000 units).

References

Baruth, B., Royer A., Klisch, A. and Genovese, G. (2008) The use of remote sensing within the MARS crop yield monitoring system of the European Commission. In *Proceedings ISPRS*, Vol. 37, Part B8, pp. 935–940.

Besag, J. (1974) Spatial interaction and the statistical analysis of lattice systems. *Journal of the Royal Statistical Society, Series B*, 36, 192–236.

Bethel, J. (1989) Sample allocation in multivariate surveys. *Survey Methodology*, 15, 47–57.

Carfagna, E., Giovacchini, A., Ragni, P., Narciso, G. (1992) SAM: Spectro-Agro-Meteorological yield forecasting model. In *The Application of Remote Sensing to Agricultural Statistics*, pp. 361–364. Luxembourg: Office for Official Publications of the European Communities.

Cerny, V. (1985) A thermodynamical approach to the traveling salesman problem. *Journal of Optimization Theory and Applications*, 45, 41–51.

Chambers, R. and Skinner C.J. (2003) *Analysis of Survey Data*. Chichester: John Wiley & Sons, Ltd.

Cochran, W.G. (1977) *Sampling Techniques*, 3rd edition. New York: John Wiley & Sons, Inc.

Cressie, N. (1993) *Statistics for Spatial Data*, revised edition. New York: John Wiley & Sons, Inc.

Dobrushin, R.L. (1968) The description of a random field by mean of conditional probabilities and condition of its regularity. *Theory of Probability and its Applications*, 13, 197–224.

Fotheringham, A.S. and Rogerson, P.A. (1993) GIS and spatial analytical problems. *International Journal of Geographical Information Systems*, 7, 3–19.

Fotheringham, A.S., Brunsdon C. and Charlton M. (2002) *Geographically Weighted Regression: The Analysis of Spatially Varying Relationships*. Chichester: John Wiley & Sons, Ltd.

Gelfand, A.E., Kim, H., Sirmans, C.F. and Banerjee, S. (2003) Spatial modeling with spatially varying coefficient processes. *Journal of the American Statistical Association*, 98, 387–396.

Geman, S. and Geman, D. (1984) Stochastic relaxation, Gibbs distributions and the Bayesian restoration of images. *IEEE Transaction on Pattern Analysis and Machine Intelligence*, 6, 721–741.

Geman, D., Geman, S., Graffigne, C. and Dong, P. (1990) Boundary detection by constrained optimization. *IEEE Transaction on Pattern Analysis and Machine Intelligence*, 12, 609–628.

Givens, G.H. and Hoeting, J.A. (2005) *Computational Statistics*. Hoboken, NJ: John Wiley & Sons, Inc.

Kirkpatrick, S., Gelatt, C.D. and Vecchi, M.P. (1983) Optimization by simulated annealing. *Science*, 220, 671–680.

Langford, M., Maguire, D., Unwin, D.J. (1991) The areal interpolation problem: estimating population using remote sensing in a GIS framework. In Masser I. and M. Blakemore (eds), *Handling Geographical Information: Methodology and Potential Applications*, pp. 55–77. Harlow: Longman.

Loader, C. (1999) *Local Regression and Likelihood*. New York: Springer.

Matheron, G. (1963) Principles of geostatistics. *Economic Geology*, 58, 1246–1266.

Matheron, G. (1971) *The Theory of Regionalized Variables and its Application*, Les Cahiers du Centre de Morphologie Mathematique de Fointainebleu, 5. Fontainebleau: École Supériere des Mines.

Metropolis, N., Rosenbluth, A., Rosenbluth, M., Teller, A. and Teller, E. (1953) Equation of state calculations by fast computing machines. *Journal of Chemical Physics*, 21, 1087–1092.

Neyman, J. (1934) On the two different aspects of the representative method: the method of stratified sampling and the method of purposive selection. *Journal of the Royal Statistical Society*, 97, 558–606.

Pfeffermann, D. (1993) The role of sampling weights when modeling survey data. *International Statistical Review*, 61, 317–337.

Särndal, C.E., Swensson, B. and Wretman, J. (1992) *Model Assisted Survey Sampling*. New York: Springer.

Sebastiani, M.R. (2003) Markov random-field models for estimating local labour markets. *Applied Statistics*, 52, 201–211.

Strauss, D.J. (1977) Clustering on coloured lattices. *Journal of Applied Probability*, 14, 135–143.

Tobler, W.R. (1970) A computer movie simulating urban growth in the Detroit region. *Economic Geography supplement*, 46, 234–240.

Valliant, R., Dorfman, A.H. and Royall, R.M. (2000) *Finite Population Sampling and Inference: A Prediction Approach*. New York: John Wiley & Sons, Inc.

van Laarhoven, P.J.M. and Aarts, E.H.L. (1987) *Simulated Annealing, Theory and Applications*. Dordrecht: Reidel Publishing.

Wackernagel, H. (1995) *Multivariate Geostatistics*. Berlin: Springer.

Wheeler, D. and Calder, C.A. (2007) An assessment of coefficient accuracy in linear regression models with spatially varying coefficients. *Journal of Geographical Systems*, 9, 119–144.

Wheeler, D. and Tiefelsdorf, M. (2005), Multicollinearity and correlation among local regression coefficients in geographically weighted regression. *Journal of Geographical Systems*, 7, 161–187.

Author Index

Agricultural Survey Methods Edited by Roberto Benedetti, Marco Bee, Giuseppe Espa and Federica Piersimoni

Subject Index

Agricultural Survey Methods Edited by Roberto Benedetti, Marco Bee, Giuseppe Espa and Federica Piersimoni